# Meat and Meat Products

# Meat and Meat Products

## Technology, chemistry and microbiology

**Alan H. Varnam**
Consultant Microbiologist
Southern Biological
Reading
UK

and

**Jane P. Sutherland**
Head of Food and Beverage Microbiology Section
Institute of Food Research
Reading Laboratory
Reading
UK

**CHAPMAN & HALL**
London · Glasgow · Weinheim · New York · Tokyo · Melbourne · Madras

Published by Chapman & Hall, 2–6 Boundary Row, London SE1 8HN, UK

Chapman & Hall, 2–6 Boundary Row, London SE1 8HN, UK

Blackie Academic & Professional, Wester Cleddens Road, Bishopbriggs, Glasgow G64 2NZ, UK

Chapman & Hall GmbH, Pappelallee 3, 69469 Weinheim, Germany

Chapman & Hall USA, 115 Fifth Avenue, New York, NY 10003, USA

Chapman & Hall Japan, Thomson Publishing Japan, Hirakawacho Nemoto Building, 6F, 1-7-11 Hirakawa-cho, Chiyoda-ku, Tokyo 102, Japan

Chapman & Hall Australia, 102 Dodds Street, South Melbourne, Victoria 3205, Australia

Chapman & Hall India, R. Seshadri, 32 Second Main Road, CIT East, Madras 600 035, India

First edition 1995

© 1995 Alan H. Varnam and Jane P. Sutherland

Typeset in 10½/12½ pt Garamond by Acorn Bookwork, Salisbury, Wilts
Printed in Great Britain by St. Edmundsbury Press, Bury St. Edmunds, Suffolk

ISBN  0 412 49560 0

A catalogue record for this book is available from the British Library

Library of Congress Catalog Card Number: 94-74685

∞ Printed on permanent acid-free text paper, manufactured in accordance with ANSI/NISO Z39.48-1992 and ANSI/NISO Z39.48-1984 (Permanence of Paper).

# Contents

# Preface

Meat has formed a part of the human diet since prehistory and the development of hunting skills. Subsequently, the rearing of meat animals became an important part of agriculture, although hunting remains important today in some societies. Despite the existence in the modern world of wholly vegetarian cultures, restrictions on meat consumption are still largely imposed by economic considerations. Economic limitation of meat consumption is most obvious in developing countries but is also a factor of life in the generally far more affluent developed world. In the developed world, however, dietary choice is much greater and a feature of recent years has been a rejection of meat-based diets. This stems largely from health concerns, although ethical considerations are also involved and pose a growing problem to the meat industry.

The perishable nature of meat and, in earlier years, its highly seasonal availability led to the early development of preservation techniques, such as drying and curing. At a later date, the relatively high cost of meat and the demands of a growing population resulted in the development of products, including sausages and meat pies, that permitted utilization of virtually every part of the animal. These two factors, in turn, resulted in the development of a large meat products industry, which is of considerable economic importance today. This industry has responded to general concerns over meat consumption and more specific concerns over levels of sodium and preservatives, especially nitrites, in meat products by the development of new products designed for the 'health and quality conscious consumer'. In some cases, however, this has led to concerns over the microbiological safety of some modern meat products.

For a number of years, technological developments in the meat and meat products industry have been underpinned by parallel developments in the understanding of the related chemistry and microbiology. For this reason it is apparent that a full appreciation of the science of meat and meat products requires a knowledge of the three basic disciplines: technology, chemistry and microbiology. The intention of this book is to provide a fully comprehensive understanding of meat science for both undergraduate-level students and those entering the meat industry. This book follows the precedent of earlier books in the *Food Products* series in providing a detailed discussion of manufacturing processes, set in the context of the fundamental concepts of technology, chemistry and microbiology.

The fast-changing nature of the meat industry is fully recognized and reflected in the contents of the book. The strong link with commerce and the wider world is further strengthened by the use of information boxes and * points, while exercises encourage the reader in the application of knowledge to unfamiliar yet realistic situations.

A.H.V.
J.P.S.

# A note on using the book

## EXERCISES

Exercises are not intended to be treated like examination questions and in many cases there is no single correct, or incorrect, answer. The main intention is to encourage the reader in making the transition from an acquirer of knowledge to a user. In many cases the exercises are based on 'real' situations and many alternative solutions are possible. In some cases provision of a full solution will require reference to more specialist texts and 'starting points' are recommended.

# Acknowledgements

The authors wish to thank all who gave assistance in the writing of this book. Special appreciation is due to:

Debbie and Phil Andrews for providing both hand drawn and computer-generated illustrations.

Mr G. French and Mr B. Duddy of J. Sainsbury plc and Dr W. Morgan of Northern Foods plc for providing information concerning current industrial practice.

The libraries of the Institute of Food Research, Reading Laboratory and the University of Reading for their assistance in obtaining information.

Our colleagues in Reading and elsewhere for their help and interest during the preparation of this book.

# 1

# INTRODUCTION

## 1.1 MEAT IN THE DIET OF MAN

Although a high meat content is taken for granted in the energy rich diets of the western industrial nations, this situation is not typical and in other areas the meat content of the diet is, of necessity, very low. The situation in many countries today is probably similar to that of the poor in the UK and other industrial countries only a little over 100 years ago.

Although meat eating remains at a high level, there have been distinct changes in the type of meat eaten. The most striking is the rise in consumption of poultry, especially chicken, at the expense of red meats. Initially the driving factor was economic: the poultry industry exploited intensive rearing methods and recently developed processing technology to produce chickens at significantly lower prices than other meats, finally fulfilling the hope of King

Henry IV of France that there should be no peasant in his kingdom so poor as to be unable to have '*tous les dimanches sa poule au pot*'. More recently chicken and other poultry have been recognized as 'healthier' (or possibly less harmful) meats. In the US in particular this has resulted not in an overall change in meat-eating habits, but in an increased utilization of chicken and turkey in processed meats as well as for home cooking. At the same time the success of fast-food outlets means that increasing quantities of beef and, to a lesser extent, other meats are eaten as burgers and similar products, rather than as traditional home-cooked joints. Recipe dishes (ready-meals) also account for an increasing proportion of meat consumption.

Vegetarianism is a long established feature of religious groups such as Hindus and Buddhists. In the western world vegetarianism (by choice rather than necessity) is largely a 20th century phenomenon. Vegetarianism was mainly the result of moral and ethical concerns. These remain today and have deepened with publicity concerning the less acceptable aspects of intensive animal rearing. Vegetarianism is now also driven by health concerns, including claims that meat eaters are at greater risk from some types of cancer as well as by the fact that non-meat dishes may actually be preferred, irrespective of moral or health issues. Substitute meat dishes, including the much maligned 'nut cutlets' were developed to meet demand from vegetarians and products such as burgers made with soya protein and tofu are now available. Many vegetarians, however, consider that it is unnecessary for vegetarian foods to mimic meat dishes and a large variety of non-meat dishes are now commercially available.

It is thought that a relatively high proportion of persons avoiding meat for reasons of health concern, or simple preference, rather than for moral reasons, do not follow a totally vegetarian regimen, but eat meat either as a minority component of the diet, or on particular occasions. This has been referred to as 'quasi-vegetarianism' and is effectively a middle way between a predominantly meat-based diet, in which vegetables are secondary, and a fully vegetarian diet.

---

* Epidemiological evidence presented in 1994 has demonstrated a positive link between meat eating and certain types of cancer. Some aspects of the evidence have, however, been questioned. It is not clear whether meat eating as such is a predisposing factor, or other aspects of the 'typical' meat eater's life-style.

---

**BOX 1.1   Dined on one pea and one bean**

Quasi-vegetarianism is seen as creating a market for 'halfway' meals. These are effectively vegetable based side-dishes, such as noodles and sauce, or combinations of rice and different types of bean, to which meat is added at low level by the consumer. The individual ingredients for 'halfway' meals are already readily available, but preparation can be time consuming. The ideal commercially prepared 'halfway' meal is seen as being shelf stable, easy to use and to allow for individualism. (Sloan, E.A., 1994, *Food Technology*, **48 (2)**, 38.)

---

Attempts have been made over a number of years to produce meat analogues, intended as a direct replacement of meat for a significantly wider market than that represented by vegetarians. Many of the earlier products were of poor quality but some success has now been achieved.

## 1.2 ANIMALS AND BIRDS AS SOURCES OF MEAT

In the world as a whole a very large range of animals, birds and even reptiles are consumed as meat. In commerce, however, cattle, pigs, sheep and (to a much lesser extent) horse and goat are the meat animals of overwhelming importance. The most important birds are chickens, turkeys, ducks and geese. All of these animals and birds are domesticated and reared specifically for meat production. In addition animals, birds and reptiles are still hunted for meat. In underdeveloped countries, hunting is often a matter of necessity and subsistence, but in the industrialized nations the shooting of a diverse range of creatures, including wild boar, bears, deer, birds (such as pheasant, grouse and even larks) and reptiles (such as alligators and crocodiles) is undertaken as a matter of 'sport'. In some cases, however, such activities are semi-commercialized, pheasant, partridge, grouse and deer all entering commerce in the UK for example. In the case of deer, significant numbers are now farmed and killed in conventional abattoirs, while there is increased farming of more exotic creatures including kangaroos, ostriches and alligators. Some animals which have only limited food use at present are also considered to have longer-term potential as sources of farmed meat, these include the Giant Amazon River turtle. Although farming of exotic species is most common in the countries of origin, some,

---

BOX 1.2   **Strong the tall ostrich on the ground**

Ostrich farming was established in the mid 19th century to provide feathers for the hats then in high fashion. The more recent revival of the industry has primarily been to supply meat, imports from African ostrich farms being banned by the UK government because of inhumane practices. Exotic meat such as ostrich can also cause legislative difficulties. According to the American Ostrich Association, the US government is unable to decide whether ostrich (and the other ratite species, emu and rhea) should be classed as poultry or as livestock. (Anon, 1994, *Food Chemical News*, **July 11 1994**, 58).

---

including ostriches, are increasingly being farmed in the UK and US.

Importation of exotic meat is also increasing, although the total volume remains very small. In the past there has been concern over meat such as kangaroo being illegally used in meat products as a substitute for more expensive meats. The focus of concern has now shifted to possible public health risks. Conditions in producing countries are often unsatisfactory and, especially in the case of meat from reptiles, there is the risk of introducing new, atypical serovars of *Salmonella*.

Despite developments concerning use of exotic creatures as meat, cattle, pigs, sheep and poultry remain of overwhelming importance as meat animals in the industrialized nations. For this reason, this book is largely concerned with meat from these animals, although mention will be made of meat from other species where appropriate.

## 1.3 MEAT PRODUCTION AT FARM LEVEL

With the exception of pigs, meat may be obtained from animals specifically raised for meat production, or from animals surplus to other requirements. Thus beef may be obtained either from specially raised animals, male and other surplus calves, or cows at the end of their milk-producing career. Similarly chickens may be specially raised broilers, broiler breeders or hens from egg production ('spent' hens). In general meat from older animals and birds is of low quality and widely used for manufacturing.

[Production of any meat is concerned with maximizing the yield of 'saleable meat' while providing a satisfactory financial return to producer] and processor. Factors such as growth rate and feed efficiency are of major importance to the producer at farm level, while the processor is concerned with meat yield and unit price. The latter is ultimately determined by quality factors, such as colour and tenderness (see page 9). 'Success' in meat production (i.e. satisfying the needs of both producer and processor) depends largely on animal breed, gender and sex status, nutrition and slaughter weight. To some extent, however, there is conflict between the producer and the processor in that breeding programmes in recent years have stressed growth rate and yield at the expense of quality. Use of porcine somatotrophin and beta-androgenic agonists as feed supplements to increase lean meat content are also considered to lead to meat of lower quality. Supplements of this type are used in the US, but not to any great extent in Europe. The importance of quality is increasingly recognized by supermarket retailers, however, and integrated production systems incorporating quality assurance procedures to ensure quality have been developed for poultry and pork (Table 1.1), although only to a limited extent for beef. These require co-operation and contractual agreement between feed compounder, producer, processor and retailer. Systems of this type often also take account of food safety and animal welfare.

## 1.4 THE COMPOSITION AND NUTRITIONAL VALUE OF MEAT

Although the consumer may choose meat primarily for its aesthetic appeal, or through habit, it is important not to overlook its nutritional value. The composition of lean meat is relatively constant over a wide range of animals (Table 1.2). Variation is most marked in the lipid content, which may be evident as different degrees of 'marbling'.

* The UK multiple retailer, Marks and Spencer, has undertaken what appears to be the most intensive exercise ever for identifying factors determining meat quality and designing integrated systems spanning primary production to retail sale. In the case of beef, Aberdeen-Angus was the preferred breed, followed closely by beef sired by Charolais or Limousin bulls. Steers were found to produce better beef than heifers or bulls and beef from suckler herds was superior to that from dairy herds. Aberdeen-Angus steers are generally 50–75% Aberdeen-Angus, having a purebred sire and a crossbred mother to promote hardiness and disease resistance in the calf. Meat from these animals is retailed as 'Aberdeen-Angus', while that from steers with a Charolais or Limousin sire is retailed as 'Traditional'. Marks and Spencer has undertaken similar exercises for pork and lamb. (*The Guardian*, July 23, 1994).

**Table 1.1** Integrated pork production: Factors specified in quality assurance schemes

| | |
|---|---|
| *Feed* | No antibiotics |
| | No growth promoters |
| *Breed* | Emphasis on meat quality and hardiness |
| | (outdoor pigs) |
| *System* | No sow tethers |
| *Transportation* | Keep rearing groups together |
| | No mixing groups of different origin |
| | Restrict travel to short distances |
| | Permit recovery in lairage |
| *Quality measurement* | Absence of bruising |
| | No pale or dark meat |
| *Carcass processing* | Use of hip suspension or electrical stimulation |
| | Longer conditioning times |

*Note*: Data from Truscott, T.G. *et al.* (1983) *Journal of Agricultural Science (Cambridge)*, 100, 257–64.

**Table 1.2** Composition of lean muscle tissue of meat animals (%)

| Species | Water | Protein | Lipid | Ash |
|---|---|---|---|---|
| Beef | 70–73 | 20–22 | 4.8 | 1.0 |
| Chicken | 73–7 | 20–23 | 4.7 | 1.0 |
| Lamb | 73 | 20 | 5–6 | 1.4 |
| Pork | 68–70 | 19–20 | 9–11 | 1.4 |

Note: Data from Fennema, O.R. (1985) *Food Chemistry*, Marcel Dekker, New York.

Meat is considered, justifiably, as a high protein food. Of the total nitrogen content of muscle *ca.* 95% is protein and *ca.* 5% smaller peptides, amino acids and other compounds. The quality of the protein is very high, the types and ratios of amino acids being similar to those required for maintenance and growth of human tissue (Table 1.3). Of the essential amino acids, meat supplies substantial quantities of lysine and threonine and adequate quantities of methionine and tryptophan, although the content of these amino acids in meat is relatively low. The biological value of meat protein is 0.75 (human milk = 1.0, wheat protein = 0.50) and the net protein utilization 80 (egg = 100, wheat flour = 52). The digestibility of meat protein, like that of milk and eggs, is 94–97, compared with 78–88 for plant proteins.

**Table 1.3**  Amino acid composition of meat proteins (g/100 g)

|  | Beef | Chicken | Lamb | Pork |
|---|---|---|---|---|
| Arginine | 13.7 | 12.8 | 12.7 | 12.2 |
| Cystine | 2.6 | 2.6 | 2.7 | 2.6 |
| Histidine | 7.5 | 6.2 | 6.7 | 8.9 |
| Isoleucine | 10.4 | 9.5 | 9.7 | 9.2 |
| Leucine | 16.3 | 15.4 | 15.0 | 14.5 |
| Lysine | 18.5 | 18.4 | 20.3 | 19.7 |
| Methionine | 5.5 | 4.9 | 5.3 | 5.6 |
| Phenylalanine | 9.1 | 9.2 | 8.0 | 7.9 |
| Threonine | 9.4 | 8.5 | 9.7 | 8.9 |
| Tryptophan | 2.6 | 2.3 | 2.7 | 2.3 |
| Tyrosine | 7.8 | 7.2 | 7.3 | 7.6 |
| Valine | 10.7 | 9.8 | 10.0 | 9.9 |

*Notes*: 1. Values for chicken are based on raw meat only, while those for other species are based on raw, lean (average) samples.
2. Data from Paul, A.A., Southgate, D.A.T. and Russell, J. (1980) *First Supplement to McCance and Widdowson. The Composition of Foods*, HMSO, London.

Meat is of relatively high lipid content. This is of dietary significance in provision of energy, especially for persons engaged in heavy labour, or where overall dietary intake is limited. In the energy-rich countries of the industrialized west, however, the lipid content of meat has been associated with obesity and with artherosclerosis. This is exemplified in the US by the very high energy intake, which can greatly exceed requirements, and the growing problem of obesity. The role of meat should, however, be placed in perspective. Meat consumption amongst affluent Americans is certainly high and largely unlimited by financial constraints. Further, there is a tendency to consumption of high fat products, such as burgers. Meat consumption, however, is only a part of the equation, since the non-meat energy intake is also high and

* Biological value and net protein utilization are parameters of protein quality. The biological value of a protein is the fraction of the nitrogen retained in the body for growth and maintenance. It is determined by nitrogen balance experiments:

$$BV = \frac{IN - UN - FN}{IN - FN}$$

where IN = nitrogen intake, UN = urinary nitrogen output, FN = faecal nitrogen output.

Net protein utilization is the ratio of nitrogen retained and total protein nitrogen intake and is thus influenced by biological value and digestibility of the protein.

contributes to the laying down of fat as a long-term reserve. The problem is further exacerbated in many cases by the low energy requirements associated with sedentary work and an almost total lack of exercise. It is concluded that overall dietary habits and lifestyle are the underlying causes of obesity rather than the lipid content of meat *per se*.

The cholesterol and saturated fatty acid content of meat have both been associated with a predisposition to heart disease. The cholesterol content of *ca*. 65–75 mg/100 g (lean muscle) is not excessively high, although consumption of large quantities of meat obviously leads to a high total intake. Cholesterol content of kidney (*ca*. 400 mg/100 g) and liver (*ca*. 430 mg/100 g) is significantly higher than that of the meat itself. Degree of saturation varies with species. Beef and mutton fat are more saturated than pork fat, which in turn is far more saturated than poultry fat (see pages 24–25). Extraction of cholesterol from meat using supercritical $CO_2$ is technically feasible, but is not carried out on a commercial scale.

Muscle tissue in general is an excellent source of some of the B-complex vitamins, especially thiamine, riboflavin, niacin, $B_6$ and $B_{12}$. The B-vitamin content of meat, however, varies according to a number of factors, including species and muscle type. Secondary factors influencing B-vitamin content of meat within a species are breed, age, sex and general health of individual animals. Vitamin A is the most important fat-soluble vitamin in meat and provides *ca*. 23% of the average intake in the US. Contents of vitamins D, E and K are generally rather low in meats, although levels of vitamin E are elevated where animals are fed high tocopherol diets.

Lean meat is recognized as a good source of iron and phosphorus, but is usually low in calcium. An exception is mechanically recov-

---

* Iron bioavailability is recognized as being the proportion of the Fe in a food, or ingested as a supplement, which is absorbed and utilized by the human body. Iron deficiency is the most prevalent nutritional deficiency in the world today and is particularly marked in developing countries. This stems partly from a deficient diet, or from a diet in which bioavailability of Fe is low. Total dietary Fe can be divided into haem and non-haem pools, meat (including fish) being the only source of haem Fe. Haem Fe is of much higher bioavailability than non-haem. Meat exerts a dose-dependent effect which enhances the absorption of non-haem Fe. This is known as the 'meat effect' and is unique, other animal foods including eggs and milk either having no effect or inhibiting absorption. The mechanism of the 'meat effect' is not fully understood, but appears to be closely associated with the digestive process.

ered meat and some types of deboned meat, where small particles of bone are present.

## 1.5 DETERMINANTS OF MEAT QUALITY

There are three main determinants of meat quality at consumer level: colour, juiciness and toughness (tenderness). Flavour is usually of importance only in a negative sense when taints are present. Colour is the most important factor with respect to initial selection. In red meats a bright red colour associated with a high content of oxymyoglobin is a positive determinant of quality, while metmyoglobin content is a negative determinant (see page 28). Two specific defects – pale, soft, exudative meat (PSE) and dark, firm, dry meat (dark cutting; DFD), both of which are due to abnormal post-mortem pH values – are also recognized (see Chapter 2, pages 76–83).

Poultry meat differs from red meat in that whole birds are retailed with the skin attached. Skin colour is thus a factor influencing perceptions of quality. Strain of bird and diet, as well as processing conditions (scalding), are often manipulated to produce a range of skin colours to suit particular markets. At the same time poultry portions and deboned meat may be retailed with the skin removed and colour of the meat itself strongly influences perceived quality. Some sectors of the poultry industry have experienced problems due to wide variation in the colour of both chicken and turkey meat. The reason is not fully understood, although there may be an indirect relationship between stress and haem content. The role of haem pigments in colour variation is not, however, established and cytochrome *c* content of the muscle may have a greater influence on colour (see Chapter 5, page 267).

The importance of colour as a quality determinant should be seen in the context of overall appearance. Perceptions of quality related to colour can be modified by other visual factors. The most important of these, in red meats, is the extent of marbling, the adipose tissue located between muscle fibre bundles in the perimysial connective tissue. Marbling is positively associated with good eating quality and can be an important factor influencing consumer choice. At the same time the amount of fat surrounding major muscles influences the appearance or 'finish' of the meat. Excessive fat has always been associated with poor quality, although a certain quantity is expected on some cuts. Concern over dietary fats now means that quality can be associated with a virtual lack of visible fat. Colour of fat is also important and a yellow appearance, which is common with some dairy breeds such as Jerseys, is a negative determinant of quality.

Perceptions of quality can also be affected by defects, some of which are essentially cosmetic. Defects include blood splash and staining of fat with blood from drip. Physical defects resulting from poor butchery or inadequate plucking of poultry are also considered to be indicative of poor quality.

Juiciness is related to the water-holding capacity of the meat (see page 26) and also to marbling. There is interaction with appearance in that, while dry meat is undesirable, excessive drip and exudation, as occurs in the PSE condition, is a specific quality defect. Juiciness together with tenderness accounts for the overall eating quality and consumers may confuse the two factors when making assessments or comparisons.

Tenderness is a consequence of inherent factors, such as the type of muscle, and post-mortem events involving onset and resolution of rigor (tenderization). Post-mortem events are discussed fully in Chapter 2, pages 74–86. In general terms the degree of tenderness is directly related to quality. There are, however, circumstances where a degree of toughness, or texture, is desirable. The firmer meat of 'Label' poultry, produced from slow-growing breeds fed high cereal diets, compared with the standard broiler is considered to be a positive feature. Some studies with very well aged beef have found that tougher pieces were preferred by a significant number of consumers and similar considerations probably apply with other meats. In the case of both juiciness and tenderness, method and extent of cooking may strongly affect perceptions of quality and can be of greater importance than the intrinsic properties of the meat.

## 1.6 NATURE OF MUSCLE TISSUE

### 1.6.1 Types of muscle and muscle fibre

Muscle may be classified in a number of ways. The simplest is as 'red' or 'white', colour reflecting the different myoglobin contents. 'Red' muscles are characterized by a high myoglobin content, a highly developed vascular system and copious supplies of oxygen.

They are consequently adapted to oxidative metabolism and are thought to be involved in sustained, repetitive activity. As a further consequence 'red' muscles have limited glycolytic activity and a relatively high content of mitochondria. 'White' muscles have a lower myoglobin content, relatively few mitochondria and a less well developed vascular system than 'red'. They have, however, greater glycolytic capacity and are thought to be involved in short bursts of violent activity, during which metabolism becomes anaerobic. The simple differentiation between 'red' and 'white' is of value in meat science since there is a broad correlation with post-mortem behaviour (see Chapter 2, pages 77–78) and functional properties in meat products (see Chapter 3, page 144). For this reason the simple 'red'–'white' classification is used in the present volume, although it is necessary to be aware of other systems.

As knowledge of muscle types has evolved, other classification schemes have been proposed. A system based on muscle contraction speed is now more widely used for research purposes than the simple 'red' and 'white' differentiation. Muscles in this system classified as fast twitch respond quickly to stimulation and resemble white muscle in physiological terms. Slow twitch muscles exhibit the physiological characteristics of 'red' muscles and respond slowly to stimulation. This classification may be refined by taking account of respiratory capacity. In this system muscles are classified into three types: slow twitch oxidative, fast twitch oxidative and fast twitch glycolytic.

Muscle fibres may be classified according to similar criteria to whole muscles. With a few exceptions, however, skeletal muscles are not homogeneous but are composed of a mixture of different fibre types in accordance with the overall muscle function. Like muscle, fibres may be classified simply as 'red' or 'white', but a more common system is to divide into 'red' (predominantly oxidative activity), 'intermediate' (oxidative and glycolytic activity) and 'white' (predominantly glycolytic activity). Fibres have also been classified into types I, IIA and IIB corresponding directly to the muscle classifications slow twitch oxidative, fast twitch oxidative and fast twitch glycolytic. In a similar system these are described as $\beta$-red, $\alpha$-red and $\alpha$-white respectively. More complex systems using intermediate classifications have been devised, using subdivisions which take account of variants of the main fibre types. The advantages of more complex classification systems in a practical context are doubtful.

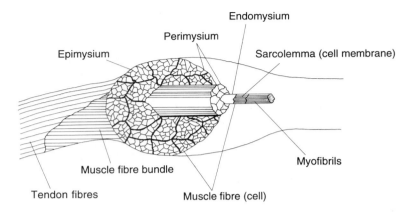

**Figure 1.1**  The overall structure of muscle.

### 1.6.2 The structure of muscle

Each muscle (Figure 1.1) is surrounded by a thick sheath of connective tissue, the epimysium, which is continuous with the tendon. The muscle itself is divided into bundles of fibres by a coarse, primary network of perimysial connective tissue. This primary network, which is obvious to the unaided eye, is further subdivided by thinner sheets of perimysial connective tissue. These typically define a bundle of muscle fibres in the order of 1 mm in cross-section. Individual muscle fibres have diameters of only 10–100 μm but, in meat animals, vary in length from a few millimetres to over 30 cm. Muscle fibres may run parallel to, or at an angle to, the length of the muscle. This is determined by the size of the muscle and its anatomical location. The perimysium contains collagen fibres which are crimped and arranged in a lattice formation. The orientation of these in relation to the muscle fibre axis varies with shortening and lengthening.

Within the bundle, individual muscle fibres are separated by the endomysial connective tissue network. The blood capillaries and nerve connections necessary for *in vivo* muscle function are contained in this connective tissue sheath. The surface of the muscle fibres is known as the sarcolemma. This probably consists of three layers: an outer network of collagen, an amorphous middle layer and an inner plasma membrane. The plasma membrane is invagi-

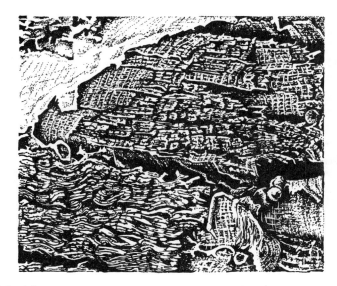

**Figure 1.2** The arrangement of myofibrils in muscle.

nated to form the transverse (T) system. The T-system effectively extends the plasma membrane into the interior of the muscle cell and is responsible for the transmission of nerve impulses for muscle contraction through T-tubules. All of the cellular components are bathed in the semi-fluid sarcoplasm, which contains soluble components, such as myoglobin, some enzymes and some metabolic intermediates.

The fibres are primarily composed of myofibrils closely packed side by side (Figure 1.2). Myofibrils account for *ca*. 80% of the muscle cell volume. These are the contractile apparatus of the muscle and are roughly cylindrical organelles, 1–2 μm in width, but of significantly greater length. 'White' fibres contain myofibrils separated only by the calcium-storing, membrane-lined channels of the sarcoplasmic reticulum. Only a few mitochondria are present. In contrast, 'red' fibres contain mitochondria intruding between adjacent myofibrils.

Whole muscle fibres have a striated appearance (Figure 1.3), which results from the packing of the striated myofibrils. The wide,

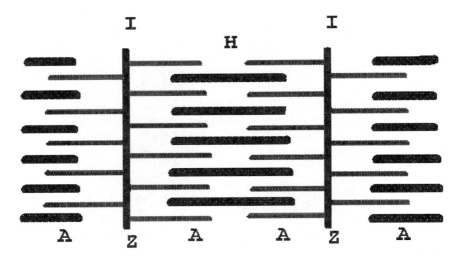

**Figure 1.3**   The structure of muscle fibres.

protein-dense bands of the muscle fibre form the A-band, each of which contains a lighter, central H-zone, while the M-line forms a dense line in the middle of the A-band. The light bands are known as I-bands, each of which is bisected by a Z-disc. The contractile unit is known as a sarcomere and lies between two Z-discs, sharing the I-band with adjacent sarcomeres. The striated appearance of the myofibril is due to the presence of two types of filaments, thick and thin, which are arranged in overlapping arrays forming a repeating pattern. Water is held in the spaces between the two types of filament, the filaments lying in a structured hexagonal lattice.

Thick filaments are composed of *ca.* 300 molecules of the filamentous protein myosin and are 1.6 µm in length. The myosin molecule has a long tail (156 nm × 2 nm), which has two curved, pear-shaped heads, 19 nm in length, attached flexibly to one end. The molecule is composed of two large subunits of molecular weight *ca.* 200 000 (the heavy chains) and four subunits of variable molecular weight in the region of 20 000 (the light chains). The heavy chains form the tail portion of the myosin molecule, *ca.* 50%

of the length being arranged as α-helices and coiled in a rope-like configuration. The remainder of each heavy chain folds separately into the head region. Each of the head regions is able to bind two light chains. In the thick filament the tails of the myosin molecules pack to form the shaft. The tails in each half point in opposite directions resulting in a smooth shaft and accounting for the manner in which three or four pairs of myosin heads protrude from the surface of the thick filaments at 14.3 nm intervals in a helical pattern. In this configuration the heads can readily interact with the thin filament.

At least seven other proteins are associated with the thick filaments. A possible structural role has been suggested for C-protein, which is situated on either side of the M-line region in evenly spaced strips. The C-protein is closely associated with the tails of myosin molecules and may be involved in controlling the formation of thick filaments. M-protein, myomesin and creatine kinase are also closely associated with the M-line, but the nature of any structural role has not been elucidated. Three proteins are associated with the A-band. No structural role has been established for the minor F- and H-proteins, while I-protein, situated close to the end of the I-band, inhibits myosin ATPase.

The major component of the thin filament is *ca.* 400 molecules of F-actin, the number varying according to the species. F-actin molecules are formed by condensation of G-actin monomers, are shaped like a dumb-bell and can bind one myosin head. The molecules are packed in a helical configuration, with the long axis of each dumb-bell perpendicular to the filament axis. The F-actin filament is sensitized to calcium by the proteins tropomyosin and troponin, which thus act as regulators. Tropomyosin is a filamentous protein, 41 nm in length, consisting of two polypeptide α-helices coiled in a rope-like structure. Tropomyosin forms tightly bound end-to-end aggregates, which run along each of the helical grooves for the entire length of the F-actin filament. Troponin is a globular protein complex containing three subunits T, I and C. The complex attaches to molecules of tropomyosin at 38.5 nm intervals along both sides of the F-actin double helix. Three minor proteins are associated with the thin filaments: β-actinin, which caps each end of the thin filaments, γ-actinin, which inhibits actin polymerase, and paratropomyosin, which is located at the A–I band junction.

### 1.6.3 Muscle contraction

The energy required for muscle contraction is derived from hydrolysis of ATP catalysed by ATPase in the myosin heads. The process
is initiated by the release of $Ca^{2+}$ ions from the sarcoplasmic reticulum (see below) in response to a nerve impulse. Calcium ions subsequently bind to the thin filament activating a switch which
permits the myosin heads of the thick filament to bind to the thin
filament and, probably through a change in shape, generate force.
The muscle is shortened by successive cyclic interactions of the
myosin heads, which leads to increased overlap of the thick and
thin filaments. This shortening is described by the sliding filament
theory. As the muscle shortens the width of the I-band decreases as
the thin filaments are drawn into the spaces between the thick filaments in the centre of each sarcomere. The A-band, however,
maintains its thickness throughout contraction. Sarcomere length is
thus dependent on the extent of overlap between the thick and
thin filaments which themselves usually remain of constant length.
Maximum reversible contraction varies from 20 to 50% of the
length of the sarcomere at rest (*ca.* 3.6 µm). Relaxation results
from removal of $Ca^{2+}$ and the myofibrils losing their ability to
hydrolyse ATP (at concentrations below *ca.* 0.5 µm). Adenosine triphosphate then acts as a plasticizer allowing separation of actin
and myosin and relaxation of the sarcomeres to the rest length.

### 1.6.4 The cytoskeletal framework

Cytoskeletal proteins are involved in maintenance of the structural
framework within which the contractile proteins function (Table
1.4). Some of the proteins which constitute the Z-disc are associated with actin, but are considered to be cytoskeletal. Connectin
(titin) is the dominant protein of the cytoskeleton and is the third
most abundant protein of muscle. It is characterized by a very high
molecular weight and highly elastic properties. Connectin is
located primarily at the A–I band junction. It extrudes as thin filaments on either side of the centre of the myosin filaments through
the actin filaments to the Z-disc. Connectin appears to be the structural component of 'gap filaments', which are observed when
muscle fibres are stretched beyond the natural overlap length of
the thick and thin filaments (Figure 1.4).

The N-lines are thin structures which run transversely across the
myofibrils parallel to the Z-discs. N-lines are present at three loca-

**Table 1.4** Myofibrillar proteins associated with the cytoskeletal network

| Location | Protein | Major function |
| --- | --- | --- |
| GAP filaments | Connectin | Links myosin to Z-disc |
| $N_2$-line | Nebulin | Not known |
| By sarcolemma | Vinculin | Links myofibrils to sarcolemma |
| Z-disc | α-Actinin | Links actin to Z-disc |
| | Eu-actinin, Filamin | Links actin to Z-disc |
| | Desmin, Vimentin | Peripheral structure of Z-disc |
| | Synemin, Z-protein, Z-nin | Lattice structure of Z-disc |

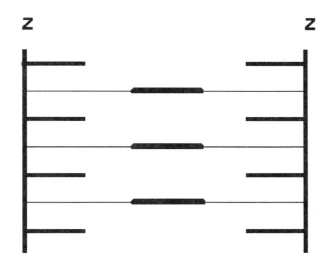

**Figure 1.4** Gap filaments.

tions: close to the Z-disc, in the centre of the I band and at the A–I band junction. The protein nebulin, which comprises up to 5% of the total myofibrillar proteins, has been positively identified as a structural component of the $N_2$-lines in the centre of the I band. The $N_2$-lines control the geometrical organization of the thin filaments, changing the configuration from a hexagonal lattice at the A–I band junction to a square lattice at the Z-disc. It has also been suggested that the N-lines are attached to connectin filaments to form a three-dimensional framework within each sarcomere, but evidence is incomplete.

The Z-disc contains a number of cytoskeletal proteins of which α-actinin is of greatest importance. This protein is thought to have a structural role in attaching actin filaments to the Z-disc. The minor proteins eu-actinin and filamin are also present in the Z-disc, while the internal lattice structure of the Z-disc consists of Z-protein, Z-nin and zeugmatin. The Z-disc is surrounded by a peripheral network of intermediate filaments consisting of desmin (skeletin) and vimentin. The intermediate filaments appear to have a major structural role in maintaining adjacent myofibrils in lateral register across the width of the muscle fibre. A further protein, synemin, is also located at the periphery of the Z-disc, but is present in only very small quantities. Latticed structures (costameres) run transversely across the sarcomeres on either side of the Z-disc and are firmly attached to the myofibril at the sarcolemma. This consists of the protein vinculin.

### 1.6.5 The sarcoplasmic reticulum

The sarcoplasmic reticulum consists of a series of transverse and longitudinal vesicles, which form a complex membranous network around each myofibril. The sarcoplasmic reticulum releases the $Ca^{2+}$ necessary for muscle contraction in response to detection of a nerve impulse and actively readsorbs $Ca^{2+}$, permitting the muscle to relax when the nerve impulse is relaxed. Transverse elements of the sarcoplasmic reticulum (the terminal cisternae) are located parallel to the Z-lines and cover much of the I-band. Longitudinal vesicles emanate from each side of the transverse elements and meet at the centre of the sarcomere, forming a hollow collar. The collar is penetrated with holes and termed a 'fenestrated collar'. The terminal cisternae usually exist in close association with the tubules of the T system. The meeting of the T system and the sarcoplasmic reticulum (triadic joint) occurs at different intrafibre locations in different muscles but, in mammals and birds, is most commonly at the A–I band junction. Irrespective of the position of the meeting, the common function is detection of the nerve impulse transmitted through the T-tubules, which stimulates $Ca^{2+}$ release from the entire sarcoplasmic reticulum.

### 1.6.6 Connective tissue

Connective tissue holds and supports the muscle through the component tendons, epimysium, perimysium and endomysium. A highly intimate relationship exists between muscle cells and connective tissue and it is probable that all substances passing into or out of the muscle cell diffuse through some type of connective tissue.

Connective tissue consists of various fibres embedded in an amorphous ground substance rich in proteoglycans and glycoproteins. Various different types of cell are also present in the ground substance. These include fibroblasts (from which the fibrous proteins collagen and elastin are synthesized), macrophages, mast cells and fat cells. The structural stability imparted by connective tissue is largely determined by the properties of the collagen fibres, although elastin is involved to a lesser extent. Collagen is a major determinant of the texture of cooked meat, the quality of the collagen rather than the quantity being of major importance (see Chapter 5, pages 268–270).

### (a) Collagen

Collagen comprises one third or more of the total protein of mammals. It is not a single substance but a family of closely related molecules consisting of at least 15 genetically distinct types. Of those of greatest importance in muscle, types I, III and V are fibre-forming (Table 1.5) and, although composed of different α-chains, share a similar basic structure and a common molecular length of 300 nm. These types of collagen are of considerably greater importance in raw meat structure than type IV, which has a molecular length of 420 nm and which does not aggregate in fibrous form. Type IV collagen is a major constituent of the amorphous basement membrane of the endomysium.

The basic structure of types I, III and V collagen consists of three α-helical protein chains wound together to form the characteristic

**Table 1.5**  Collagens of greatest importance in muscle connective tissue

| Type | Molecular length (nm) | Aggregation | Main location |
|------|-----------------------|-------------|---------------|
| I    | 300 | Striated    | Epimysium |
| III  | 300 | Striated    | Perimysium |
| IV   | 420 | Non-fibrous | Endomysium[1] |
| V    | 300 | Striated    | Epimysium |

[1]in basement membrane

triple helix of the tropocollagen molecule. The structure is a consequence of the amino acid content and sequence of the component α-chains. With the exception of short non-helical (teleopeptide) end regions of the tropocollagen molecule, every third amino acid residue in the chains is glycine in a repeating Gly–X–Y sequence where X and Y are often proline and hydroxyproline. The abundance of proline and hydroxyproline interferes with the α-helical structure, restricting the rotation of each polypeptide chain and causing it to twist into a left-handed helix with three amino acid residues per turn. The three α-chains are then wound to form the right-handed super helix of tropocollagen. This molecule is stabilized by hydrogen bonds arising from the regular sequence of glycine along each constituent chain.

Collagen molecules are linked end to end to form fibrils. The charge distribution on the outer surface of the tropocollagen molecules is asymmetric contributing to aggregation in a quarter-stagger parallel array, with a small end-overlap of 25 nm. This precise arrangement of collagen generates the characteristic 67 nm axial repeat pattern along the fibril (Figure 1.5).

The fibrous structure of collagen is stabilized against mechanical stress by enzyme-induced cross-linking. The aldimine and keto-imine chemically reducible bifunctional cross-links contribute con-

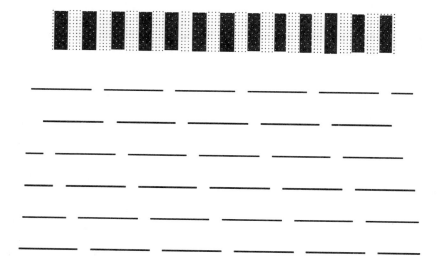

**Figure 1.5** Organization of collagen fibres.

siderable tensile strength to the collagen fibres. The overall reaction is the oxidative deamination of specific lysine or hydroxylysine residues in the non-helical end regions of each tropocollagen molecule. The reaction is mediated by the enzyme lysyl oxidase. Aldimine cross-links are formed when the $\epsilon$-NH$_2$ group of a lysine residue is converted to an aldehyde, which condenses with a hydroxylysine residue on the triple helical portion of an adjacent molecule. Formation of keto-imine cross-links involves oxidative deamination of a hydroxylysine residue in the non-helical region of tropocollagen. In this case, however, the aldimine initially formed undergoes an Amadori rearrangement to form a keto-imine cross-link. Keto-imine cross-links are heat and acid stable, while aldimine cross-links are heat and acid labile.

The tensile strength of collagen increases with ageing, but the proportion of aldimine and keto-imine correspondingly decreases. It appears that these cross-links undergo further reactions, producing trifunctional cross-links in place of bifunctional ones. This results in the gradual development of a network of cross-links within the fibre. The mechanism of formation of some of the trifunctional cross-links is now fairly well understood. The aldimine cross-link can react with histidine from the adjacent molecule to form histidino-hydroxylysinonorleucine, while the keto-imine cross-link can react with a hydroxylysine-aldehyde from another molecule to form the ring compound hydroxylysyl-pyridinoline (Figure 1.6).

### (b) Elastin

Elastin is a minor component of connective tissue, present in most muscles at less than 5% the level of collagen. In such cases, elastin appears to be mainly associated with the capillary system. There are exceptions, and the latissimus dorsi and semitendinosus muscles both contain up to 30% of the collagen content (*ca.* 2% of the muscle dry weight). The additional elastin in these muscles is associated with the perimysium and may well have a role in determining texture. Elastin has a molecular weight of 70 000 and a similar amino acid composition to collagen. Hydroxylysine, however, is absent and histidine present at only very low levels. The molecular structure is a random coil. Elastin fibres are composed of a glycoprotein-rich myofibrillar component surrounding an amorphous protein fraction which consists of elastin itself.

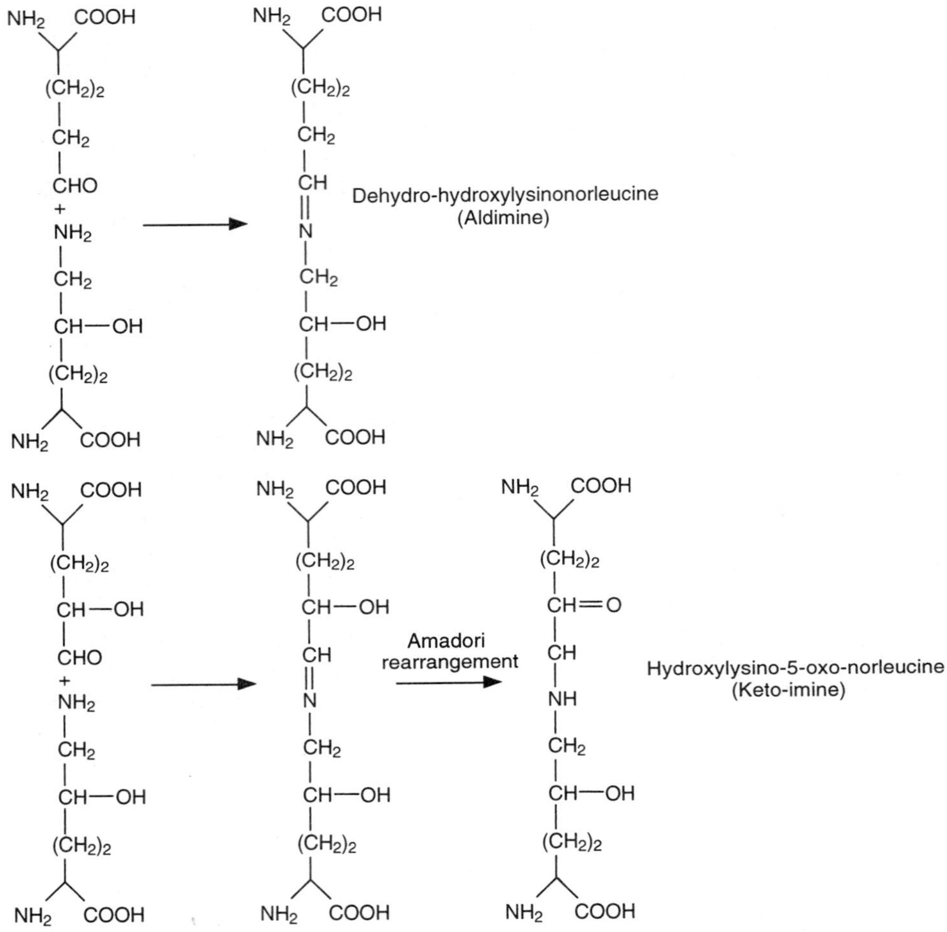

**Figure 1.6**　Cross-linking of collagen.

## 1.7 THE LIPIDS OF MEAT

Lipids in meat are of three discrete types: subcutaneous, intermuscular and intramuscular. Lean mutton, beef or pork usually contains 5–10% fat; chicken contains *ca*. 4%. Trimming all visible fat from beef can reduce the fat content to as low as 2%, while trimmed mutton has a fat content of *ca*. 4%. Fatty tissue of carcasses usually contains *ca*. 70% triacylglycerol fat.

Some fat is laid down in all animals fed on an adequate diet. The amount that accumulates in an animal depends on a number of factors including genetic predisposition, age, gender and sex status, level of nutrition and exercise. The current trend is towards leaner animals but, where this has been achieved by genetic selection, it is sometimes considered that overall quality is adversely affected (see page 5). Entire males usually have a higher lean-to-fat ratio than castrates as well as a better feed conversion ratio. In the past the use of entire males has been limited by their aggressive nature and, in the case of boars, the obnoxious odour of boar taint. Boar carcasses have a higher incidence of bruising as a result of fighting before slaughter, while glycogen depletion leads to a high incidence of DFD meat. Use of entire male animals in meat production, therefore, has implications for animal welfare as well as quality.

Fat generally begins to accumulate abdominally or under the skin, fat in muscle being the last to be laid down. Accumulation of fat around the muscle is known as 'marbling' and is often considered indicative of good eating quality and a factor in choice of meat (see page 9). As this is the last fat to be deposited a high plane of nutrition is necessary to produce marbling. The amount of fat surrounding and within muscles is not constant, but reflects the food intake and other factors. Fat is laid down in fat cells, which probably arise from undifferentiated mesenchyme cells, development usually commencing around blood vessels. Development is accompanied by accumulation of lipid droplets, which grow by coalescence until a single large drop of fat exists in the mature fat cell. The fat cell is surrounded by cytoplasm containing subcellular organelles. Fat cells are located outside the muscle bundle, either subcutaneously or in perimysial spaces, but not in the region of the endomysium. Lipid in the latter area is usually present as very small droplets within the muscle fibres, or is associated with membranes.

The lipids of meat are essentially triacylglycerols. Lipids from different animal muscles vary in properties, this variation being related

---

* Fighting and bruising *per se* can lead to abnormal ultimate pH values and poor quality meat. Increased bruising has been related to both the PSE and the DFD condition, depending on the time at which fighting occurred prior to slaughter and the duration of the fight. (Moss, B.W. 1992, in *The Chemistry of Muscle-based Foods*, (eds D.E. Johnston, M.K. Knight and D.A. Ledward), Royal Society of Chemistry, London.)

to the composition of the constituent triacylglycerols and, ultimately, the fatty acids. These vary according to animal species and also according to the position of the muscle in the body, diet, age, sex and gender status and weight. Diet is of particular importance with non-ruminant (monogastric) animals and the composition of the lipids in the body fat tissues tends to reflect that of the dietary fats. This is particularly marked with pigs and animals fed on a diet high in unsaturated fatty acids tend to have softer body fats. In ruminant (polygastric) animals, unsaturated fats are subject to hardening by hydrogen produced by rumen micro-organisms. For this reason the fat of ruminants tends to be harder than that of non-ruminants.

There can also be marked variation according to the location of the fat in the body. Internal body fats are significantly harder than those near the skin. This is thought to reflect the fact that the lower temperatures at the outside of the body mean the fat must have a lower melting point to permit mobilization. Conversely internal fats must have some structural rigidity and thus have a higher melting point. Differences reflect the fatty acid composition, the major compositional characteristic. The fatty acid composition of fats from major meat animals is listed in Table 1.6 and variations in the fatty acid content in various parts of a sheep carcass in Table 1.7.

## 1.8 THE WATER OF MEAT

Water is quantitatively the most important component of meat comprising up to 75% of weight. The water content is inversely related to fat content, but is unaffected by protein content except in young animals. Water in meat is associated with muscle tissue and proteins have a central role in the mechanism of water binding. In the living animal, muscle proteins impart a gel structure to the tissue and there is little loss of water from tissue cut immediately after slaughter. This is attributed to the water molecule behaving as a dipole and binding strongly to surfaces by a number of non-covalent forces.

In meat it was previously thought that up to 60% of water was bound by the myofibrils, but this is now recognized to be an overestimate and a figure of *ca.* 10% appears to be more realistic. The majority of water (*ca.* 85%) is bound between the thick and thin myofibrils. Binding is looser than in the living animal and some loss, as drip, from freshly cut surfaces is inevitable, if undesirable.

**Table 1.6** Fatty acid composition of fats of the major meat animals (m/m%)

| Fatty acid | Mutton | Beef | Pork | Chicken |
|---|---|---|---|---|
| 14:0 | 2.0 | 2.5 | 1.5 | 1.3 |
| 14:1 | 0.5 | 0.5 | 0.5 | 0.2 |
| 15:0 | 0.5 | 0.5 | | |
| 16:0 | 21.0 | 24.5 | 24.0 | 23.2 |
| 16:1 | 3.0 | 3.1 | 3.5 | 6.5 |
| 17:0 | 1.0 | 1.0 | 0.5 | 0.3 |
| 18:0 | 28.0 | 18.5 | 14.0 | 6.4 |
| 18:1 | 37.0 | 40.0 | 43.0 | 41.6 |
| 18:2 | 4.0 | 5.0 | 9.5 | 18.9 |
| 18:3 | | 0.5 | 1.0 | 1.3 |
| 20:0 | 0.5 | 0.5 | 0.5 | |
| 20:1 | 0.5 | 0.5 | 1.0 | |
| Others | 2.0 | 2.5 | 1.5 | 0.3 |
| P/S ratio[1] | 0.07 | 0.11 | 0.25 | 0.64 |
| Iodine value[2] | 42.6 | 48.7 | 60.3 | 78.3 |

[1]polysaturated/unsaturated ratio; [2]by calculation from fatty acid composition.
*Note*: Data from Rossell, J.B. (1992) In *The Chemistry of Meat-based Foods*, (eds D.E. Johnston, M.K. Knight and D.A. Ledward), Royal Society of Chemistry, London.

**Table 1.7** Variation of fatty acid content of fats from different parts of the sheep carcass

| | Major fatty acids (mass %) | | | | |
|---|---|---|---|---|---|
| | 14:0 | 16:0 | 18:0 | 18:1 | 18:2 |
| Perinephric[1] | 2.0 | 26.0 | 34.0 | 30.0 | 1.0 |
| Mesenteric[2] | 2.0 | 24.0 | 33.0 | 32.0 | 1.0 |
| Chest | 2.0 | 26.0 | 21.0 | 42.0 | 1.0 |
| Rump | 2.0 | 23.0 | 11.0 | 53.0 | 2.0 |
| Leg | 1.0 | 17.0 | 3.0 | 69.0 | 1.0 |
| Intramuscular[3] | 5.0 | 24.0 | 14.0 | 38.0 | 4.0 |

[1]kidney fat; [2]inter-intestinal fat; [3]all visible fat trimmed off.
*Note*: Data from Rossell, J.B. (1992) In *The Chemistry of Meat-based Foods*, (eds D.E. Johnston, M.K. Knight and D.A. Ledward), Royal Society of Chemistry, London.

The extent of drip loss from meat is largely a function of post-mortem changes, predominantly those affecting the ultimate pH value of the meat and the extent of changes in myofibrillar volume. These are discussed in Chapter 2, pages 84–85.

Meat is sold by weight and drip loss must be minimized for economic reasons. Drip is also unsightly to the consumer and excessive drip is a negative determinant of meat quality (see page 10). Water-holding capacity is therefore an important quality parameter and is defined as the ability of meat to retain the tissue water present in its structure. Water-holding capacity is also important with respect to the manufacturing properties of the meat, together with a second parameter, water-binding capacity, which is defined as the ability of meat to bind added water. It should be appreciated that while meat of low water-holding and water-binding capacity is undesirable for both retail consumption and manufacturing, the converse does not necessarily hold true. Dark, firm, dry meat has very good water-holding and binding capacity, but is unacceptable to most consumers. Dark, firm, dry meat is, however, perfectly acceptable for some, but not all, manufacturing purposes.

## 1.9 THE COLOUR OF MEAT

The characteristic colour of meat is a function of two factors: the meat pigments and the light-scattering properties.

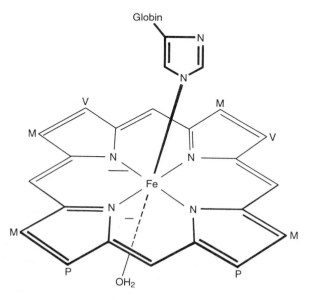

**Figure 1.7**  Structure of myoglobin.

### 1.9.1 Meat pigments

The basic pigment of fresh meat is myoglobin (Figure 1.7). Haemoglobin, which is very similar in chemistry, is also present in small quantities, especially if bleeding has been inefficient. The myoglobin molecule consists of a protein, the globin, and a haem group, which comprises the flat protoporphyrin IX ring system with a central iron atom. The iron atom has six co-ordination links, one of which is attached to the globin and four to nitrogen atoms. The other co-ordination link may bind to other substances providing these are of the correct electronic configuration and able to occupy the haematin pocket in the protein. The nature of the sixth ligand (usually water or oxygen) and the oxidation state of the central Fe will affect the electronic arrangement of the *d*-electrons of the iron. This arrangement will in turn affect the spectral characteristics and thus the colour.

The basic structure of myoglobin from all animals is similar, but small differences exist which may account for observed differences in visual appearance and colour stability of different meats. Myoglobin from cattle and sheep, for example, has been stated to have different stabilities in acid environments to that from pigs. The amino acid composition and isoelectric point of pig myoglobin also differs from that of cattle or sheep.

The content of myoglobin in muscles, and thus the colour of the meat, varies considerably according to species. A dark meat such as beef, for example, contains 4–10 mg myoglobin/g wet tissue (as much as 20 mg/g in 'old' beef), while in contrast pork and veal contain no more than 3 mg/g wet tissue. Differences also occur between animals of the same species and between muscles from the same animal. Myoglobin levels vary according to breed and age, concentration increasing with age. Meat from male animals also usually contains more myoglobin than that from females. The function of myoglobin in the living animal is oxygen storage and levels are accordingly higher in muscles with the higher work loads. Leg muscles contain more myoglobin and are of darker colour, for example, than loin. Equally myoglobin levels in muscles of animals reared under free-range conditions are likely to be higher than those reared intensively. The meat of free-range poultry, for example, is often darker than that of poultry reared indoors with restricted movement.

Myoglobin in fresh meat usually exists in three forms, the colours of which differ. All three forms are present in fresh meat and are in equilibrium with each other. Reduced myoglobin has reduced iron ($Fe^{2+}$) and $H_2O$ at the sixth co-ordination point. The pigment is purple-red in colour and found in the centre of a piece of meat, where $O_2$ is absent, and also in vacuum packs. Oxymyoglobin is the oxygenated form of myoglobin and while the iron is in the reduced ($Fe^{2+}$) form, $O_2$ occupies the sixth co-ordination link. This pigment is bright red in colour and is the desired pigment of fresh meat. Myoglobin has a great affinity for $O_2$ and formation of oxymyoglobin (blooming) is rapid at the surface of the meat where the $O_2$ surface tension is high. With time the narrow oxymyoglobin layer spreads downwards into the meat, the depth depending on the $O_2$ tension and activity of $O_2$-utilizing enzymes. At $0°C$, the oxymyoglobin layer is typically 1–3 mm thick after 2 hours exposure to air and 7–10 mm after 7 days. Oxymyoglobin formation is favoured by conditions which increase $O_2$ solubility (e.g. low temperature) and decrease enzyme activity (e.g. low temperature, low pH value). Loss of enzyme activity during prolonged storage in vacuum packs also enhances oxymyoglobin formation. Oxygen penetration in dark, firm, dry meat is limited and enzyme activity high. Formation of oxymyoglobin in DFD meat is therefore very limited.

There is variation in rates of oxymyoglobin formation between different muscle, but this is considerably less marked than variation between species. Fresh lamb, for example, has a high rate of enzyme activity and 'blooms' much less readily than beef under comparable conditions. Formation of oxymyoglobin appears to be very limited in turkey muscle, especially legs and breast, the colour of which is probably due almost entirely to reduced myoglobin.

Metmyoglobin has iron in the oxidized state ($Fe^{3+}$) and $H_2O$ at the sixth co-ordination point. The pigment, which is brown in colour, is unable to form an oxygen adduct. Metmyoglobin is usually present in a zone of low $O_2$ concentration between the anaerobic interior of the meat and the oxygenated zone at the surface. As meat ages the metmyoglobin layer widens and becomes visible below the thinning layer of oxymyoglobin. This results in loss of perceived quality and is a serious problem during retail display of meat. (Factors influencing formation during storage are discussed in detail in Chapter 2, pages 92–97.)

Other myoglobin-derived pigments can be present in fresh meat under specific, and usually uncommon, circumstances. These include the green-pigmented sulphmyoglobin, formed by combination of myoglobin with $H_2S$ of bacterial origin. This is a common problem with New York dressed (uneviscerated) poultry due to production of $H_2S$ in the intestine and its diffusion into the meat. Significant bacterial $H_2S$ production can also occur in vacuum-packed meat, especially (but not exclusively) if the pH value is in excess of 6.0. The precise structure of sulphmyoglobin has not been elucidated, but it is known that each haem molecule contains one S atom which is bound to the ring system. Similarly the pathway of formation has not been fully elucidated, but may involve a ferryl intermediate (Figure 1.8). Production would require the presence of $H_2O_2$, derived from the oxidation of oxymyoglobin to metmyoglobin, the activity of bacteria or enzymes, lipid peroxidation, or flavins. Under extreme circumstances, the pigment may be broken down. The haem portion becomes detached from the protein, the porphyrin ring is ruptured and finally the iron atom is lost from the haem structure. These changes result in the formation of green choleglobin and colourless bile pigments. This is rare in fresh meat.

The chemistry of cooked meat pigments is discussed in Chapter 5 (pages 266–268) and that of cured meat pigments in Chapter 4 (pages 202–203) and Chapter 6 (page 304).

### 1.9.2 Light-scattering

The extent of light-scattering is related to the structure of the muscle and appears to depend on myofibrillar volume. Pale, soft,

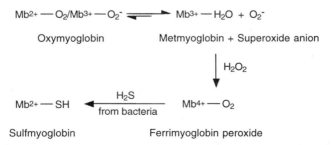

**Figure 1.8** Formation of sulphmyoglobin.

exudative meat, which is of low myofibrillar volume, has a high light-scattering ability. Light is unable to penetrate a significant distance into the meat without being scattered. This means that there is relatively little absorption by myoglobin and the meat appears pale. Dark, firm, dry meat has only very limited light-scattering ability, permitting incident light to penetrate for a considerable distance. Considerable absorption by myoglobin occurs and the meat appears dark. The PSE and DFD conditions represent extremes and variation in the extent of light scattering will occur in 'normal' meat, modifying the visual appearance to a greater or lesser extent.

### 1.10 THE FLAVOUR OF MEAT

Raw meat has very little flavour and effectively tastes of blood. Development of the characteristic flavour is dependent on heating, when a large number of chemical reactions occur between the many non-volatile compounds of meat. Many hundreds of volatile compounds are produced by these reactions, but probably only a relatively small number play a significant role in determining flavour and aroma (see Chapter 5, pages 258–260, for a more detailed discussion).

Although the characteristic flavour of meat develops during heating, the precursors are present in the raw muscle. Precursors are derived from the minority components of muscle, lipid (*ca*. 2.5%), carbohydrates (*ca*. 1.2%) and other water-soluble non-protein compounds (*ca*. 2.3%). This latter fraction includes amino acids, peptides, reducing sugars, vitamins and nucleotides.

Flavour precursors in raw meat are affected by the course of post-mortem glycolysis and conditioning (see Chapter 2, pages 85–86 and 91–92). There are also inherent differences due to species, which are reflected in the different flavour characteristics of the

---

* The generation of undesirable odours in meat by lipid oxidation is well known, but the role of lipids in desirable flavour production is a matter of contention. Many traditionalists attribute poor flavour in meat to the lack of fat. Certainly cooked fat has a characteristic flavour, liked by some and loathed by others, but lean meat from a lean animal appears to be indistinguishable in flavour from lean meat from a fat animal.

cooked meat. It is suggested that the effect of species on flavour stems from the genetic control of lipid metabolism and composition. This contention appears to be justified in the case of lamb and chicken, but the situation with beef and pork is less clear. The characteristic 'sweaty sour' flavour of lamb and mutton (and also goat) is derived from branched chain fatty acids, primarily 4-methyloctanoic and 4-methylnonanoic, arising from metabolic processes in the rumen. These fatty acids are not present to any significant extent in beef. The characteristic flavour of chicken has been attributed to unsaturated aldehydes, produced by oxidation of linoleic acid, which is present at a high concentration in the triacylglycerols.

Pork may be identified by 'porky' or 'sour' aromas, but beef has only a very weak species-specific odour. The origin of the specific odours in these animals is not known. Some have a characteristic off-odour, 'boar taint', which occasionally occurs in females and castrates as well as entire boars. This is distinct from the species-specific odour and is due to two compounds, $5\alpha$-androst-16-en-3-one and $5\alpha$-androst-16-en-3-$\alpha$-ol, which have odours of urine and musk respectively.

Diet can affect the flavour precursors present in raw meat. Feed taint was once a widely described problem, although in most cases the underlying cause was unsuitable or poor quality feed. Deliberate attempts are made to enhance desirable flavour attributes by dietary modification but it appears that it is easier to produce poor flavour through inappropriate diets than to improve flavour.

In both beef and lamb, it has been demonstrated that animals fed on high-energy concentrates are of superior flavour to those pasture-fed. Improvements have been associated with changes in the fatty acid composition of lipids. Forage grass contains relatively high levels of ($\omega$-3) fatty acids and the ratio of these to ($\omega$-6) fatty acids is significantly higher in grass-fed animals than in concentrate-fed. The oxidation products of these different classes of fatty acids differ as a consequence of differences in the position of double bonds. This has a marked effect on the final flavour, that of grass-fed animals being described as 'milky', 'oily' and 'sweet–gamey' compared with the clean flavour of concentrate-fed animals. Terpenoids derived from chlorophyll via fermentation in the rumen may also contribute to the flavour of grass-fed animals. Flavour may also vary according to the nature of the pasture and, with lamb, distinct differences have been recorded between ryegrass, vetch, clover and lucerne.

It has been claimed that the flavour of poultry is amenable to improvement by dietary modification. Improved flavour scores have been obtained by feeding a diet containing a high proportion of whole wheat and by allowing access to green vegetables. It has also been claimed that the superior flavour of free-range poultry is due to compounds derived from nettles and other wild plants. This does not, however, appear to have been confirmed on a scientific basis.

## 1.11 MEAT ANIMALS AS SOURCES OF PATHOGENS

Meat animals are considered to be important reservoirs of pathogenic micro-organisms. Human pathogens from animal sources constantly enter the food supply chain, where cross-contamination during processing can lead to a high rate of contamination of retail meat and meat products. This in turn is contributory to human morbidity through undercooking or recontamination of cooked products. There has been widespread discussion of the relative merits of elimination at farm level and decontamination of the carcass by, for example, irradiation. The two approaches should not be considered mutually exclusive and pathogen control is best achieved by an integrated programme. There are, however, grave doubts concerning the feasibility of eradication of pathogens such as *Salmonella* and *Campylobacter* at farm level, although levels of infection must be minimized (see below). It should also be emphasized that knowledge of the ecology of pathogens on farms is often far from complete. This is illustrated by *Escherichia coli* O157:H7, which is considered to be predominantly of bovine origin. Although some management practices have been found to be

---

* *Bacillus anthracis* is the causative organism of anthrax, an often fatal disease of humans as well as cattle and other animals. Intestinal anthrax of humans, which has a very high fatality rate, results from consuming flesh of infected animals. Consumption of infected meat is usually the consequence of severe economic hardship and large outbreaks of intestinal anthrax have occurred in developing countries, where animal immunization programmes were interrupted by civil war. A large outbreak of intestinal anthrax also occurred in the Siberian city of Sverdlovsk during the 1980s. It is likely that this was due to consumption of illegally butchered meat from condemned carcasses, although allegations were made in the US that the source was a biological warfare plant.

**Table 1.8** Association between management factors and on-farm prevalence of *Escherichia coli* 0157:H7.

| Practice | Association |
| --- | --- |
| Use of computerized feeders | + |
| Irrigation of pastures with manure slurry | + |
| Sharing of unwashed feeding utensils in calves | + |
| Feeding oats to calves in starter rations | + |
| Feeding grain to calves in first week of life | + |
| Feeding clover to calves as first forage | 0 |
| Feeding whole cottonseed to cows and heifers | 0/– |
| Feeding milk replacer to calves | 0/– |
| Feeding ionophores to calves | 0/+ |
| Small herd size | 0/+ |
| Grouping of calves before weaning | 0/+ |

*Note*: + significant association with increased prevalence; 0/– no association or decreased prevalence; 0/+ no association or increased prevalence

---

### BOX 1.3 **In farm and field**

Publicity has tended to be focused on the role of intensive farming practices in the spread of foodborne zoonotic disease. There is no doubt that some practices are, by any standards, undesirable. It must, however, be appreciated that zoonotic disease is neither new, nor associated exclusively with intensive agriculture. Tuberculosis and brucellosis were previously endemic in the national herd, which was managed almost entirely using traditional practices.

---

significantly associated with increased prevalence of the bacterium (Table 1.8), the current state of knowledge of its ecology is insufficient to propose an approach to on-farm eradication. It is considered that risk should be managed, with attention being paid to fast-food outlets, which have been associated with a number of outbreaks of *E. coli* O157:H7 infection.

### 1.11.1 Control at farm level

Control of pathogens at farm level is seen as being of considerable importance in reduction of morbidity due to meatborne zoonotic pathogens. In some cases, notably chickens raised under intensive conditions, the level of contamination with pathogens such as

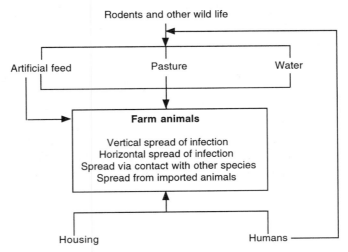

**Figure 1.9**   Possible pathways of infection of meat animals at farm level. (After Varnam, A.H. and Evans, M.G., 1991, *Foodborne Pathogens: An Illustrated Text*, Wolfe Publishing, London.)

*Salmonella* is very high at farm level and there is considerable interest in competitive exclusion as a means of reducing infection. Farm animals, however, may be infected via a number of routes (Figure 1.9) and specific procedures, such as competitive exclusion, are unlikely to be successful in the absence of general good practice and attention to hygiene.

From the viewpoint of food safety, every stage in the rearing of the animal, from birth to arrival at the slaughterhouse, is a critical point at which spread of infection should be controlled. It should be stressed that there is no conflict between requirements for food safety and good husbandry. Equally both producer and consumer ultimately benefit from good practice and the observation of basic precautions.

### (a) Feedstuffs

Feedstuffs are an obvious source of infection, especially when waste animal protein is used without adequate heat treatment. Feed can also play an important secondary role in other routes of infection. In recent years in the UK, much attention has been given to the connection between the use of potentially scrapie-infected sheep brain and nervous tissue, as cattle-feed, and the emergence of BSE (see page 39). At an earlier date, however, the practice of 'recycling' protein (such as feathers and other animal parts which cannot be utilized as human food) as animal food was recognized as being at least partly responsible for the high incidence of *Salmonella* infection in poultry and some other meat animals. Young chicks are particularly susceptible and can be colonized by *Salmonella* within a few hours of hatching. A technologically straightforward, albeit energy-expensive, means of reducing the scale of the problem is ensuring that all protein foods are adequately heat-treated. The feasibility of adequately decontaminating poultry feed using formic acid has also been demonstrated. Irrespective of the method used, the benefits of feed decontamination can easily be negated by infection from other sources.

Animals may be infected from feeds other than processed protein. Poorly made silage, especially of the 'big bale' type is recognized as a source of *Listeria monocytogenes*, which may also be acquired from naturally contaminated pasture. Pasture may also be contaminated by fertilization with animal manure or sewage sludge, by irrigation with contaminated water, or by flooding. Use of animal manure and sewage sludge has been of particular concern with respect to *Salmonella*, but while the potential importance of these routes is difficult to assess, there are some indications that the overall risk is low.

### (b) Water

Water from contaminated streams and ditches can be an important source of contamination and may well be chosen by farm animals in preference to supplies of clean water. In the case of adult cattle, such sources are thought to be of particular importance in infection with *Aeromonas* and *Campylobacter*. Infection may, however, also be acquired from sources such as boreholes, which are apparently of high quality. Widespread contamination of a broiler rearing unit with *Salmonella* was traced to water from a borehole, which had been subject to contamination of a remote, and unidentified, origin. Water supplies from boreholes should be monitored on a regular basis.

Water distribution systems within rearing units may also become contaminated. It has been suggested that *Campylobacter*, possibly introduced by birds, can persist for extended periods in distribution systems and that a regular programme of cleaning and sanitizing tanks and pipelines is effective in reducing the incidence of infection due to this bacterium.

## (c) Animal-to-animal spread of infection

Animal-to-animal spread of infection is of particular importance where housing is intensive and hygiene poor. The faecal–oral route is of greatest importance and usually involves indirect routes via contaminated litter, bedding, food or water. Infection may be either vertical or horizontal, the vertical route being of particular importance among poultry. Birds may be infected before hatching either by trans-ovarian infection or by transmission through a damaged or porous egg.

Uninfected herds or flocks are at risk from animals brought in from outside. Different sites are often used for different stages of rearing of poultry and, to a lesser extent, pigs and strict precautions are required in these cases. These include quarantine of incoming stock and ensuring that purchase is made only from a known and reputable source.

Purchase of animals at auctions, stock sales, etc. carries an inevitable risk of introducing infection. Conditions at livestock markets can be unhygienic and stressful for animals with a high risk of cross-infection. Where purchase of stock at open markets is considered necessary, strict quarantine should be observed before the brought-in animals are allowed contact with the existing herd or flock.

Cases are known where infection, including *Salmonella*, has been transmitted between different species of meat animal reared on the same farm. Control requires strict segregation of species and good hygiene. Domestic cats and dogs have also been implicated in infection of meat animals. In one example, puppies suffering *Campylobacter* enteritis allegedly infected both farm staff and cattle. Such cases, however, are probably rare and attract disproportionate attention.

## (d) Infection by rodents

Rodents may be carriers of a number of zoonotic diseases and have been implicated in infection of meat animals. Precautions against rodents in housing and feed facilities are a matter of good practice.

It should be appreciated that infection can be acquired from dead, as well as living, rodents. The omnivorous pig has a taste for dead rats and is at particular risk. Dead rats are also an important source of infection with the nematode parasite *Trichinella spiralis* under some circumstances.

## (e) Infection by other wild animals and birds

Faeces of wild animals and birds may contaminate pastures or water supplies and infect meat animals. Evidence is largely circumstantial, although direct evidence exists for transmission of *Salmonella* and *Campylobacter* by seagulls. In the UK badgers are considered to be a reservoir of *Mycobacterium bovis* and a controversial eradication programme is enforced. The source of infection of a cow with *Salmonella agona*, which led to abortion, was also badgers. Grazing animals are considered to be at greatest risk from this route of infection, but birds may contaminate water and feed supplies of animals housed indoors. The overall importance of infection by wild animals and birds is probably low but the route may be significant under local conditions. Insects have also been implicated in infection of animals but the relative importance is not known.

## (f) Infection by humans

Passive transmission of infection on hands, clothing and footwear is probably of greatest significance. This normally involves transmission within a farm, but spread of infection from farm to farm may also occur, especially if basic hygiene precautions are neglected. Direct transmission from humans to animals does occur but is rare under most circumstances.

## (g) Buildings

Contamination of the fabric of buildings is an important source of infection for successive batches of animals. Pathogens may persist for extended periods even when rigorous attempts are made at sanitation. *Salmonella typhimurium*, for example, persisted for many months in a calf-rearing unit of good construction, despite depopulation, cleaning and disinfection. Some older buildings are very difficult to disinfect and problems are generally considerably less in buildings constructed to modern standards. Even in these cases difficulties may arise due to damage to surfaces and poor maintenance.

## 1.11.2 Competitive exclusion

Competitive exclusion (the Nurmi principle) was developed as a means of controlling the incidence of *Salmonella* in poultry at farm level. The underlying principle is the establishment of an adult gut microflora at the earliest opportunity, thus providing the young bird with a degree of protection normally only available to the adult. Early work involved feeding of a saline suspension of the gut content of an adult *Salmonella*-free bird, but application of this process was limited due to fears that the suspension could transmit other human or avian pathogens. Later work has therefore been concentrated on use of cultures of micro-organisms derived from the adult gut. These cultures may either be defined, consisting of 30 to 50 strains of known identity, or undefined, in which case the number and identity of strains is not known. The exact mechanism of competitive exclusion is not yet known but the most likely mechanisms appear to involve competition for binding sites, production of volatile fatty acids and possibly other inhibitory metabolites in the caecum and competition for limiting nutrients. The establishment of the adult gut microflora does seem to involve true colonization and it is likely that a combination of mechanisms is involved.

Cultures for competitive exclusion are available on a commercial basis, the most widely used defined culture being Broilact® and the most widely used undefined being Aviguard®. Broilact® is an anaerobic culture of selected caecal microflora from screened adult birds, while Aviguard® consists of undefined intestinal bacteria from healthy specific-pathogen-free birds.

Studies have shown both defined and undefined cultures to be effective in reducing the prevalence of infected chicks following challenge with *Salmonella*. In some cases there has, however, been considerable variation between surveys, possibly because of variation in experimental conditions. Results have generally been better under experimental circumstances and the protective effect is diminished when repeatedly challenged under commercial conditions. There appears to be no benefit in repeated dosing with competitive exclusion cultures. There is, however, considerable synergy with strict hygiene, including protection of the very young bird by feeding *Salmonella*-free feed. Competitive exclusion should therefore be used as a part of an integrated anti-*Salmonella* strategy.

Despite a large number of studies of various approaches to competitive exclusion, there have been few systematic comparisons of defined and undefined cultures. A review made during 1993 suggested that defined cultures are less effective overall than undefined, especially when subject to repeated challenge, but these findings are not universally accepted.

Although originally applied to control of *Salmonella*, the competitive exclusion principle has been extended to *Campylobacter*. Defined cultures have been found to prevent colonization of 40–100% of treated birds and to reduce the population in birds which were colonized. Various defined cultures have been used, including a combination of Broilact® and 'K-bacteria' and a combination of *Citrobacter diversus*, *Klebsiella pneumoniae* and non-pathogenic *E. coli*.

An alternative approach to competitive exclusion is the feeding of carbohydrates to create intestinal conditions which favour colonization by the indigenous caecal microflora. This approach has also been combined with the feeding of competitive exclusion cultures, although the value has been doubted. Mannose and lactose have been most commonly fed, the mechanism of action being inhibition of adherence by pathogens and a lowering of the pH value by fermentation favouring the caecal microflora and increasing the inhibitory effect of volatile fatty acids. There has also been interest in the feeding of fructooligosaccharides (FOS), a mixture of tri-, tetra- and pentasaccharides (glucose and fructose), prepared by the action of β-fructofuranosidase from *Aspergillus niger* on sucrose. Caecal bacteria, especially *Bifidobacterium*, are stimulated by FOS at the expense of *Salmonella*.

### 1.11.3 Bovine spongiform encephalopathy

Bovine spongiform encephalopathy (BSE) is a relatively 'new' disease of cattle, which has attracted continuing publicity since 1990 as a result of fears that humans may acquire the disease through consumption of infected meat. Although BSE has only been recognized in recent years, spongiform encephalopathies are a well established group of infections which include scrapie of sheep and goats, chronic wasting disease of mule, deer and elk and the human infections kuru, Creutzfeldt–Jacob disease and Gerstmann–Straussler syndrome. All of these diseases are chronic, progressive and invariably fatal infections of the central nervous system. The infectious agent appears to replicate at a very slow rate, apparently with no eclipse phase, and requires many months to reach high titres. Spongiform degeneration of the central nervous system is the most important histopathological feature, neurons being the only cells which appear directly injured.

Bovine spongiform encephalopathy is caused by the agent responsible for scrapie in sheep and is believed to have crossed the species barrier to infect cattle as a result of incorporating by-products from sheep, containing nervous tissue, into cattle feed. This practice was restricted to the UK, where the disease is of greatest prevalence. Subsequently the agent has been found to be capable of infecting cats. There is no evidence that humans can contract BSE through consuming meat containing nervous tissue of infected animals. The ability of the causative agent to cross the species barrier is, however, a cause of concern. Similar spongiform encephalopathies affecting humans are also well known. Direct experimentation on infectivity to humans is obviously not possible and the long incubation period makes epidemiological evidence difficult to assess.

Bovine nervous tissue, such as brain, was previously used in a wide range of meat products. The causative agent will be present in material from infected animals and is not inactivated by conventional cooking procedures. As a precaution, and in response to public concern, the UK government banned the use of potentially infected offal (see Chapter 2, page 70) in meat products. Restric-

---

* The nature of the agents of spongiform encephalopathies is not known, although distinct differences from conventional viruses are recognized. Electron microscopy has consistently revealed two structures in tissues of infected animals: scrapie-associated fibrils (SAF) and tubulovesicular particles. Proponents of conflicting theories of the nature of the agent have attempted to relate the hypotheses to the presence of the SAF structures. At present there are two main hypotheses. The **virino** hypothesis considers the agent to be a virus-like structure constructed of specific low molecular weight (small) nucleic acid protected by host-encoded protein. The nucleic acid is not thought to have coding capacity for protein, but induces the disease. Host enzymes are responsible for replication of the infective nucleic acid. It has been implied that the SAF structures are the virinos. The **prion** hypothesis suggests that the infective agent is entirely proteinaceous and contains no nucleic acid. Prions may be seen as consisting of self-replicating protein, but the mechanism of transcription and replication is unknown. Confirmation of the prion theory would mean recognition of an entirely new class of infective agents and may violate the central dogma of molecular biology, which holds that genetic information is transferred from nucleic acids to proteins. The prion hypothesis has, however, attracted a considerable amount of support.

tion was placed on material from animals more than 6 months in age, it being considered that animals of less than this age were too young to have developed the disease. It has subsequently been shown that infectivity in the ileum may develop more rapidly than was previously thought and during 1994 the ban was extended to the intestine and thymus of calves under 6 months of age. The infectious agent may, however, be present in lymphatic tissue and removal is possible only by careful deboning. This is a European Union requirement for imported meat. In general old dairy cows are most likely to be infected and up to half the dairy herds in the UK have had one or more cases of the disease.

The emergence of BSE has led to considerable acrimony between the UK Ministry of Agriculture and consumer groups. There have also been economic implications in that the export trade in UK beef has been affected both by unofficial concern over BSE and through embargoes placed by importing governments. It is difficult to escape the conclusion that political and economic factors have served to obscure scientific facts.

---

### BOX 1.4    **Mad cows and Englishmen**

During the summer of 1994 the German government threatened a total ban on imports of UK beef. High-level negotiations followed which resulted in a withdrawal of the threat of a unilateral ban. With predictable chauvinism this was presented as a 'victory' by the then Minister of Agriculture, Gillian Shepherd. It subsequently transpired that, as part of a compromise, the UK government had accepted a European Union proposal that only cattle known not to have been in contact with BSE-infected animals could be exported. The high level of infection in British herds and the inability to trace the movements of individual animals sold through cattle markets, etc. effectively means an embargo on all British beef exports. A pyrrhic victory indeed!

---

## 1.12 MEAT ANIMALS AS SOURCES OF ANIMAL PARASITES

Meat animals are historically recognized as important sources of human infection with parasites such as *Trichinella spiralis*. In some cases, including *Tr. spiralis*, the incidence of infection of both humans and animals has fallen over the years. At the same time, meat has been recognized as a source of infection with more recently identified parasites, such as *Toxoplasma gondii*. Meat may also be a vehicle of *Cryptosporidium parvum* and *Giardia lamblia*, but the role is secondary to water and possibly other routes of infection. Animal parasites are not heat-resistant and problems normally result from consumption of undercooked or raw meat.

### 1.12.1 *Trichinella spiralis*

*Trichinella spiralis* is a nematode parasite that infects humans and a number of other hosts, including food animals and domestic pets. The most important food animal is the pig. Larvae encyst in the skeletal muscles of infected animals and if ingested cause a serious and long-lasting illness with a mortality up to 30%. Larvae are inactivated by heating to 58.3°C and such a treatment is a statutory requirement for semi-dried fermented meat products in the US (see Chapter 7, page 333). Larvae of *Trichinella* are also destroyed by holding at -25°C for 20 days. Infection of meat animals is much reduced by control of rats. The incidence of *Trichinella* infection is high in wild pigs in central Europe; and in areas such as Canada, where bears are hunted for food, the bear is a major source of infection.

### 1.12.2 *Taenia* spp. (tapeworms)

Two species of *Taenia*, *T. solium* (pork tapeworm) and *T. saginata* (beef tapeworm), are common human parasites. Cystercerci encyst in the muscle of infected secondary hosts (cows or pigs), giving the muscle a 'measly' appearance which may be detected by visual inspection of meat. Liver and other organs are occasionally infected. Ingestion of live cystercerci by humans leads to infection and the pathological condition, taeniasis. Cystercerci are killed by heating to a temperature of 60°C, or by holding at -18°C for 7 days. Man is the primary host and the parasite can be controlled by preventing cattle and pigs having contact with human faecal contamination. This is readily achieved in countries with efficient sewage systems, but can present difficulties elsewhere.

### 1.12.3 *Toxoplasma gondii*

*Toxoplasma gondii* is a protozoan parasite that can cause serious congenital infections, which may result in blindness, mental instability, abortion and death of the foetus. The domestic cat has been identified as the definitive host and many infections result from contact with feline faeces. Other domestic animals, including dogs, are also sources of human infection. Toxoplasmosis is also associated with consumption of raw or undercooked meat. Carriage rates in farm animals may be as high as *ca.* 22% of cattle, 30% of sheep and 11% of pigs. *Toxoplasma gondii* is inactivated by heating to 70°C, or by holding at −20°C for 2 to 3 days. There may, however, be a sub-population which is more resistant to freezing.

### 1.12.4 *Cryptosporidium parvum*

The importance of *Cr. parvum* as a cause of human disease has only been recognized relatively recently. The organism is a protozoan parasite that causes acute and often protracted enteritis. This is self-limiting in normal persons, but is often chronic and life-threatening in the immunocompromised.

Cryptosporidial infection may be acquired from a wide range of sources, including direct contact with infected animals. Water has been identified as the most significant source of epidemic cryptosporidiosis and the organism can survive some water treatment processes. There is no direct evidence of transmission via meat, but epidemiological evidence has identified raw or undercooked meat and offal as vehicles of infection. Infection resulting from consumption of meat and offal is likely to be sporadic rather than epidemic.

### 1.12.5 *Fasciola hepatica* (liver fluke)

Liver flukes are shared between the primary host (usually sheep but occasionally cattle or rarely pigs) and the secondary host, the water snail (*Limnaea trunculata*). Flukes infect the liver and, less

---

* *Cryptosporidium parvum* is one of a number of micro-organisms which present a major risk to persons suffering from AIDS. Infection usually develops into a prolonged and life-threatening illness, resembling cholera in symptoms and severity. Diarrhoea frequently becomes irreversible and fluid loss excessive, passage of 3–6 litres of watery fluids per day being common and as much as 17 litres per day having been reported. Extraintestinal symptoms may also occur. Cryptosporidial infection in the immunocompromised host often carries an ominous prognosis. (Current, W.L. and Owen, R.L., 1989, in *Enteric Infection*, (eds. M.J.G. Farthing and G.T. Keusch), Chapman & Hall, London.)

commonly, lungs and peritoneum of the host, causing serious and possibly fatal disease. Heavy infections are readily detected by visual inspection of meat, although an occasional fluke may still be found during consumption of liver from a lightly infected animal. The adult fluke, which is readily destroyed by heating, is not infectious to humans; infections arise indirectly from drinking contaminated water containing the cecarial stage of the fluke, or by eating plants such as watercress grown in contaminated water. In countries such as the UK commercially grown watercress is protected from contamination with *F. hepatica* and the incidence in animals has been significantly reduced by precautions at farm level.

---

### BOX 1.5   **Messing about in boats**

Snails cultivated for human consumption are produced under conditions protected from any possibility of infection with *Fasciola hepatica*. In some countries, however, wild snails are consumed with a resulting risk of fascioliasis. During the 19th century fascioliasis was a common complaint amongst boatmen navigating the Kennet and Avon Canal between Reading and Bath. This resulted from the practice of consuming, as a delicacy, a water snail which thrived on lock walls and other canal-side structures.

---

### 1.12.6 Other animal parasites

A number of other animal parasites can cause disease in humans, although direct evidence for meat as a vehicle of infection is often lacking.

#### (a) Giardia lamblia

A waterborne route of transmission of the protozoan parasite *G. lamblia* is well established and there is some indication that foods may be more important than previously recognized. A case is known where infection was attributed to the 'accidental' tasting of tripe being prepared for a cat. *Giardia lamblia* is inactivated by a heat treatment equivalent to boiling for 10 minutes but may, under some circumstances, survive freezing.

## (b) *Sarcocystis* spp.

Sarcocystes are primarily parasites of cattle and, to a lesser extent, pigs. The incidence among cattle can be very high. Occasional illness, involving mild diarrhoea, has been reported in humans, in whom infection is associated with consumption of heavily infected beef or pork.

## (c) *Balantidium coli*

*Balantidium coli* can cause a disease resembling intestinal invasive amoebiasis. Ulceration of the gut epithelium occurs and stools may be bloody and mucoid. *Balantidium coli* is primarily a parasite of pigs and many cases of human infection appear to result from direct transmission from animals. A foodborne route of infection exists where hygiene is poor and *B. coli* is now recognized as an occasional cause of travellers' diarrhoea.

# 2

# CONVERSION OF MUSCLE INTO MEAT

---

OBJECTIVES

After reading this chapter you should understand
- Pre-slaughter handling of animals and birds
- Slaughterhouse techniques
- Electrical stimulation
- Post-slaughter processing of meat
- Added-value processing
- Packaging of meat
- Meat quality and its controlling parameters
- Relationship between technology and *post-mortem* changes
- Biochemical changes during maturation
- Colour of meat during retail display
- Meat as a source of pathogenic micro-organisms
- Spoilage of meat
- Grading, quality assurance and control

---

## 2.1 INTRODUCTION

Conversion of muscle into meat marks the transition from the live animal to a food commodity. The central process, from both a philosophical and technical viewpoint, involves slaughter of the animal, although this stage should not be viewed in isolation from pre-slaughter handling or post-slaughter processing.

Conversion of muscle into meat is fundamentally a straightforward process and in less developed countries slaughter, butchery and sale are carried out at the same premises over a very short time span. In developed countries, there has been a continuing trend away from small slaughterhouses serving a local area, or even a single shop, towards very large, centralized operations. This trend has been reinforced both by the development of high-throughput, capital-intensive, automated equipment and by the increase of meat sales through supermarkets at the expense of specialist butchers. Developments in packaging technology permit extensive central butchery and preparation of meat on the same site as the slaughtering operation. Modified atmosphere packaging has led to central preparation of meat in retail packs, a development which has allowed supermarkets to reduce costs by almost entirely dispensing with in-store butchery. At the same time, centrally prepared retail packs of fresh meat are now available to small supermarkets, with no specialist knowledge of meat or facilities for butchery. To some extent, this has permitted small supermarkets to replace the neighbourhood butcher, with respect to serving those unable or unwilling to travel to an out-of-town supermarket.

As in all areas of the food industry, there is considerable interest in developing added-value variants of basic retail cuts of meat. The scope is relatively limited compared with formulated products and many added-value fresh meats, such as pre-diced stewing steak, offer the consumer nothing but marginally less preparation. Concern over the high fat and cholesterol content of meat has also led to products such as 'extra-lean' mince and cuts of meat from which most visible fat has been trimmed. These command a premium price, although the dietary benefit is dubious.

## 2.2 TECHNOLOGY

### 2.2.1 Pre-slaughter handling of animals and birds

Pre-slaughter handling of animals and birds has a profound effect on the quality, and therefore the value, of the meat. At its simplest, agitation leads to injury. This is a particular potential problem with poultry, where broken legs and wings lead to a downgrading of the carcass. Some pigs are highly stress-prone and fatalities can occur, although a more common consequence of poor handling is onset of the pale, soft, exudative (PSE) condition. This can occur even with less stress-prone genotypes of pig, while stress in other animals leads to other quality problems (Table 2.1), including a high incidence of dark cutting meat in cattle. At the same time, stress leads to increased defaecation and contamination of the bodies of animals and birds, resulting ultimately in an increased risk of contamination of the meat. The incidence of *Salmonella* in pigs

**Table 2.1**   Examples of the effects of pre-slaughter stress on meat quality

| Animal | Effect on quality |
| --- | --- |
| Cattle | High incidence of dark cutting (DFD) meat |
| Pig | High incidence of PSE meat |
| Poultry | Increased toughness of breast tissue |
| | Increased darkening and toughening of breast tissue |

and poultry can increase considerably under these conditions. It has also been suggested that stress can lead to shedding of *Escherichia coli* O157:H7 by cattle which had previously tested as non-carriers.

Problems can be minimized by maintaining, as far as possible, a calm atmosphere; and certainly the gratuitous cruelty still seen at some cattle markets can only be condemned. Animals and birds should only be transported in correctly designed vehicles and overcrowding should be avoided. Adequate ventilation is required during warm months to prevent overheating. This is again particularly important with pigs which are prone to heat stress. In some

---

### BOX 2.1   **Freedom Food**

The welfare of meat animals has long been a concern of many meat-eaters as well as vegetarians. In the UK during 1994, the Royal Society for the Prevention of Cruelty to Animals, which for many years has monitored conditions at cattle markets, launched an initiative, Freedom Food™ to give farm animals five basic freedoms:

1. Freedom from fear and distress
2. Freedom from pain, injury and disease
3. Freedom from hunger and thirst
4. Freedom from discomfort
5. Freedom to behave naturally

Farmers, hauliers and abattoir owners wishing to join the scheme have to agree to stringent conditions and accept both regular inspections and spot checks. Meat produced under conditions approved under the Freedom Food initiative is identified by a special sticker on the retail pack.

countries, such as Denmark, transporters must be fitted with mechanical ventilation systems for use during warm weather.

### 2.2.2 Slaughter and primary processing

#### (a) Cattle

The procedures in conventional slaughterhouse operations are outlined in Figure 2.1. In large-scale operations a chain system is used, in which the entire procedure is organized as a series of unit operations carried out by a number of workers. In some cases a

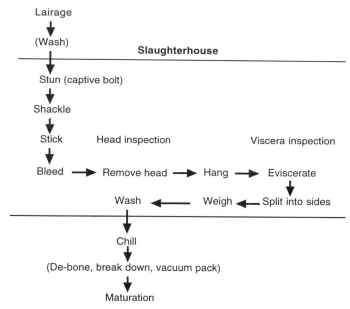

**Figure 2.1** Conventional procedures for slaughter and dressing of cattle.

* Entry into the slaughterhouse is a critical control point with respect to food safety and control of pathogens would be enhanced by preventing infected animals entering the plant. Obviously sick animals can be identified and excluded, but in many cases of infection with potential pathogens, symptoms are not obvious. According to Dr S.M. Gendel of the United States Food and Drugs Administration, DNA fingerprinting technology offers the potential to detect cattle carrying pathogens such as *E. coli* O157:H7 and thus prevent the animals entering the slaughterhouse. The value of rapid tests in this context is, however, debatable, Dr D. Hancock of the University of Oregon stating that 'thinking that a rapid test will solve the *E. coli* O157:H7 problem is believing a myth'. (Anon, 1994, *Food Chemical News*, **July 4**, 4; **July 25**, 23.)

high level of automation has been adopted. In contrast, in small operations animals are killed and dressed on a one-man:one-animal solo operation either in a slaughterhouse or, in some countries, on the farm. Conditions and facilities are generally of a higher standard in large-scale operations and, in theory, should lead to improved standards of dressing and a lower level of microbial contamination. In practice standards may well be higher in small operations. This is a consequence of the lower level of operator skill in unit operations and a higher level of handling. In some plants, the very high line speeds tend to negate good practice.

Chilling is a critical control point in determining the microbial quality of meat. The general principle is that chilling should be as rapid as possible. In the European Community, regulations require a deep muscle temperature of 7°C within 24 hours of slaughter and similar regulations exist elsewhere. Rapid cooling of beef, however, can lead to 'cold shortening' and consequent toughness (see pages 78–80). It is generally accepted that if the meat is cooled to a temperature of 10°C by the time the pH value falls to *ca.* 5.5 and rigor commences, cold shortening is probable. The conventional solution is to reduce the cooling rate, seeking a balance between the conflicting requirements of rapid cooling and prevention of cold shortening. Such a balance can be difficult to achieve and an alternative is use of electrical stimulation (see pages 79–80) to accelerate the onset of rigor.

Air chillers are commonly used for cooling beef sides. Refrigeration capacity must be sufficient to meet peak loadings under the hottest conditions and air flow must be carefully designed to permit even cooling. Design of airflow is also a compromise between the need for efficient cooling and the need to avoid excessive dehydration and weight loss. The humidity must, however, be carefully controlled to reduce bacterial growth on the meat surface. Operation of chillers must be carefully supervised and it is imperative to avoid overloading.

Beef undergoes maturation and should be held for at least a week (preferably longer) before butchery into retail joints. In traditional operations, sides remained intact up to the point of butchery, but it is now common practice to break down the carcasses into deboned primal joints, which are vacuum packed. Preparation of primal joints in packing plants adjacent to the slaughterhouse reduces refrigeration and transport costs. Supermarkets in particular also favour use of primal joints for convenience in pre-packing operations. The demand for premium, high quality meat, however, has led to a resurgence in maturing beef as sides. Field stretching, which involves suspension of the side by the hip bone, is now used by some retailers. The theory of field stretching is that beef is more tender due to the muscle being in the natural position adopted in the live animal. Field stretching is most effectively used as part of an integrated process.

### (b) Sheep

The slaughter of sheep involves a similar process to that of cattle. In older practice killing involved 'gash' cutting of the throat, but electric stunning is now common followed by sticking. Sheep are prone to cold shortening and careful control of cooling is required. A common protocol is to reduce the temperature rapidly to 12–15°C, holding at this temperature for up to 18 hours, followed by a slower reduction in temperature to below 5°C. Deboning and maturation as vacuum-packed primal joints is increasingly common practice. Electrical stimulation is also used, especially in New Zealand, where rapid freezing of whole carcasses is common.

### (c) Pigs

Pigs differ from cattle and sheep, not only in their propensity for the PSE condition, but also since the skin is not normally removed after slaughter. Pigs are stunned, either electrically or by $CO_2$ anaesthesia, before killing by sticking. The method of stunning can affect quality and electrical stunning has been stated to have an incidence of PSE meat three times that of $CO_2$ stunning. The onset of rigor is also earlier with electrical stunning, this and the higher incidence of PSE meat being attributed to an increased rate of metabolism in the muscle and vigorous spasms when the current is disconnected. Spasms can result in 'blood splash' due to rupture of minor blood vessels.

Following stunning, pigs are killed by sticking. The carcass is usually scalded shortly after sticking. The most common means of scalding is immersion in water at *ca.* 80°C. Scalding reduces the microbial load, but the scald water can be a source of thermoduric micro-organisms, which can be drawn into the carcass through the sticking hole. An alternative process to scalding involves singeing the carcass by passage through gas flames. The hair is then removed by rotating brushes and water sprays and the carcass is rinsed.

Control of cooling rate is an important factor in determining both the microbial and the organoleptic quality of pork. Chilling rate affects meat toughness, colour and water holding capacity. Cold shortening can occur as a result of cooling to a core temperature of 10°C, while the pH value is greater than 6.2. There may also be a diminution of enzyme activity contributing to toughness. Where post-mortem glycolysis is relatively rapid, quick chilling is effective in reducing drip loss and paleness. Quick chilling is not possible in all cases, however, due to the slow rate of post-mortem glycolysis in stress-resistant genotypes. Electrical stimulation is not widely used because of the risk of inducing the PSE condition. The skin of pig carcasses becomes dehydrated during chilling and, while this is minimized to reduce weight loss, the dryness of the skin is an important factor in controlling microbial growth. Following cooling, pork is often broken into deboned primal joints for distribution. The primal joints may be vacuum packed, but this remains relatively uncommon.

### (d) Poultry

Although some poultry is killed and dressed on an individual basis, the majority is processed in large-scale operations. A high throughput is required to justify the capital expenditure and there may be a considerable degree of automation. The stages of processing of chickens and turkeys are similar (Figure 2.2), although there may be variation from plant to plant.

Scalding is an important stage and is necessary to permit complete removal of feathers. Tanks of circulating hot water are the most common means of scalding, although limited use has been made of hot water sprays and steam. It is customary to use a 'hard' scald at *ca.* 60°C for birds which are subsequently to be frozen and a 'soft' scald at *ca.* 50°C for birds which are to be retailed chilled. A number of processors adopted a higher temperature for all birds, with the intention of preventing a build-up of *Salmonella* in the scald tank. It has subsequently been found, however, that a temperature of 56°C, or below, which avoids removal of the epidermis, is effective in reducing attachment of *Salmonella* to the skin, at both this and subsequent processing stages. Spoilage micro-organisms are also affected, which explains why soft-scalded birds

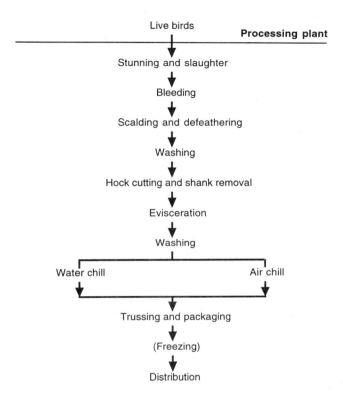

**Figure 2.2** Conventional processing of chickens and turkeys.

tend to have a longer shelf life. A number of attempts have been made to combine scalding with decontamination (see page 61).

Defeathering is now almost invariably mechanical and involves a series of rubber flails. There is considerable aerial spread of micro-organisms at this stage and defeathering should be physically separated from cleaner processes. The fingers themselves are prone to colonization by micro-organisms (see page 98) and cross-contamination can occur at this stage. The design of the defeathering plant can also affect the microtopography of the skin and the extent of attachment by *Salmonella* and other pathogens. Spray washing follows immediately after defeathering and, in addition to removing debris, is important in removal of micro-organisms. To minimize attachment by bacteria, delays in washing should be avoided.

Evisceration may be carried out manually or mechanically. It is important to minimize contamination of the carcass with intestinal contents. Equally, incomplete evisceration can lead to rapid spoilage and taints and a high level of control and inspection is necessary at this stage. Giblets are handled separately and are often placed in paper or Cellophane packets and packed with the dressed carcass. Handling of giblets can be a weak point in some plants, leading to rapid spoilage.

In the past, it was common practice to inject freshly eviscerated poultry with solutions of polyphosphates or, less commonly, NaCl. It was claimed by processors that these procedures improved quality by increasing succulence. A major economic factor, however, was the weight gain due to water uptake. This became the subject of much critical opinion from consumer organizations and injection is now less common. A further practice is the injection of the freshly eviscerated carcass with butter, vegetable oil or a mixture of fats to produce a self-basting bird.

Carcasses receive a second spray washing after evisceration. Even efficient washing brings about no more than a log-cycle reduction of microbial numbers, but the stage may be of greater importance in removal of *Salmonella*. Washing should follow directly after evisceration to minimize attachment of bacteria to the chicken skin.

The deep muscle temperature of freshly eviscerated poultry is invariably above 30°C and rapid cooling is required to prevent premature spoilage. There are two main types of chilling: water chilling and chilling in cold air. Water chilling systems consist of one or a series of large tanks, each of which can hold several hundred carcasses. Water flows continuously through the tanks, the last one of which in a series (and possibly earlier tanks) contains mechanically cooled water, or is cooled by the addition of ice. The carcasses are agitated mechanically and, in some designs, by compressed air, a common arrangement being a large Archimedean screw, mounted horizontally, which serves both to agitate the birds and to move them through the chiller (spin chiller). Systems may be through-flow in which the water and the carcasses move in the same direction, or counter-flow in which the water and the carcasses move in opposite directions. The counter-flow type is considered to have significant advantages in terms of hygiene, since carcasses leaving the system meet the cleanest water. Only the counter-flow type of chiller is permitted in the European Community (EC).

A considerable washing of carcasses may occur in the chiller, but equally micro-organisms including *Salmonella* may accumulate, leading to high levels of cross-contamination. It seems likely that, under poor management and control, water chillers may be important sources of contamination, but the potential problem can be minimized by implementation of stringent precautions during operation and by good sanitation at the end of each working period. Several codes of practice have been developed, that of the EC requiring an overall water usage of 2.5–6.0 l per carcass, depending on size, with 1–2 l per carcass in the last part of the system. The water temperature at the carcass entry and exit points must not exceed 16°C and 4°C respectively. Disinfection of water in the chillers by chlorine (added as hypochlorite), chlorine dioxide or ozone is common practice. Chlorine is most widely used, high levels (45–50 mg/l total residual) being necessary to overcome the reduction of activity due to low temperature and formation of chloramines in the presence of high levels of organic matter. Chlorination has little, if any, effect on *Salmonella* on the carcass, but is effective in preventing a build-up in the chiller water. Chlorination must, however, be used as part of an overall hygiene strategy and not as a means of attempting to correct earlier bad practice. In-plant chlorination of surfaces and equipment at levels of 10–20 mg/l residual chlorine is sometimes used as part of the strategy, but the benefit is doubtful.

In some small plants poultry is chilled by holding overnight in static tanks of slush ice. With turkeys and large chickens, this procedure may be applied as a secondary chilling stage after the spin chiller. Holding in slush ice for up to 24 hours may also be used as an 'ageing' process and allegedly improves flavour and tenderness. Numbers of psychrotrophic bacteria can increase *ca.* 5-fold during a 24-hour hold at 0°C.

Air chilling is favoured in Europe for birds which are to be retailed as fresh, unfrozen products and is also used in the UK, where one retailer will only accept air chilled birds. Primary cooling takes place in a chill room, although there may be a preliminary period in an air-blast tunnel. Risk of cross-contamination is lower, but the washing effect associated with water chilling is absent. Unless hygiene standards are high the chill room and its equipment may also be a significant source of psychrotrophic bacteria.

Liquid nitrogen sprays have been used as a means of dry chilling but have not been widely adopted. Solid carbon dioxide 'snow' has also been used as a secondary chilling method. Use of $CO_2$ snow had some popularity as a means of extending the shelf life of turkeys at Christmas and other times of peak demand. To avoid surface discolouration, the snow is added after packaging. The surface of the bird is crust-frozen, the carcass then equilibrating to *ca*. $-2°C$, giving an additional 3–4 days of shelf life. To some extent use of $CO_2$ snow has been superseded by bulk storage in a 100% $CO_2$ atmosphere (see page 68).

Following chilling, carcasses may be wrapped for sale unfrozen, or frozen (see Chapter 8, page 370), or broken down into portions. The flesh may be removed from the bones and deskinned either for retail sale or for further processing.

Processing of ducks is basically similar to that of other poultry but feather removal is much more difficult. The usual procedure is to scald the birds at *ca*. 60°C, pluck mechanically and then remove remaining feathers by dipping in molten wax at 80–120°C. The wax is set by spraying with, or immersion in, cold water and is then removed, usually by hand, together with remaining feathers.

### 2.2.3 Hot boning

Hot boning involves butchery of the carcass in the pre-rigor condition, up to 3 hours after slaughter, when the muscles are still warm (15–20°C). A number of advantages are claimed for hot boning including an increased water-holding capacity and, possibly, a more uniform colour. Refrigeration costs are lower than those for chilling whole, bone-in carcasses and there is less evaporative weight loss. Either conventional butchery or muscle seaming techniques may be used. Hot boning has been used for preparation of primal joints of beef and lamb and for pork destined either for curing or for other manufacturing. Some use has also been made of hot boning in preparation of poultry meat.

The technique is not used as widely as was originally predicted, partly because some of the theoretical advantages are difficult to demonstrate in practice. Hot boning may also be accompanied by a small amount of toughening, due to limited muscle shortening occurring just before depletion of ATP. This is partly offset by the beneficial effect on tenderness which occurs during holding at 15–20°C. There is also concern that the process favours the growth of micro-organisms, including pathogens, due to the favourable conditions existing on moist, warm, contaminated surfaces. It has been stated that problems are not insuperable but that, in many cases, little attempt has been made at resolution. This may stem from the further difficulties of operating both conventional boning procedures and hot boning in the same plant. In dedicated plant it is possible to design cooling procedures as effective as those used for conventionally butchered meat. A practical point, however, is that difficulties in handling a pliable, pre-rigor carcass may result in a higher initial level of contamination.

### 2.2.4 Electrical stimulation

Electrical stimulation of the carcass may be applied up to 1 hour after slaughter to ensure rapid onset of rigor mortis. The main purpose is to permit rapid chilling, but in some cases electrical stimulation is now applied routinely to control fall of pH value. The general principle is to apply a pulsed voltage to the carcass. Either high voltages of *ca.* 500–1000 V or low voltages of *ca.* 90 V may be applied, a typical electrical stimulation process involving 600 V at a current of 5–6 A applied in 16–20 pulses of *ca.* 2 seconds duration. Electrical stimulation is used during processing of beef and lamb, but not usually pork where there is a significant risk of the PSE condition developing. Electrical stimulation has also been applied to poultry destined for further processing.

On a number of occasions, it has been claimed that electrical stimulation enhances tenderness and may improve colour. Electrical stimulation exerts its greatest tenderizing effects when carcass cooling is sufficiently rapid to provoke appreciable cold shortening. When cooling is too slow to elicit a shortening response, there may still be an influence on toughness but the direction of the influence depends on cooling rate. When electrical stimulation is applied, chilling at a relatively rapid rate still results in more tender meat but delayed or very slow cooling results in significantly less tender meat. In an absolute sense, maximum tenderness of meat is associated with an intermediate rate of glycolysis and not the very high rate often obtained using electrical stimulation. It is suggested, in beef at least, that the goal of maximizing early post-mortem pH fall

is misguided and can lead to sub-optimal tenderness. The underlying biochemistry is discussed in pages 79–80.

### 2.2.5 Decontamination of carcasses

Decontamination of carcasses has been of interest for a number of
years. There are two main approaches. The first is a general reduction of the microbial load, with the prime object of increasing
storage life but also to reduce the incidence of pathogens. In this
context it should be stressed that decontamination must not be
seen as a means of compensating for shortcomings in slaughterhouse hygiene and poor handling practice. The second approach is
targeted at pathogens, especially *Salmonella*, and is currently of
greatest interest for application to poultry. This approach is most
effective as part of an overall strategy for reduction of pathogens.

A wide range of methods of decontamination has been devised.
These may be placed into three categories: physical inactivation or
removal, chemical inactivation or removal and ionizing radiation.

### (a) Physical methods

The simplest physical method involves high pressure water sprays.
This method is used to reduce the overall microbial load. Many
bacteria adhere tightly to red meat muscle and poultry skin and
pressures are considerably greater than those used for simple
washing, typically 100 lb/in$^2$. Temperatures of up to 75°C are used,
permanent discolouration occurring above 80°C. Evaluations of efficiency have produced variable results, but the general effect on the
microbial load is at best limited. As with conventional sprays, water
remaining on the carcass after washing enhances subsequent
microbial growth. There is also concern over the creation of
aerosols and resulting dissemination of micro-organisms. Steam has
been proposed as an alternative to water sprays, exposure for 10
seconds being effective in reducing load without discolouration.
Steam generation is, however, expensive and distribution requires
special precautions.

Use has been made of both ultrahigh frequency (ultrasonic) and
very low frequency sound to remove bacteria from the surface of
poultry. The major purpose is removal of pathogens. Poultry is
passed through baths of turbulent water incorporating a sound
source, which is intended to loosen the adherence of bacteria to

the skin. Removal is completed by the turbulent water flow. Equipment of this type is little used and appears to be of low effectivity. Cross-contamination can occur unless the operation is carefully controlled.

Use of very high pressure for decontamination of meat is a new technique, which is applied to retail cuts and mince rather than to whole or half carcasses. Experimental work has shown that pressures of 200–300 mPa are moderately effective and delay microbial growth by 2–6 days, largely through inhibition of the Gram-negative microflora. Higher pressures of 400–450 mPa reduced counts through 3–5 log cycles and apparently eliminated *Pseudomonas*. After a lag period, dependent on the intensity of processing, it was again possible to isolate *Pseudomonas*. It appeared that a sub-population comprising *ca.* 0.01% of cells survived and resumed growth after a repair phase of 3–9 days at 3°C. The effectiveness of pressure treatment can be increased by combination with other treatments, such as heating to 50°C, but it is obvious that further work is necessary to determine the value of the process in the decontamination of meat.

## (b) Chemical inactivation

Chemical inactivation of micro-organisms on meat involves use of sprays or, in the case of poultry, dips. National legislation varies and chemicals are not permitted in a number of countries. A wide and sometimes bizarre range of chemicals has been used, including chlorine, hydrogen peroxide, β-propiolactone, glutaraldehyde and various organic acids. Chlor-Chill™, a patented method of treating beef, pork and mutton, has been developed; it involves use of chlorine at 100–200 mg/kg. The method is reported to be effective both in lowering total bacterial numbers and in reducing the incidence of *Salmonella*, without causing taints.

Organic acids have attracted greatest interest for decontamination of red meats. A wide range has been used, either singly or in

---

* The inhibitory effects of organic acids are largely related to the quantity of undissociated acid present, undissociated acid being 60 to 100 times more inhibitory than dissociated acid. Undissociated acids enter the cell and dissociate after diffusion through the cell membrane. This has profound effects, including acidification of the cytoplasm, disruption of the proton-motive force, inhibition of substrate transport, energy-yielding processes and macromolecular synthesis.

combination, including acetic, adipic, ascorbic, citric, fumaric, lactic, malic, propionic and sorbic acids. Use has also been made of the inorganic acids, hydrochloric and phosphoric. Acetic and lactic acids are the most widely used and appear to be most effective.

Organic acids are applied before chilling, by spray or, in the case of poultry, by dips. Dips are recognized as being more effective but are not practical with large carcasses, though some use of dips has been made with primal joints of hot-boned beef. The effectiveness of organic acids is enhanced by application at 50°–55°C. The procedure is generally considered effective in reducing microbial load, although the extent of reduction appears variable. In general, however, decontamination is most effective when microbial numbers are high. In such a situation it has been claimed that application of 3% acetic acid at 10–15°C reduced aerobic bacteria by 99% and increased the storage life by 18 to 21 days. In the case of poultry, it has been argued that the same microbiological status can be obtained without decontamination, by improved hygiene and temperature control. Similar considerations may apply to red meat.

Initial interest in use of organic acids lay in reducing the microbial load. The effect on pathogens was extrapolated from the effect on spoilage micro-organisms and it was assumed that the incidence of pathogens would be significantly reduced. Studies on decontamination of beef, however, have shown that *E. coli* O157:H7 is remarkably resistant to acetic, citric and lactic acids, while *S. typhimurium* is resistant to a lesser extent. This may appear to be unimportant where the function of decontamination is reduction of microbial load, but there is concern that removal of the spoilage microflora may provide an opportunity for growth of pathogens.

Organic acids are very aggressive and are likely to affect the colour of meat and impart flavour notes. The extent of discolouration varies but, on some occasions at least, the appearance of treated meat is unacceptable for retail sale. This is of little consequence where the meat is to be used for further processing, but acidic flavour notes may persist and be detectable in the final manufactured product.

With poultry, reduction of the incidence of pathogens has always been of prime concern, although any possibility of extending shelf life is, of course, welcome. The nature of poultry processing offers greater scope for different approaches to decontamination and efforts have been made to incorporate a decontamination stage into existing processing operations. Attempts to reduce contamination by either raising or lowering the pH value of water in the scald tank have been made for a number of years. Such a procedure is effective in reducing the numbers of *Salmonella* and *Campylobacter* in the circulating water (and thus in lessening cross-contamination), but was no effect on contamination of the carcass. Attempts have also been made to enhance the effect of chlorination in water chillers. Chlorination is undoubtedly a valuable adjunct to hygiene, but should not be viewed as a means of decontamination of the carcass. Ozone has been used as an alternative to chlorine but overall results have been unimpressive.

Many attempts to reduce the incidence of pathogens on poultry parallel methods are used with red meat. A patented method, AvGard™ shows promise as a means of reducing the incidence of *Salmonella* and possibly other pathogens. This process, which is permitted in the US and some other countries, employs trisodium phosphate ($Na_3PO_4$; TSP). Trisodium phosphate reduces numbers of *Salmonella* by *ca.* 2 log cycles, which results in a contamination rate on treated carcasses of <5%. It is probable that TSP operates by minimizing attachment of bacteria to the skin, rather than through bactericidal properties. The AvGard™ process requires special equipment (Figure 2.3), but efforts are being made to incorporate treatment into a normal processing line. The most appropriate stage is scalding, a soft scald temperature of 50°C enhancing activity, especially against *Campylobacter*, which appears unaffected by TSP at lower temperatures. The effect of TSP on flavour and colour is minimal. Trisodium phosphate has only a limited

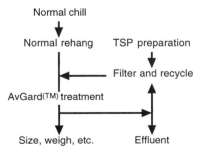

**Figure 2.3** The AvGard™ process for *Salmonella* reduction in poultry.

effect on the spoilage microflora, but is more effective against *E. coli* (including serovar O157:H7) than *S. typhimurium*. For this reason there is interest in extending the process to beef.

A further recently developed decontamination process is also attracting considerable interest in the US. This makes use of a peroxidase-catalysed reaction between iodine and $H_2O_2$. The reaction is based on use of peroxidase, donor molecules (iodine ions), low concentrations of peroxide and accelerators (sugars). In the first stage of the reaction peroxidase transfers electrons from donor molecules to $H_2O_2$. This reduces $H_2O_2$ to $H_2O$ and generates the powerful antimicrobials tri-iodide and iodine.

### (c) Ionizing radiation

Ionizing radiation at doses of *ca.* 3 kGy has been proposed for a number of years as a means of eliminating *Salmonella* and other pathogens from poultry. Such a dose, applied at *ca.* 2°C, reduces numbers of *Salmonella* through 5 log cycles and effectively eliminates the organism. *Campylobacter* is rather more sensitive than *Salmonella* and is also effectively eliminated. The technology is well established, but use has so far been limited (see below). Either radioisotopes or electron machines, such as a linear accelerator, may be used as source, although radioisotopes have been preferred for foods because of limiting electron penetration. The radio-isotope cobalt 60 has been very widely used as a source and is the basis of many current radiation facilities. Cobalt 60 is a gamma emitter, with a half-life of 5.3 years, which is produced in large quantities in some nuclear reactors by neutron bombardment of cobalt 59 over extended periods. There is doubt over the availability of cobalt 60, should irradiation become a more widely used process. Caesium 137 may be used as an alternative source but is likely to be very expensive. Despite the low penetration, electron machines are attractive in many circumstances because of their high operating rate. The performance of electron machines has been significantly improved and modern types are capable of producing electrons of high beam-energy as well as X-rays, which give the same penetration as gamma-rays at the same energy level.

Irradiation-treated poultry has been marketed on a small scale in the US and elsewhere. Birds are irradiated after packaging to avoid problems of post-irradiation contamination. Irradiation effectively eliminates the vegetative spoilage microflora and an increase in storage life is a collateral benefit. Endospores, including those of *Clostridium botulinum* survive and growth of this bacterium is a possibility. During a recent trial marketing of irradiated poultry in the US, the use of $O_2$-permeable packaging film was mandatory to obviate the possibility of growth of *Cl. botulinum*; at the same time it is necessary to ensure that the film is impermeable to both moisture and bacteria.

Although poultry is the prime concern, some investigations have been made into the possibility of eliminating pathogens from red meat. Most work has involved beef, where *E. coli* O157:H7 is causing considerable concern. Studies in the US have shown that a dose of 2.5 kGy applied to beef would inactivate $10^{8.1}$ *E. coli* O157:H7, $10^{3.1}$ *Salmonella* and $10^{10.6}$ *C. jejuni*. Although there is variation in sensitivity to irradiation associated with temperature and fat content, 2.5 kGy may be considered to effectively eliminate the main vegetative pathogens.

Despite the availability of technology for irradiation decontamination of poultry and other meat, widespread use does not appear likely in the immediate future. Consumer reaction is largely distrustful of anything related to radiation and nuclear technology, while not all scientists share enthusiasm for irradiation decontamination. This is based not on fears of induced radioactivity or production of toxic radicals, but on the use industry will make of

---

### BOX 2.2 **Radiation fever**

It is unfortunate, although probably inevitable, that the debate about use of irradiation decontamination has been hijacked by out-and-out enthusiasts on the one hand and those seeking 15 minutes of fame on the other. It is necessary to be aware, however, that underlying public mistrust is a genuine concern and that fear is none the less real for being irrational. Public distrust of the food industry and its scientists is, regrettably, partly due to a general anti-science attitude. At the same time the handling of concern over *Listeria*, *S. enteritidis* and bovine spongiform encephalopathy is very far from exemplary. It is also salutary to remember that respected food scientists once supported the use of tetracyclines to suppress bacteria on poultry carcasses and argued publicly in favour of food colours now recognized as potential carcinogens.

irradiation. 'Cleaning up' badly contaminated foods and permitting excessively long storage lives may well be the major attraction for industry rather than reducing morbidity due to elimination of pathogens.

### 2.2.6 Conditioning (maturation) and tenderization of meat

It has long been recognized that conditioning of meat by storage increases post-cooking tenderization. The rate of tenderization varies considerably amongst animal species. At temperatures of *ca.* 5°C at least 14 days is required for 80% of tenderization to be achieved for beef, in contrast to only *ca.* 5 days for pork. The tenderization period for lamb is intermediate between these two extremes, while that of poultry is very rapid and usually complete within *ca.* 48 hours at 5°C. Conditioning requires little positive action by the processor but it is necessary to ensure that the temperature is controlled at a maximum of 5°C. The relative humidity must also be controlled unless the meat is packed in vacuum packs or other impermeable film. A high humidity leads to rapid microbial growth, while a low humidity is economically unacceptable due to excessive drying and weight loss.

Tenderness is an important factor in consumer perception of meat quality. In countries such as the US where meat is a major dietary constituent, there is a long-standing preference for steak over ground beef and roasts. Less than 25% of carcass weight, however, produces high quality steaks, with resulting high costs. Attempts have been made to use lower grade material, such as beef chuck, traditionally marketed as a low priced steak or roast and described as the most underutilized wholesale cut. Various approaches have been taken to upgrading lower grade meat. Proteolytic enzymes of plant origin such as papain, ficin and bromelain have been used, either injected into the animal immediately before slaughter, or applied to the meat surface. Such enzymes have greatest activity at higher temperatures and most tenderization takes place during the early stages of cooking, the enzyme subsequently being inactivated. Proteolytic activity can be demonstrated, but tends to lead to a softening distinct from 'natural' tenderization. There is also no effect on collagen. An effective collagenase is available but its source, *Clostridium perfringens*, makes it unacceptable for use in foods. Despite a number of attempts to market enzyme-tenderized joints and steaks under brand names, the process is not widely used.

Attempts have also been made to enhance tenderness of beef and lamb by injection of calcium chloride into the muscle, either pre-, or post-rigor. The rationale is the stabilization of calpain activity (see pages 88–91) but the procedure is generally considered ineffective. Problems also arise due to bitter or salty notes, which persist to consumption.

Marinading in a mixture of acetic and lactic acid has been used in tenderization of particularly recalcitrant meat such as the drumsticks of spent layers, which are of very high collagen content. This limits use even for manufacturing. Marination is effective but imparts acidic flavours, which are detectable in the end-product. Marination is only practical, therefore, when the acidic flavours are compatible with the nature of the end-product.

A variety of mechanical methods have been proposed for meat tenderization, most of which are ineffective. Blade tenderization involves the penetration of the muscle by thin blades, which disrupt the structure. The effectiveness of blade tenderization has been demonstrated with post-rigor muscles, such as longissimus, after 7 days storage. The life of blade tenderized muscle, however, is likely to be very short due to high drip losses.

Tenderization of meat is of particular interest in the use of high pressures in food processing. It has been suggested that tenderness of both pre-rigor and post-rigor meat can be improved by pressure treatment. The effect appears to be greatest in pre-rigor meat, where weakening of myofibrils can be demonstrated by both optical and electron microscopy. A combined treatment of high pressure and heat treatment to 45–60°C has been proposed to improve tenderness in beef subjected to wide variation in post-slaughter handling. Pressure tenderization, if practical on a commercial scale, is likely to be used for up-grading of poor quality meats. As such, the substantial additional cost must be balanced against the higher value of upgraded meat and it is considered unlikely that the consumer will benefit rather than the profits of the processors.

## 2.2.7 Long-term refrigerated storage of fresh meat and poultry

Although commonly used methods, such as vacuum packing of primal joints, provide a sufficiently long storage life for domestic needs in the UK and many other countries, there is continuing pressure for longer storage lives. The most obvious means is freezing (see Chapter 8, pages 368–370) but frozen meat is perceived as being of low quality and is either sold at a lower price or defrosted and sold as fresh. In the latter case, it is often 'forgotten' that the meat was once frozen. This practice is considered unethical, if not dishonest. A considerable amount of vacuum-packed chilled meat does enter international commerce, although there can be problems with spoilage.

---

BOX 2.3   **The townsman's prayer**

Although canned meat, of very dubious quality, was imported from the late 1860s, large-scale importation of meat into the UK had to await the development of steamships with refrigerated holds. The first really successful shipment of frozen beef and mutton was made, from Melbourne to London, by the SS *Strathleven* in 1880. Within a few years large-scale importation of New Zealand lamb, Argentinian beef and American pork was an established feature of trade and resulted in an effective halving of the cost of meat. This led to a revolution in the working class diet and a greatly increased meat consumption. In 1902 the value of imported meat reached £50 million and consumption per head was more than 56 lb per year. The working class diet also benefited from a significant fall in bread prices due to cheap wheat, imported from the New World and Australasia. Although a demand for high-quality home-produced meat continued amongst the wealthy, UK farmers inevitably suffered. There was, however, little sympathy for the farmer amongst the working class, who had long memories of high prices and poor diet. This is quite apparent from the townsman's prayer: 'God speed the plough in every land but our own.'

---

Packaging in an atmosphere of 100% $CO_2$ together with storage at *ca.* $-1.5°C$ is now recognized as a means of extending the storage life of fresh meat to 16 weeks, or more. Successful use of 100% $CO_2$ atmospheres requires use of a high purity gas and a very low residual $O_2$ concentration to prevent metmyoglobin formation. This requires that the $O_2$ concentration at sealing should be a maximum of 0.1% and preferably no more than 0.05%. Sufficient $CO_2$ must be added to saturate the tissues. The packaging must be gas impermeable to maintain the $CO_2$ concentration throughout the prolonged storage period and flexible film packaging incorporating a metal foil gas barrier is used.

To obtain the best results it is necessary to develop an entire integrated process. This may be illustrated by reference to the Captech® process developed by the New Zealand Meat Research Institute. The process was originally developed to permit a minimum 16-week storage life for lamb but may equally be applied to other meats. A specially designed combined chamber–snorkel machine, operating under computer control, is used for evacuation of air and gassing with $CO_2$. Phased evacuation of the chamber and pack avoids air entrapment and obviates the need for flushing. Mechanical stresses on the meat and pouch are avoided. The packaging machine is equipped with on-line sensors, which abort the operating cycle if problems occur. The meat is packed in triple laminate pouches (Figure 2.4) containing an aluminium foil gas barrier layer. These pouches are of sufficient mechanical strength to avoid splitting or puncture during transport and seal well even when the sealing surfaces have been contaminated with meat. Additional protection is provided by strong cardboard outer cases. A certain amount of drip is inevitable and an adsorbent is required to prevent staining of fat. In the case of lamb, the adsorbent is combined with bone guard to prevent damage to the pouch. At present Captech® and similar processes are used for primal joints of meat, but it is probable that the principle will be extended to pre-packed, branded consumer packs.

Fresh turkeys present a different problem in that a high percentage of total annual sales are made in the days immediately before Christmas. A longer than normal life is required to permit sufficient stocks to be built up to meet demand. This is a matter of days

(a)
(b)
(c)

**Figure 2.4** Structure of triple laminate pouch used in the Captech® packaging system: a = high-strength plastic film; b = aluminium foil; c = low-strength plastic film.

rather than weeks, however, and less rigorous conditions are required than those for beef, lamb, etc. A number of approaches have been taken, including crust freezing with $CO_2$ snow, but these represent only a partial solution. Use is now made of bulk storage at *ca.* $-1.5°C$, in an atmosphere of 100% $CO_2$, economics dictating that the cold store must be filled and emptied in single operations. After removal from $CO_2$ storage, the birds enter the normal distribution and display system in air-permeable packaging.

### 2.2.8 Meat at retail level

*(a) Red meat*

In traditional small butchery operations, meat is purchased, maturation is completed on the premises and the meat is butchered into retail portions largely to customer requirements. No packaging is required other than a simple paper or film wrap. The growth of meat retailing through supermarkets led to a greatly increased use of pre-preparation. Cuts and joints are prepared in advance, placed on expanded polystyrene trays and wrapped in an air-permeable film for sale from refrigerated cabinets. A certain amount of centralized preparation was introduced, but the relatively rapid deterioration in appearance of prepackaged meat stored in air meant that a significant amount of preparation continued in individual stores.

Modified atmosphere packaging (MAP) permits a significant extension of retail shelf life in terms of appearance and also markedly delays spoilage. Modified atmosphere packaging (which may also be referred to as controlled atmosphere packaging) involves packaging retail cuts of meat in an atmosphere containing $CO_2$ to inhibit microbial growth and $O_2$ to maintain the desired colour of meat. Many gas mixtures have been tested, but the most common for packaging retail cuts is 20–40% $CO_2$ with 60–80% $O_2$. Meat is packaged in highly impermeable material, which usually consists of a pre-formed tray heat-sealed to a lid. Drip can be a problem and it is usual to include an absorbent pad.

The scope for 'adding value' to red meat is limited, but not beyond the capabilities of the imaginative technologist. The simplest means is increasing the extent of butchery operations and thus minimizing the amount of meat handling in the home. Stewing meat, for example, may be cut into cubes before packaging, while central preparation of pieces of meat for kebabs enables the retailer not just to add value, but also to exploit the barbecue market and counter sales from take-away operations. A demand for variety can be met, especially with pork and beef, by retailing suitable cuts after marination in various flavoured sauces. An alternative approach is to retail meat co-packaged with a sachet of marinade.

Concern over heart disease has led to an increased demand for lean meat. This is reflected in the increased demand for chicken at the particular expense of beef and lamb. Amongst beef eaters, however, a demand exists for joints with most fat trimmed off and for mince of lower fat content. Such cuts may be viewed as a type of added value product, since the removal of fat is reflected in a proportionate increase in price. Trimming obviously reduces the overall fat content, but has little effect on cholesterol levels.

### (b) Poultry

With the exception of New York dressed turkeys and ducks, the great majority of whole poultry is packed in oxygen-permeable film at the processing plant. Attempts have been made to use vacuum packaging but have been unsuccessful due to the unpleasant appearance of the packaged bird. Poultry pieces and poultry meat are also prepared on a centralized basis, packed on polythene trays and covered in an air-permeable overwrap. As with red meat, value may be added by extending centralized butchery to simple operations such as cubing, usually carried out in the home. Whole birds are available which have been pre-prepared for cooking by addition of herbs, lemon slices, etc. and poultry pieces are available which have been flavoured by marinading, or by addition of spice mixtures. This type of product, however, is less common than its cooked equivalent.

## 2.3 HANDLING OF BY-PRODUCTS

In some cases no more than 50% of liveweight of a meat animal is converted into carcass meat. This value, which is often known as the killing-out percentage, is fairly common in sheep. The killing-out percentage for cattle is slightly higher (*ca.* 55%) than for sheep, but significantly lower than for pigs (*ca.* 75%). This is a consequence of the smaller size of the pig's (non-ruminant) digestive tract and the edible nature of the skin and head meat.

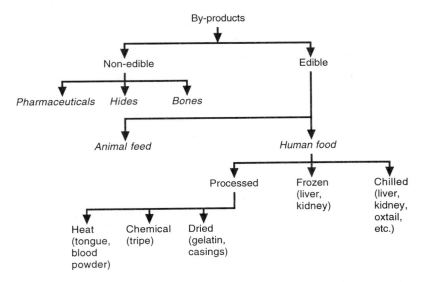

**Figure 2.5**　Types of meat industry by-products.

By-products are usually classified as non-edible and edible (Figure 2.5). Non-edible by-products include bones and (in the case of cattle and sheep) hide. Edible by-products include organ meats and tissues, such as liver, kidney and sausage casings (made from the gastrointestinal tract of cattle, pigs and sheep), as well as blood. Use of edible by-products (offal) is subject to restrictions in many countries. Mammalian liver, kidney, oxtail and heart, as well as avian giblets (gizzard, liver, heart, neck) may be sold for consumption without processing. These offals, as well as diaphragm, head, etc., may also generally be used in any meat products without processing and, in the UK, count toward meat content. Other offals, including feet, lungs, intestines, etc., must be processed before use and do not count towards meat content. In the UK concern over bovine spongiform encephalopathy (BSE) has led to severe restrictions on the use of offal in meat products for humans (see Chapter 1, page 40).

## 2.4 MEAT INSPECTION AND QUALITY ASSURANCE

Inspection of meat is commonplace in industrialized countries. In those such as the UK, meat inspection was introduced to prevent the sale of unfit meat and this function is still an element of meat inspection today. The historic purpose of meat inspection varied, however, from country to country. In several European countries, where inspection is by veterinarians, the prime function was animal welfare, while elsewhere it was seen as a means of preventing fraudulent practice. In many countries, especially those with major export interests, meat is also inspected as a quality assurance function to ensure that the expectations of the customer are met. This function is distinct from that of ensuring that the meat is fit for consumption and is invariably linked to a carcass grading scheme.

Historically meat inspection to ensure the fitness of meat for sale involved visual detection of meat from animals suffering diseases, such as tuberculosis, or which were infected with parasites. Inspection takes place immediately after slaughter in the early stages of dressing. It has been argued, particularly from within the industry, that meat inspection in its current form is not justified. This stems from the virtual elimination of diseases, such as tuberculosis, which can be detected by examination of the carcass. Irrespective of this, it may be argued that the presence of a meat inspector is a disincentive to bad practice. Further, there is growing pressure in many countries for a widescale introduction of HACCP, currently little used in the meat and poultry industry, which could result in a broader if rather different role for the meat inspector.

Inspection of meat for grading purposes is a means of maintaining quality and is considered essential for the export trade. Grading schemes are operated on a national basis, but there are many

---

### BOX 2.4  A lawyer forced to go to law

In July 1994, a group of Iowa farmers followed the example of a private cattle feeder in filing a suit against US Agriculture Secretary Mike Espy. This concerned the zero tolerance policy towards faecal matter applied in interim guidelines for cattle, but not poultry. Unlike the cattle feeder the Iowa farmers are not seeking compensation, but that the court enters an order 'declaring the Secretary's regulatory scheme and requirements for the control and removal of pathogenic contaminants, including faeces and ingesta, from meat and poultry, arbitrary, unreasonable, and an abuse of discretion and not in accordance with law'.

---

BOX 2.5  **The Laws of Robotics**

In Denmark, the ability to grade pig carcasses accurately is considered essential to the prosperity of the meat industry. Grading has been automated for a number of years, with robotically positioned fibre-optic probes being used to determine a reflectance profile relating to the meat content of each carcass. This value was previously converted to a percentage meat content using a computer-based expert system. The Danish Meat Research Institute at Roskilde has subsequently developed a neural network (a computer system that mimics the human brain by 'learning') to control the pig grading system. Using the network to determine the positioning of probes and calculate the percentage meat content has resulted in 98% of gradings being correct using 9 probes. This compares very favourably with the expert system which graded 90% of carcasses correctly using 17 probes. Over a two-year period *ca.* £500 000 has been saved in maintenance costs, while it is estimated that the potential benefit to sales, through accurate carcass grading, is *ca.* £2 million per year. (Geake, E., 1994. *New Scientist*, **15 January**, 17.)

---

similarities. Meat is usually inspected when the carcass is fully dressed and often after breaking down into primal joints. Parameters include carcass conformation and correct ratio of lean meat to fat, meat content, colour and overall appearance and standard of butchery. Grading is primarily the work of expert graders and may include a considerable amount of visual assessment. There is, however, substantial interest in grading carcasses on the basis of tenderness. Various instruments based on fibre optics appear to be the most promising for routine use.

At present microbiological analysis of carcasses and primal joints of meat is not considered necessary on a routine basis. As noted above, however, this may be changed by developments in methodology. A possible exception is meat that has been stored for very long periods in $CO_2$, where microbial counts have been used as predictors of retail shelf life, but the validity of such analysis is uncertain.

Very few quality assurance procedures have been applied to meat at retail level, other than informal inspection to ensure standards of butchery and presentation are acceptable. In some cases butchery and presentation standards of centrally prepacked meat remain low, butchery is of a poor standard and bone dust (produced during use of saws in butchery) as well as bone chips are common contaminants. The adoption of MAP-packed retail cuts by multiple retailers has, however, led to a general improvement of standards. An improvement may also be expected if use of brand names in fresh meat retailing is extended. Microbiological analysis of MAP-packed meat at point of packing is common practice as part of general quality control. Chemical analysis is not normally a part of quality control at this level, an exception being the determination of the fat content of mince to ensure that legislative requirements are met.

Despite the existence of formal meat inspection and grading schemes, these are largely imposed on the production process and, in many cases, no structured quality assurance scheme exists. Quality assurance systems, such as HACCP, are not generally used in the processing of red meat and, while there has been some application, use also remains limited in poultry processing. Introduction of HACCP is being widely promoted in the US as a means of improving hygiene standards and reducing contamination with pathogens. Similar principles can be applied to assuring the quality of meat.

Buying in meat from wholesalers can cause difficulties. The most satisfactory means of quality assurance is to purchase meat only where the originating processor is known to operate under good conditions. Ideally specifications based on HACCP should be agreed. In other circumstances, it may be necessary to use microbiological analysis to assess conditions at slaughter and during subsequent handling. Where doubt exists a method of determining meat species should be available.

* A fuller knowledge of changes in beef quality with respect to time and process conditions would enable accurate prediction of the effects of processing on quality. Selection of optimum processing conditions would then allow beef of the highest quality to be obtained with minimum chilling time. Kinetic models of changes in quality indices during chilling have been developed and used to obtain criteria of chilling for accelerated beef processing. Optimum chilling conditions were obtained using a multi-parameter optimization programme. (Mallikarjunan, P. and Mittal, G.S., 1994, *Journal of Food Science*, **59**, 291–302.)

## 2.5 CHEMISTRY

### 2.5.1 Post-mortem glycolysis

The death of the animal at slaughter initiates metabolic processes in muscle which alter its *in vivo* nature. When circulation stops, muscles are no longer able to obtain energy by respiration, since mitochondrial activity ceases with depletion of internal oxygen. In response glycogen, the main energy store in the muscle, is converted to lactic acid by anaerobic, post-mortem glycolysis. This reaction provides the necessary energy for local attempts to maintain structural and functional integrity. Post-mortem glycolysis involves the Embden–Meyerhof pathway, glycogen being converted through a series of phosphorylated 6-carbon and 3-carbon intermediates to pyruvate, which is then reduced to lactate.

Glycolytic enzymes were originally thought to be localized in the cytoplasmic phase of the cell. This concept was first propounded in 1934 and persists today despite the binding of cytoplasmic enzymes to the I-band being demonstrated as early as 1962. The I-band is not, in fact, the only site at which glycolytic enzymes are bound, but it appears to be of greatest importance. The major structural protein of the I-band is F-actin and, under conditions of low ionic strength, the glycolytic enzymes have a considerably greater affinity for this protein than others. Physiological adsorption is primarily to the thin filaments of the I-band, where F-actin occurs as a complex with tropomyosin and troponin.

Binding of glycolytic enzymes is a dynamic process, in which enzyme proteins can move from the cytoplasmic phase to a particulate binding site in response to a stimulus, such as electrical stimulation. The quantity bound is directly related to the work done and a significant factor is that the ability of glycolytic intermediates to modulate the extent of enzyme binding is important in control of glycolysis. The intermediate fructose diphosphate, for example, can release aldolase. This ability reflects the relevance of binding in respect of the altered kinetic properties of the glycolytic enzymes. In general terms, movement of enzyme from cytoplasm to myofibril provides for accelerated enzyme activity. This is demonstrated by the increase in binding of the four principal glycolytic enzymes in response to electrical stimulation.

In comparison with respiration, post-mortem glycolysis is an inefficient means of providing energy; it is essentially the resynthesis of adenosine triphosphate (ATP) from adenosine diphosphate (ADP). The yield of ATP during post-mortem glycolysis is only 2-3 mol ATP per mole of glucose, compared with 36 or 37 mol during aerobic respiration. In the immediate post-mortem period some ATP is regenerated by conversion of creatine phosphate to creatine and transfer of its phosphate to ADP. Some ATP may also be generated by the adenylate kinase system, which converts two molecules of ADP to one molecule of ATP and one molecule of adenosine monophosphate (AMP). None of the mechanisms of post-mortem ATP generation are able to maintain ATP levels for more than a limited period. Adenosine triphosphate is gradually depleted by various ATP-ases, some derived from the contractile proteins, but most from the membrane systems. Glycolytic activity finally ceases either due to exhaustion of glycogen reserves or, more usually, due to the fall in pH value from *ca.* 7.2 to *ca.* 5.5, which accompanies glycolysis.

The fall in pH value can be closely correlated with lactate production, although the hydrogen ions generated come from hydrolysis of ATP and not from the production of lactate. The enzymes responsible for glycolysis are progressively denatured as the pH value falls towards 5.5 and the isoelectric point of the proteins. At the same time, events such as the deamination of AMP remove essential cofactors from the muscle system.

A dramatic consequence of the cessation of post-mortem glycolysis and the fall in ATP (and ADP) levels is the onset of rigor mortis. This occurs when ATP levels are no longer sufficient to serve as plasticizers for actin and myosin. At this point actin and myosin interact to form inextensible actinomyosin, while stiffness develops due to the tension developed by antagonistic muscles. Not all muscles stiffen, but all become inextensible when ATP is depleted. In contrast to the situation in living muscle, ATP levels fall effectively to zero and there is no plasticizing effect whatsoever. At this point attachment between the myosin heads and the thin actin filaments becomes very strong and the filaments are drawn closer together. This results in further stiffening of the muscles.

Post-mortem glycolysis has other effects, several of which have consequences for the quality and properties of the meat. The lack of energy prevents resynthesis of protein molecules. Those present begin to denature and become susceptible to attack by endogenous proteinases, including the calpains and the cathepsins. This leads to tenderization and is discussed in greater detail on pages 86–92. There is also a tendency to oxidation of the meat pigments to the undesirable metmyoglobin and concomitant oxidation of fats (see pages 93–94), while there is an increasing concentration of the precursors of cooked meat flavour.

### (a) Factors affecting the rate of post-mortem glycolysis

The rate of post-mortem glycolysis is affected by a number of intrinsic and extrinsic factors (Table 2.2). The main intrinsic factors are species of animal, genotype, age, type of muscle and location within it. These factors cannot be controlled by manipulation at the slaughterhouse. Extrinsic factors are, however, amenable to manipulation. The most important are: pre-slaughter drug administration; temperature during post-mortem glycolysis; and contrived factors, such as comminution, salting, imposition of pressure and oxygen tension.

### Intrinsic factors

For a given muscle, held under controlled conditions, the rate of post-mortem glycolysis is generally faster in the pig than in cattle or sheep. In the pig particularly, the breed and genotype within the breed are important determinants of rate of glycolysis. Certain breeds of pig, notably Landrace, Hampshire and Piétrain, are prone to production of PSE pork. The PSE condition results from a high level of denaturation of meat proteins. This can be a consequence of two phenomena: rapid post-mortem glycolysis leading to a

**Table 2.2**  Factors affecting the rate of post-mortem glycolysis

| Intrinsic | Extrinsic |
| --- | --- |
| Animal species | Pre-slaughter drug administration |
| Genotype | Post-mortem pithing and shackling |
| Age of animal | Post-mortem temperature |
| Type of muscle | Use of electrical stimulation |
| Intramuscular location | Post-mortem comminution |
| | Post-mortem salting |
| | Post-mortem pressure |
| | Post-mortem oxygen tension |

*Note*: Data from Lawrie, R.A. (1992) In *The Chemistry of Muscle-based Foods*, (eds D.E. Johnston, M.E. Knight and D.A. Ledward), Royal Society of Chemistry, London.

muscle pH value of *ca.* 5.5 while the carcass temperature is still high, or post-mortem glycolysis leading to an abnormally low ultimate pH value of *ca.* 4.8. The attainment of a low pH value while the muscle is still warm leads to a high level of protein denaturation. When myosin is denatured the head length shortens from *ca.* 19 nm to *ca.* 17 nm. This relatively small change appears to be sufficient to induce a very marked contraction of the space between the thick and thin fibres of the myofibril. Water-holding capacity is much reduced by conditions in PSE muscle and exudation significantly increased.

The PSE condition is controlled by at least two genes. One gene promotes a high rate of post-mortem glycolysis and is associated with sensitivity to the anaesthetic halothane. Sensitivity to halothane can be used as a marker for potential development of the PSE condition. The other gene, which is not associated with halothane sensitivity, promotes an abnormally low pH value, although the rate of pH fall is not usually excessively rapid. The PSE condition in Landrace and Piétrain pigs, which are often halothane-sensitive, stems from a high rate of post-mortem glycolysis while that in Hampshire pigs, which are not usually halothane-sensitive, stems from an abnormally low ultimate pH value. The PSE condition is not an all-or-nothing phenomenon, however, and there can be considerable variation in post-mortem glycolysis in pigs of the same breed and age, held under standardized conditions. This may be attributed to other genotypic differences.

Poultry meat enters rigor extremely rapidly. Breast muscles of commercially processed chickens, for example, may enter rigor within an hour of death, although there can be wide variations within a single flock. It is recognized that rate of glycolysis can vary and that both the PSE and DFD conditions can occur. The situation with poultry differs to that with mammals in that post-mortem changes such as ATP catabolism and lactic acid accumulation do not appear to be related to rigor mortis development.

The muscle type is of considerable importance in determining the rate of post-mortem glycolysis. This stems from evolutionary differences resulting from the different roles of 'red' and 'white' muscles. 'Red' muscles, adapted to carry out slow, prolonged activity, have a much higher level of respiratory enzymes than 'white' muscles which, adapted to fast intermittent activity, have the prerequisites for efficient anaerobic metabolism. This involves high levels of ATP-ase activity, a high content of creatine phosphate and glycogen and a highly developed sarcoplasmic reticulum to control glycolytic activity through rapid release and recapture of $Ca^{2+}$ ions. For these reasons the rate of post-mortem glycolysis can be significantly greater in 'white' than in 'red' muscles. Differences in rate of pH fall between muscles may, however, be partly obscured by location of the muscle within the carcass and consequent differences in cooling rates. In cattle, at least, there are also likely to be higher rates of post-mortem glycolysis in older animals.

*Extrinsic factors*

The most important extrinsic factor, especially with respect to the relationship between meat quality and post-mortem processing, is temperature during glycolysis. The rate is high at *in vivo* temperatures of 37–39°C, decreasing with falling temperature, but increasing again as the temperature approaches 0°C. Shortening of unrestrained muscle at *ca.* 37°C has been recognized for many years ('high temperature rigor'; 'warm rigor'), but shortening and associated toughening also occurs at low temperatures and is commonly known as 'cold shortening'. Cold shortening can equal shortening at *in vivo* temperatures and was recognized as a problem in the meat industry during the 1960s, when the rapid chilling of lamb carcasses became common practice. While the fall in rate of glycolysis between *in vivo* and ambient temperatures is readily explained in terms of enzyme kinetics, it is obvious that a further factor must be involved in the apparently anomalous situation at low temperatures. In the 'red' muscle of cattle, sheep, pigs and rabbits, the rate of ATP turnover is as high at 2°C as at 15°C. This is explained by the release of $Ca^{2+}$ ions from the tubes of the sarcoplasmic reticulum, leading to a 30–40-fold increase in the concentration of the ions in the sarcoplasm and a massive stimulation of contractile ATP-ase of the myofibrils and their consequent contraction. Release of $Ca^{2+}$ ions from the sarcoplasmic reticulum actually commences when the temperature is *ca.* 25°C, but the concentration only reaches a sufficiently high level to cause contraction below 11°C. Discharge of $Ca^{2+}$ can be reversed fairly readily in 'white' muscles, which have a well developed sarcoplasmic reticulum. For this reason, 'white' muscles are less prone to cold shortening than 'red' muscles, which have a less developed sarcoplasmic reticulum. The whiter muscle of pork, for example, is less likely to undergo cold shortening than the redder muscles of beef and lamb. Cold shortening and associated toughening does, however, occur in some pork muscles, where glycolysis is accelerated by very rapid chilling.

Cold shortening has been described as a biphasic process. In meat of high ATP level and pH value, cold shortening of excised muscle occurs very rapidly, the shortened condition being maintained to the onset of rigor. The shortened muscle may be induced to relax for a time by application of loading, but gradual shortening recommences as the second phase commences, leading to rigor. The initial rapid phase of shortening can be prevented by increasing the mechanical load on the excised muscle, although this has no effect on the final degree of shortening in the second phase. More work is done during the second gradual phase than during the initial rapid phase of shortening. In carcass meat, skeletal attachment places restraint on the muscle and the initial shortening phase is not usually apparent. In practice this means that the second phase is of greater importance in determining the extent of cold shortening of carcass meat. Even where cold shortening is prevented by mechanical restriction, however, severe crimping of individual myofibrils often occurs. This is a consequence of localized cold shortening occurring in some, but not all, individual muscle fibres.

Electrical stimulation is used to prevent cold shortening, by accelerating post-mortem glycolysis until the pH value has fallen to *ca.* 6.2. Cold shortening does not occur below this pH value and rapid chilling may then be applied. Electrical stimulation is associated with sarcomere shortening and with rapid ATP breakdown but, under most conditions, shortening is transitory. This is because at the prevailing *in vitro* temperature of the carcass the sarcoplasmic reticulum is able to recapture the $Ca^{2+}$ ions released

---

* The effect of temperature on post-mortem glycolysis is not simple. A number of factors are potentially involved, the balance, and often a very fine balance, between these determining the observed outcome. An example lies in the observation that while increasing the intensity of prerigor chilling, in the absence of conditions predisposing to cold shortening, usually reduces the extent of denaturation of myosin, there is little effect if the intrinsic rate of post-mortem glycolysis is high and there is a rapid onset of rigor mortis. Further the extent of denaturation of myosin at *in vivo* temperatures is greater in muscles with low rates of post mortem glycolysis. These observations have been explained by hypothesizing that the denaturation of myosin is markedly inhibited by actin when the two combine to form actinomyosin. (Offer, G., 1991, *Meat Science*, **30**, 157–64.)

by electrical stimulation. This removes the contraction stimulus at a stage when ATP and ADP levels are high enough to plasticize and relax the sarcomeres. At the same time it is important to appreciate that electrical stimulation cannot totally 'overcome' temperature related effects. It is for this reason that electrical stimulation is most effective at intermediate rates of chilling. The extremes of high temperature rigor and cold shortening are thus avoided, allowing the physiological benefits of electrical stimulation to show to best advantage.

An alternative to electrical stimulation as a means of minimizing the effects of cold shortening is to alter the method of carcass suspension. This redistributes tensions within the muscles so that shortening can be prevented in the most important parts of the carcass. In beef carcasses hung in the conventional position (vertically from the achilles tendon) the muscles of the outer hindquarter such as the semimembranosus, the semitendinosus, the biceps femoris and the longissimus dorsi are relatively free to shorten during rigor, while many of the forequarter muscles are passively stretched and the psoas major fully stretched. Hanging the carcass in a horizontal position from the obturator foreamen, however, reverses the situation. Muscles such as the longissimus dorsi enter rigor in a stretched position, while forequarter muscles and the psoas major are free to shorten.

It has proved possible to chill carcasses of relatively small bulk, such as those of lambs, before the onset of rigor and without use of electrical stimulation and still avoid cold shortening. The key lies in chilling in air at $-20°C$, the absence of cold shortening being attributed to skeletal restraint of the carcass and 'hardening' of the surface without freezing.

Although temperature is of major importance in determining the course of post-mortem glycolysis and related events, other extrinsic factors can be involved. Pre-slaughter administration of drugs, including hormones, can alter the course of glycolysis. Calcium salts, for example, increase the rate of glycolysis, while magnesium salts have the opposite effect. Slaughterhouse procedures can affect the rate of glycolysis. The process of pithing, which is used to destroy the central nervous system at slaughter, accelerates post-mortem glycolysis. Glycolysis can also be accelerated by electrical stunning, this being of particular significance with pigs since the PSE condition may result (see pages 76–77).

Other environmental factors can affect the rate of glycolysis, although some are of little or no practical significance. Comminution of pre-rigor meat and consequent damage to the structure leads to increased contact between $Ca^{2+}$ ions, enzymes and substrates and rapid glycolysis-associated changes. Comminution in the presence of NaCl enhances the rate of glycolysis, but cold shortening does not occur even though the ATP level is depleted. This procedure has found some application in meat for the manufacture of burgers.

Respiration can be re-established by subjecting meat to high oxygen tensions. This effectively reverses post-mortem glycolysis, lactic acid is removed from muscle tissue and ATP and creatine phosphate are synthesized. It is thought that this process occurs at the extreme peripheral edge of meat surfaces even under normal atmospheric conditions.

The rate of post-mortem glycolysis is greatly increased by subjecting meat to pressures sufficient to disrupt the sarcoplasmic reticulum. There are concomitant benefits in terms of increased tenderness due to solubilization of myofibrillar proteins and an increase in proteolysis. High pressure processing is of considerable interest to the meat industry, but application at present is limited (see pages 59 and 65).

## (b) Factors affecting the extent of post-mortem glycolysis

It is important to understand the distinction between the rate of post-mortem glycolysis and the extent, as reflected by the ultimate pH value. The same intrinsic and extrinsic factors (Table 2.3) may affect both rate and extent, but different underlying mechanisms may be involved.

### Intrinsic factors
Under strictly controlled conditions, the ultimate pH value of cattle, pigs and sheep falls in the same range of *ca.* 5.4–5.6. Under commercial conditions, the glycogen levels of pigs at death tends to be lower than that of the other species and the ultimate pH values correspondingly higher. In pigs there is also significant variation between breeds. The ultimate pH value of muscle of Large White pigs, for example, is usually significantly higher than that of other breeds. At the same time, the ultimate pH value of

**Table 2.3**   Factors affecting the extent of post-mortem glycolysis

| **Intrinsic** | **Extrinsic** |
|---|---|
| Animal species | Stress |
| Genotype | Pre-slaughter drug administration |
| Sex | Post-mortem temperature |
| Age of animal | |
| Animal temperament | |
| Type of muscle | |
| Pathology | |

*Note*: Data from Lawrie, R.A. (1992) In *The Chemistry of Muscle-based Foods*, (eds D.E. Johnston, M.E. Knight and D.A. Ledward), Royal Society of Chemistry, London.

Hampshire and, occasionally, Landrace pigs can be abnormally low and associated with the PSE condition.

Although some rather dogmatic statements have been made concerning the effect of age and sex on the ultimate pH value, hard evidence is limited. It has been demonstrated in a large survey of cattle that the ultimate pH value (based on the longissimus dorsi muscle) is higher in entire males than in castrates and females. This may be a consequence of the greater stress susceptibility of entire males, although steers have been shown to have a high anaerobic capacity. Older animals are stated to produce meat of lower ultimate pH value. With the exception of limited studies with pork, however, there appears to be no evidence to support this contention.

Pre-slaughter stress is a major factor in depletion of glycogen reserves and thus high ultimate pH value. In the past, it was considered that stress was entirely an extrinsic factor, but it is now recognized that intrinsic stress susceptibility (excitability of temperament) is a major factor affecting the ultimate pH value. In many instances, stress susceptibility is a greater influence in determining the extent of post-mortem glycolysis than fasting or enforced exercise. This may be readily demonstrated with steers (Table 2.4). Mixing batches of cattle of different origin is a common cause of stress, leading to a low equilibrium glycogen level in muscles, a high ultimate pH value and dark cutting beef. The higher incidence of DFD meat in bulls has been attributed to their greater aggression. In pigs, the PSE condition can be induced by pre-slaughter stress leading to abnormally high rates of pH fall, or abnormally low ultimate pH values.

**Table 2.4** Initial glycogen levels and ultimate pH value of the longissimus dorsi muscle of steers

|  | Glycogen (mg/100 g) | Ultimate pH |
|---|---|---|
| Normal control | 1126 | 5.50 |
| Excitable control | 390 | 6.38 |
| Forcibly exercised | 1158 | 5.50 |
| Fasted (28 days) | 768 | 5.44 |

*Note*: Data from Lawrie, R.A. (1992) In *The Chemistry of Muscle-based Foods*, (eds D.E. Johnston, M.E. Knight and D.A. Ledward), Royal Society of Chemistry, London.

The ultimate pH value of 'red' muscles tends to be higher than that of 'white'. This is a consequence of the biochemical specialization (cf. pages 77–78) of the two types of muscle, 'red' muscles having relatively low glycogen reserves and being of relatively limited glycolytic enzyme activity. Differences in the ultimate pH value of different muscles may be explained by the differing ratios of 'red' and 'white' fibres present.

*Extrinsic factors*

Although susceptibility to stress is recognized to be an intrinsic factor, stress itself is applied extrinsically. Extreme stress is usually associated with mishandling and even overt cruelty, but any factors leading to anxiety or anger are stressors. These include fasting, high and low temperatures, anoxia, and conditions during transport and at the slaughterhouse. It is probable that all stressors involve a common physiological complex, which arises in the hypothalamus and proceeds to the level of the affected tissue via the pituitary and adrenal glands. This has a number of consequences, including induction of various protective proteins and polypeptides, but the effect on meat quality stems from a lowering of stored glycogen levels. In general, high ultimate pH levels are associated with glycogen levels of less than *ca*. 0.7%.

Pigs are particularly prone to glycogen depletion resulting from pre-slaughter events. Attempts to overcome this by feeding an easily assimilable sugar to build muscle glycogen reserves have been made, but with limited success. Problems of high pH meat resulting from glycogen depletion are not, however, restricted to pigs. The sudden onset of cold, damp weather imposes stress on cattle, which is reflected in a high incidence of dark cutting beef. A similar effect, although usually less marked, occurs where cattle are slaughtered without resting after transport. The effect of several stressors affecting the animal simultaneously is cumulative and there is a much greater elevation of ultimate pH value than the sum of the individual stressors.

Cooking of pre-rigor meat inactivates glycolytic enzymes and fixes the pH at the value attained when inactivation occurred. In theory meat cooked pre-rigor is very tender, but in practice rigor may commence during cooking. Freezing meat either before rigor has commenced, or in its early stages when residual ATP is present, leads to 'thaw rigor' on defrosting. This is discussed in the context of frozen meat in Chapter 8, pages 369–370.

### 2.5.2 Relationship between post-mortem glycolysis and other characteristics of fresh meat

#### (a) Development of drip

The relationship between glycolysis and water-holding capacity is well established in relation to the PSE and DFD conditions. These conditions represent extremes but it should be appreciated that glycolysis is also involved in the development of drip in 'normal' meat. The water-holding capacity of meat falls with reducing pH value due to reduced ionic binding. At the same time lateral shrinkage of the myofibrils expels water simply because of the decrease in interfilament space. The extent of decrease is very much less than occurs in PSE meat, but is still significant. Shrinking of the myofibrils has consequences for the structure of muscle fibres, which also shrink. Moisture accumulates, initially around the perimysial network and subsequently around the endomysial network. Continuous longitudinal channels form between the fibre bundles (perimysial network; Figure 2.6), drip appearing to arise by gravity draining of fluid through these channels to the cut surface of the meat.

Degradation of protein and an increase in the concentration of peptides and amino acids results in a post-rigor increase in the intracellular osmotic pressure. This tends to increase the water-

---

* An alternative explanation for the varying degrees of drip loss lies in the contention that higher rates are due to pH-dependent changes in the muscle cell membranes. It has been shown, however, that the cell membrane becomes leaky under all conditions of rigor. Further it may be shown that damage to the cell membrane is not itself a cause of drip. (Offer, G. *et al.*, 1989, *Food Microstructure*, **8**, 151–70.)

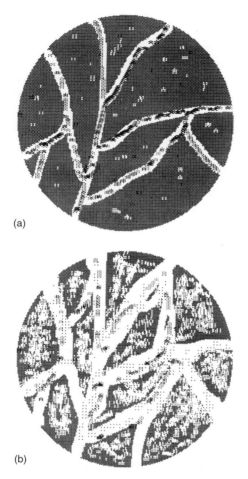

(a)

(b)

**Figure 2.6** Diagrammatic representation of post-mortem development of fluid (drip) filled channels between muscle fibre bundles. *Note*: Based on appearance by light (a) 2 hours post-mortem (b) 24 hours post-mortem microscopy

holding capacity, but the effect is not sufficient to overcome factors leading to drip.

### (b) Flavour

Although a link has been established between the pH value of meat and flavour after cooking, it is not straightforward. 'Normal' beef has a stronger meat flavour than DFD, but this may be a consequence of depletion of precursors, such as free sugars, as a result of stress, or the direct effect of pH value on formation of either precursors or final flavour compounds. Formation of some flavour compounds is known to be favoured by low pH values. The situation with pork is rather different in that high pH value pork is often considered to have enhanced and agreeable 'porky' flavour notes. In contrast PSE pork may have a predominantly sour flavour, possibly due to the high levels of lactic acid.

### 2.5.2 Meat conditioning

Tenderness is probably the most important factor affecting the perceived quality of meat and is increased by conditioning during storage (resolution of rigor). The rate of tenderization varies according to species (see page 64), but a similar mechanism appears to be involved in each case. A number of factors may be involved, but proteolysis and the post-mortem rise in osmotic pressure appear to be of greatest importance.

### (a) Changes occurring in skeletal muscle during conditioning

Skeletal muscle undergoes a number of changes during post-mortem storage (Table 2.5). An event of major importance is the weakening or degradation of the Z-disc, which is a primary process in the development of tenderness. Weakening or degradation of the Z-disc makes myofibrils more susceptible to fragmentation during homogenization and the myofibril fragmentation index may be used as an index of meat tenderness. Disappearance of troponin-T and degradation of desmin is taken as evidence that structural alterations occur in the myofibril as well as at the Z-disc. The exact relationship between troponin-T and meat tenderness is

**Table 2.5** Summary of key changes in skeletal muscle during conditioning at $2-4°C$

1. Weakening and/or degradation of Z-disc
2. Disappearance of troponin-T and simultaneous appearance of 28–32 kDa polypeptide
3. Degradation of desmin
4. Degradation of titin
5. Degradation of nebulin
6. Appearance of 95 000 kDa polypeptide

*Note*: Data from Koohmarie, M. (1992) *Biochimie*, **74**, 239–45.

not understood, but degradation of desmin leads directly to fragmentation of myofibrils, probably through disruption of transverse cross-linking. The significance of other observed changes, degradation of titin and nebulin and appearance of a 9 5000 kDa polypeptide, is not known.

A number of earlier reports claimed that changes occurred in the contractile proteins actin and myosin. For a number of years this was the subject of controversy, but it is now recognized that these proteins are not affected even during prolonged post-mortem storage at low temperatures. There is, however, evidence that some degradation of myosin occurs at storage temperatures above 15°C (see below).

Changes in the Z-disc and myofibril have obvious consequences for the whole muscle fibre. These may be illustrated by comparing unconditioned and conditioned meat (after cooking) from the viewpoint of longitudinal and lateral breaking strengths. In unconditioned meat the longitudinal breaking strength is more than 10 times that of the lateral breaking strength. This is a consequence of the fibre bundles making a substantial contribution to the longitudinal breaking strength, but none to the lateral breaking strength. During conditioning the lateral breaking strength is unchanged, despite possible changes to the perimysium. In contrast, the longitudinal breaking strength decreases markedly during conditioning. This is attributed to degradative changes in the myofibrils causing the fibre bundles to fail at smaller extensions (Figure 2.7). As a direct consequence the contribution of the fibre bundles to longitudinal breaking strength and maximum stress declines markedly.

**Figure 2.7** Effect of conditioning on the longitudinal and lateral breaking strength of meat after cooking. In unconditioned meat (a), the strength of the muscle fibres means that the longitudinal breaking strength is very much greater than the lateral breaking strength. Proteolysis during conditioning and subsequent loss of muscle integrity leads to a marked decline in longitudinal breaking strength (b). Lateral breaking strength is unaffected.

## (b) Proteolysis during conditioning

Proteolysis was empirically recognized as being of importance in changes during conditioning as early as 1917. Empirical recognition is now fully supported by experimental evidence (Table 2.6) and the role of proteolytic enzymes is now generally considered as being beyond doubt. Of the three classes of skeletal muscle (sarcoplasmic, connective tissue and myofibrillar), proteolytic degradation of myofibrillar protein appears to be the major mechanism of tenderization. Some degradation of sarcoplasmic protein does occur, but probably has no direct role in tenderization. Equally, connective tissue does not appear to be subject to proteolysis on the same scale as myofibrillar protein. The connective tissue content and extent of cross-linking, however, varies with age and this fraction plays a significant role in meat tenderness from animals of different ages. This also explains why the extent of tenderization is less in older animals.

Skeletal muscle contains a large number of proteases. A protease capable of bringing about post-mortem changes leading to tenderization, however, must have specific properties (Table 2.7). At present activity has only been demonstrated against myofibrils by two protease systems (Table 2.8): calpains (calcium-activated proteinases), which are present in the sarcoplasm, and lysosomal cysteine proteinases (cathepsins). As a consequence, research over recent years has concentrated on these two types of enzyme. Other

**Table 2.6** Experimental evidence for role of proteolysis in meat tenderization during conditioning

---

1. Proteolysis of myofibrillar proteins and fragmentation of myofibrils induced by incubation of muscle slices with calcium chloride. Proteolysis inhibited by calcium chelators (EDTA and EGTA).
2. Infusion of whole carcass with calcium chloride accelerates changes to skeletal muscle and tenderizes without storage.
3. Infusion of whole carcass with zinc chloride inhibits changes to skeletal muscle and prevents tenderization.
4. Muscle from β-adrenergic agonist (BAA)-fed lambs, which do not undergo post-mortem proteolysis, is tougher than muscle from control lambs. Toughness can be eliminated by infusion of BAA-fed carcasses with calcium chloride.

---

*Note*: Data from Koohmarie, M. (1992) *Biochimie*, **74**, 239–45.

**Table 2.7** Characteristics of proteases involved in meat tenderization during conditioning

1. Must be located within the muscle cell
2. Must have access to the substrate (myofibrils)
3. Must be able to hydrolyse the same proteins *in vitro* that are degraded during conditioning.

**Table 2.8** Proteases involved in meat conditioning

|  | Protease | Activity |
|---|---|---|
| **Sarcoplasmic** | | |
| | Calpain I | Releases α-actinin, Z-nin |
| | Calpain II | Degrades desmin, filamin, connectin, nebulin |
| | | Degrades troponins, tropomyosin |
| | | Degrades C- and M-proteins |
| **Lysosomal** | | |
| | Cathepsin B | Degrades myosin, actin, troponin T |
| | | Degrades collagen |
| | Cathepsin L | Degrades myosin, actin, troponins |
| | | Degrades tropomyosin, α-actinin |
| | | Degrades collagen |
| | Cathepsin D | Degrades myosin, actin, α-actinin |
| | | Degrades troponins, tropomyosin |
| | | Degrades collagen |

*Notes*: 1. Calpain I, cathepsins B and L are thiol proteases, calpain II and cathepsin D are aspartate proteases.
2. Cathepsin L is probably most active in tenderizing.

proteolytic systems may be directly or indirectly involved, including the multicatalytic protease of mammalian cells.

The relative importance of calpains and cathepsins has been a matter of considerable debate. Calpains have an optimum pH value above 6.0, while cathepsins are more active at lower pH values. For this reason it has been thought that calpains, activated by release of $Ca^{2+}$ ions from the sarcoplasmic reticulum, are of greatest importance during the first stage of post-mortem glycolysis, but that cathepsins, released from lysosomes at acidic pH values, become responsible for proteolysis at pH values below 6.0. Some workers have further argued that calpains are of only limited importance under any circumstances, since $Ca^{2+}$ is not released in significant quantities until the pH value is already at the lower end of the range for calpain activity. Other workers consider that even at an ultimate pH value of 5.5, residual calpain is of more significance in proteolysis than cathepsins. In support of this contention it has been noted that a significantly greater release of $Ca^{2+}$ ions occurs below pH 6.0, while the calpain-inhibitor calpastatin loses proportionately more activity than calpains at low pH values.

It has also been stated that there is no direct evidence that cathepsins are released from lysosomes at low pH values and that a comparison of proteolytic effects occurring in muscle during conditioning with the known activities of calpains and cathepsins (Table 2.9) suggests that the former are predominant. There is, however, evidence that myosin degradation occurs at temperatures above 15°C. For this reason, it has been postulated that calpains are primarily responsible for proteolysis during conditioning at low temperatures, despite the apparently hostile pH environment. At the same time, cathepsins have been considered to be the most important proteases during prolonged conditioning at high temperatures. Under most circumstances, however, it is probable that the action of cathepsins and calpains in tenderization is synergistic and many observations can only be explained in terms of such a relationship between the two types of proteinases.

The role of cathepsins has been of particular interest in relation to electrically stimulated meat, it being hypothesized that the observed increase in tenderness of such meat is due to increased release of cathepsins from lysosomes and/or stimulation of the

**Table 2.9** Comparison of the activity of calpains and cathepsins with changes in myofibrils during conditioning

|  | Conditioning | Calpains | Cathepsins |
|---|---|---|---|
| Z-disc degradation | + | + | ± |
| Titin degradation | + | + | + |
| Nebulin degradation | + | + | + |
| Myosin degradation | − | − | + |
| α-Actinin degradation | − | − | + |
| Desmin degradation | + | + | − |
| Actin degradation | − | − | + |
| Troponin-T degradation | + | + | + |

*Note*: After Koohmarie, M. (1992) *Biochimie*, **74**, 239–45.

enzymes at low pH values and high temperatures. Actual events during conditioning of electrically stimulated meat have not been fully elucidated.

It has been suggested that differences either in the levels of proteinases or in their activity can account for differences between species in conditioning rates. In addition, it has been stated that the major cause of differences in tenderness between 'tender' breeds of cattle (*Bos taurus*) and 'tough' breeds (*Bos indicus*) is the lower rate of myofibrillar degradation during conditioning. Systematic studies, however, suggest that the rapid conditioning time of chicken cannot be accounted for in terms of levels of activity of either cathepsins or calpains. It is possible that in fast-conditioning species initial cleavage sites at the Z-disc are more susceptible to proteolysis. Within a single species, proteins from slow-twitch muscles are markedly less susceptible to proteolysis than those from fast-twitch. The ratio of the levels of proteolytic enzymes to those of their inhibitors is also of major importance in determining the rate of conditioning.

Although the role in tenderization is limited, some degradation of connective tissue does occur during conditioning. This appears primarily to involve limited proteolysis of endomysial collagen. A reduction in the tensile strength of raw beef perimysium during 14 days storage at 1°C has also been observed.

## (c) Osmotic pressure of post-mortem muscle

The osmotic pressure of muscle rises during post-mortem processes and can be almost twice that of muscles in the living animal. The increase in osmotic pressure parallels that of acidity, values increasing from *ca.* 300 mOsmol to 500–600 mOsmol. At post-mortem levels, the osmotic pressure is sufficiently high to have a direct effect in weakening myofibrils. Although this effect is secondary to proteolysis, there is considerable synergy between the two mechanisms.

## (d) Effects of conditioning on flavour

Extended conditioning of beef has long been associated with an increase in flavour intensity. This involves either 'beefy' or 'fatty' notes, the increased intensity being preferred by some individuals and disliked by others. Prolonged ageing carries an increased risk of fat oxidation, which is the source of many unpleasant flavours. Increase in flavour intensity is due to an increase in concentration of precursors. Similar increases in flavour intensity have been noted with chickens and other poultry. In the case of uneviscerated birds, however, microbial activity may be partially or wholly responsible.

### 2.5.4 Biochemical basis for meat discolouration

The colour of fresh meat during retail display is of prime importance in consumer acceptability. Acceptable fresh meat colour is often short-lived and a greater or lesser degree of surface discolouration is inevitable. Discoloured meat is often considered spoiled by consumers even when microbial numbers are low and the product perfectly edible. This leads to significant economic loss through the need to withdraw discoloured packs from sale and, at best, convert the meat to lower-value mince. The colour stability of meat varies considerably and different muscles from the same beef carcass possess a wide range of stabilities (Figure 2.8) when stored in an oxygen-containing atmosphere.

The main cause of fresh meat discolouration is the accumulation of metmyoglobin (metMb) at the lean surface (see Chapter 1, pages 27–29 for a discussion of the chemistry of fresh meat pigments). The rate of formation is primarily a function of intrinsic factors, which are modified by post-mortem conditions. Continuing formation during retail storage is known to be influenced by a number of extrinsic factors and control of some of these factors, such as temperature, light and oxygen content of the atmosphere, currently offer the best means of maintaining attractive fresh meat colour.

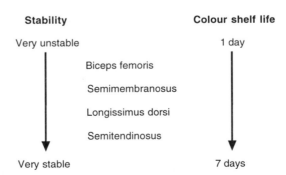

**Figure 2.8** Relative colour stability of four beef muscles stored under standard conditions.

Control of intrinsic factors affecting fresh meat colour, however, probably offers the best long-term approach to prolonging colour stability. It is important to be aware that no single factor operates independently and that the outcome of interaction between factors is of considerable importance in determining the rate of discolouration of meat. These interactions are poorly understood.

## (a) Formation of metmyoglobin

Development of metMb at the meat surface depends on the rate of autoxidation of myoglobin and the effectiveness of one or more metMb enzymic reducing systems. The oxygen consumption rate (OCR) is an important factor in determining metMb formation by favouring the reduced rather than the oxy-form of myoglobin. In the past it has been argued that differences between muscles, under standard storage conditions, can be explained in terms of different rates of autoxidation. More recently it has been demonstrated that autoxidation rates are similar over a wide range of different muscles and it appears likely that it is differences in the effectiveness of the reducing systems which are of prime importance in determining rates of metMb formation. Different muscles are subject to different temperature:pH regimes during post-mortem glycolysis (see page 78) and it is probable that reactants in the reduction of metMb and, possibly, oxidation of myoglobin are affected to a differing extent, modifying the inherent colour stability of the muscle.

The nature of the enzymic metMb reducing systems in meat is not fully understood. Specific NADH-dependent metMB reductases have been isolated from bovine skeletal muscle and other mammalian sources. Both aerobic and anaerobic systems exist, although the role of the aerobic system has been doubted. Mincing of beef destroys the aerobic metMb reductase system, which may account for the higher rate of discolouration. Metmyoglobin reductase activity continues at 1°C and the rate at 3.3°C may be as great as 2% per hour. An increase in temperature of 20°C lead to a *ca.* 10-fold increase in rate of reduction. Activity is also greatest at pH values above 5.8.

Non-specific diaphorases capable of reducing both metmyoglobin and methaemoglobin have been described, although activity is *ca.* 100 times less than that of specific metMb reductases. Methaemoglobin reductase is a component of red blood cells and can reduce either methaemoglobin or metmyoglobin. Residual methae-moglobin activity may be present in meat as a consequence of retained blood and this enzyme may contribute to control of dis-colouration. Metmyoglobin may also be reduced non-enzymatically by NADH or NADPH, but the significance to meat colour is not known.

There have been a number of reports concerning production of enzymes with metMb reducing activity by micro-organisms. A number of Gram-negative genera have been involved, although the incidence of occurrence of metMb reductase activity is not known. Reversal of discolouration of meat has been attributed to microbial metMb reductase activity, although the numbers of bacteria present suggest that undesirable organoleptic changes would occur before metMb reduction.

There has been some controversy concerning the time course of metMb formation at the meat surface. The situation is highly complex, but it seems likely that an essentially triphasic process is involved. The first phase lasts from 4 to 10 days at 1°C, during which metMb formation is rapid. This is followed by a second phase, of varying length, where metMb formation is slow. This phase may be referred to as pseudo-equilibrium and it is followed by a third phase in which metMb accumulates rapidly.

It is well established that metMb formation is much more rapid in meat up to 3 days post-slaughter than in meat aged for longer periods. The initial high rate appears to be a consequence of the high OCR in the days immediately after slaughter. The OCR subse-quently declines, accompanied by a decrease in the rate of metMb formation. Under these conditions the net rate of metMb formation is dependent on the relative effectiveness of the reducing systems. These are known to remain active for many weeks at 1°C.

Conditions in the early post-mortem period can have a profound effect on subsequent metMb formation. In general metMb forma-tion is favoured by low pH value and high temperature. There is,

---

* In some muscles, of very high oxygen consumption rate, an apparent lag in met-myoglobin formation occurs in the first few hours of storage. Such meat has only a very narrow layer of oxymyoglobin at the meat surface, permitting detection of the reduced myoglobin in the lower layers. This effectively 'dilutes' the metmyoglobin present at the meat surface and delays recognition of the typical discolouration.

however, variation between muscles and, while metMb formation is increased in longissimus dorsi and semimembranosus muscles, there is no effect in psoas major. Electrical stimulation leads to rapid pH fall while the carcass temperature is relatively high. Electrical stimulation can improve the initial colour of the longissimus dorsi muscle of beef by eliminating 'heat ring', a dark discolouration sometimes seen near the lumbo-dorsal fascia caused by rapid chilling. Electrical stimulation can also improve the colour of lean meat by producing a lighter, brighter colour on the surfaces. This is an optical effect due to the slightly lowered water binding capacity of electrically stimulated meat and the greater reflectance of light from unbound water at the freshly cut surface (see Chapter 1, page 30). In contrast there have been reports that electrical stimulation decreases the colour stability of muscles such as the l. dorsi, especially at deeper, more slowly chilled locations. Darkening of chicken muscle following electrical stimulation has also been demonstrated.

Optical effects on colour, such as those associated with electrical stimulation, are superimposed on changes due to metMb formation. The extreme cases are represented by PSE and DFD meat. In the case of DFD meat, mitochondrial activity may also be involved. Survival and function of mitochondria is enhanced at high pH levels, the increased activity possibly leading to a higher OCR and more rapid metMb formation. This hypothesis does not appear to be valid in all circumstances.

### (b) Extrinsic factors affecting colour stability during storage

#### Temperature

Temperature is the most important factor affecting continuing discolouration of meat after the immediate post-mortem period. There is both the direct effect resulting from increased OCR and indirect effects of an increased rate of lipid oxidation and increased bacterial growth. Increased storage temperature also reduces $O_2$ solubility in meat, which favours dissociation of $O_2$ from oxymyoglobin. This results in a greater proportion of reduced myoglobin in the less stable deoxy-form and consequently a greater tendency to metMb formation.

Lipid and pigment oxidation are closely coupled in beef, an increase in one resulting in a similar increase in the other. The relationship is not fully understood but, in the context of lipid oxidation promoting pigment oxidation, it may involve free radical formation. It may be postulated that free radicals produced during lipid oxidation have a direct role in oxidizing pigments and/or act indirectly by damaging metMb reducing systems.

Antioxidants, such as ascorbic acid, butylated hydroxyanisole and propyl gallate, are effective in reducing discolouration through limiting lipid oxidation. Some antioxidant systems, however, appear to promote metMb formation. In many countries, addition of antioxidants to fresh meat is illegal, although attempts have been made to overcome this restriction by injecting sodium ascorbate directly before slaughter. A procedure which may be regarded as more acceptable involves the inclusion in the animal diets of high levels of the lipid-soluble antioxidant, vitamin E (tocopherol). The original purpose of this procedure was to prevent the formation of off-flavours associated with lipid oxidation, but pigment stability was also found to be improved.

The role of bacteria in metMb formation has received relatively little attention. It is generally thought to be of minor importance compared with other factors and also of limited significance in the spoilage function. Metmyoglobin formation is usually associated with aerobic bacteria, especially *Pseudomonas*, although *Lactobacillus* has also been implicated. It is generally assumed that the role of bacteria in promoting metMb formation is the reduction of the partial pressure of $O_2$ by the high bacterial oxygen consumption but a large bacterial population ($10^8$–$10^9$/cm$^2$) is required, at which point spoilage would normally be detectable. Accounts of significant promotion of metMb formation by bacteria at lower numbers should, therefore, be treated with considerable caution.

*Partial pressure of oxygen ($pO_2$)*
It has been known for many years that metMb formation is favoured by low, non-zero $O_2$ partial pressures. It has been calculated that metMb formation in semitendinosus muscle is maximal at a $pO_2$ of 6 $\pm$ 3 mm Hg at 0°C and 7.5 $\pm$ 3 mm Hg at 7°. Minimizing metMb formation requires either total exclusion of $O_2$ from the packaging environment or presence of the gas at saturation levels. Effective vacuum packaging reduces the $pO_2$ to significantly below the maximum for metMb formation and thus delays discolouration in bulk storage. In film-wrapped retail packs discolouration can be rapid, if meat is of a high OCR, even where film of high $O_2$ permeability is used. Use of film of lower $O_2$ permeability, however, can

significantly increase the rate of metMb formation. The problem has been partially resolved in MAP-packaged meat. At higher $O_2$ concentrations of *ca.* 80%, the colour shelf life is prolonged, the effect being particularly marked with muscles such as psoas which have poor colour stability. At such high $O_2$ concentrations, the atmosphere satisfies the $O_2$ demand of residual mitochondria without affecting the oxygenation and stabilization of the reduced pigment.

*Light*

The effect of light on meat discolouration became of particular significance after the widespread introduction of prepackaging and display in brightly lit cabinets. In controlled experiments, the metMb content of meat stored in the light was shown to be *ca.* 5.5% higher than that of controls stored in the dark. Ultraviolet irradiation is particularly detrimental to fresh meat colour and UV-impermeable film has been proposed as a means of extending colour shelf life.

The mechanism of the pro-oxidative (photo-oxidative) effect of light involves modification or destruction of amino residue 'R' groups, DNA or lipids. Singlet oxygen, produced by interaction between ground state triplet oxygen and a photoenergized sensitizer molecule such as haem protein, is considered to be the active agent. It is probable that the apoprotein structure can be sufficiently modified by photo-oxidation to result in rapid pigment oxidation.

In some types of display cabinet, it is probable that direct or indirect temperature effects are partly responsible for accelerated metMb formation. A significant temperature rise can occur at the surface of the meat, the problem being particularly marked where older display cabinets have been subsequently fitted with lighting.

*Processing methods*

Mincing is known to increase the rate of metMb formation, but relatively little attention has been given to the effect of other processing methods. It has been observed, however, that colour stability of knife-cut steaks is greater than that of saw-cut steaks.

## 2.5.5 Chemical analysis of fresh meat

Only limited chemical analysis is carried out on fresh meat. Methods applicable to fresh meat for analysis of fat content, determination of meat species and rheological properties are discussed briefly in Chapter 3, pages 152–155.

## 2.6 MICROBIOLOGY

### 2.6.1 Origin of micro-organisms in meat

*(a) Pathogenic micro-organisms*

The source of most pathogens is the gut of the animal or bird. Contamination of meat may occur indirectly from faeces on hide or feathers being disseminated during slaughterhouse procedures. Alternatively poor slaughterhouse technique may permit direct contamination of meat, or the environment, with gut contents. Poultry processing in particular permits rapid dissemination of *Salmonella* from the gut of infected birds, while pigs may be contaminated during scalding.

The possibility of the processing environment or equipment permitting the long-term survival of pathogens and thus acting as a focal point for contamination must always be considered. Prevention is, of course, a function of plant hygiene but the situation is not straightforward. Contamination of the processing environment by *Listeria monocytogenes* has been reported and it is likely that wet cleaning procedures designed for control of *Salmonella* are inappropriate for *Listeria*. Biofilms which include pathogens as part of the community can develop rapidly on equipment after cleaning and serve both to protect the pathogen and act as a contamination focus. In this context problems may be increased by higher levels of automation. *Staphylococcus aureus* is capable of colonizing poultry packing plants, the rubber fingers of feather pluckers being most vulnerable. Endemic strains often have a 'clumping' phenotype when grown in liquid media. This probably aids attachment to the plucker fingers and markedly increases resistance to chlorine and other sanitizing agents.

Although much attention is still given to the role of the meat handler as a source of pathogens, it is unlikely that the carrier state is of major significance as a source of pathogens. Meat handlers may, of course, be involved in dissemination of pathogens on hands, clothing, etc.

BOX 2.6 **Scales of justice**

Concern over the role of meat handlers as sources of pathogens stems from an unthinking extrapolation from severe enteric pathogens, such as *S. typhi*, to common causes of food poisoning. This has led to prolonged exclusion from work and even dismissal from employment under circumstances which cannot possibly be justified on public health grounds. Although this effectively reflects the lay view, which places undue importance on the carrier state in spread of disease, it is likely to be upheld by law. In one case a judge upheld as fair the dismissal of a *Salmonella* excreter employed by a meat packing plant, even though his work involved no contact with food. (Bush, M.F.H., 1985, *Community Medicine*, 7, 133–5.)

Although other sources of pathogens exist, their practical importance with respect to raw meat is limited. Certainly there have been reports, in industrialized countries, of meat being contaminated by contact with vermin, bird droppings, insects and even dogs, but these have been rare examples of extreme bad practice. Contamination from water remains a possibility, especially where borehole or tanked supplies are used rather than directly drawn mains water. A high level of contamination of veal with a toxigenic strain of *Aeromonas* was attributed to use of contaminated borehole water for washing carcasses. With the exception of poultry, intestinal carriage cannot account for the high incidence of *Aeromonas* on meat (see below) and water has been identified as a possible major source.

### (b) Spoilage microflora

Members of the gut microflora are not usually involved in spoilage. In the case of cattle and sheep, the major source of the psychrotrophic spoilage microflora is the hide and fleece of the animal, which itself is contaminated by soil and water. The situation with poultry is similar, the major source of psychrotrophs being the feathers. Considerable variation in the level of contamination of poultry occurs and partly reflects conditions in the rearing houses.

The presence of large numbers of spoilage micro-organisms on hides and fleece means that hide pulling and defleecing are critical stages for meat contamination. The same consideration applies to defeathering of poultry. Considerable airborne dissemination can occur at this stage, resulting not only in direct contamination of the meat, but contamination of the environment. Growth of spoilage micro-organisms on equipment and other meat contact surfaces results in further sources of contamination. Inadequate sanitation of cold stores appears to be a common problem and permits the development of a large psychrotrophic microflora.

The situation with pigs is rather different to that with other animals, since the skin is not normally removed, but singed or scalded before dehairing. Singeing may result in a reduction in microbial numbers, although the effect is variable and significant reduction probably occurs only in localized areas. Scalding at high temperature and under good conditions of hygiene also results in a reduction of numbers and virtual elimination of the Gram-negative microflora. Where control is poor, however, the scald tank may be a source of cross-contamination with both gut organisms and organisms from the skin. The possibility also exists that contaminated water may be drawn into the carcass through orifices and the sticking wound.

Contamination during scalding largely involves Gram-positive bacteria. Psychrotrophic Gram-negative micro-organisms are primarily derived from subsequent operations. The dehairing procedure has been identified as an important stage in this context, although Gram-negative bacteria may also be derived in large numbers from general detritus in the combined slaughter/scald/dehairing facility. A considerable degree of contamination is also possible during subsequent cutting operations, even where these are undertaken at a site remote from the abattoir. Cutting operations are usually the major source of lactic acid bacteria and *Brochothrix thermosphacta*.

### 2.6.2 Meat as an environment for microbial growth

Lean tissues provide readily available sources of energy, carbon and other nutrients. The pH value, typically in the range 5.5–6.5,

---

* Determining sites of contamination can be difficult where reliance is placed on plate counts, since microbial numbers may be below the detection limit. A new analytical procedure, determination of the Contamination Index, has been designed to overcome this problem. This is calculated as the sum of bacterial counts obtained at fixed days during aerobic storage of excised meat samples. (Gustavsson, P. and Borch, E., 1993, *International Journal of Food Microbiology*, 19, 67–83.)

is also ideal for growth of most bacteria. Growth rate may, however, be slightly reduced at low pH values (PSE meat) and greater at high pH values (DFD meat). This may also exert a slight selective effect, but it is generally considered that pH effects in the range 5.5–7.0 have no effect on the gross composition of the microflora. Theoretically a very wide range of bacteria can grow on meat, including the majority of the myriad contaminants which may be present immediately after slaughter. In practice extrinsic factors are applied, which limit the extent of microbial growth and the number of strains capable of growth. Refrigeration is the major extrinsic factor and has been widely used in preservation of meat since mechanical means of chilling were first available. More recently, vacuum and modified atmosphere packaging have been widely adopted and further restrict development of micro-organisms as well as modifying the composition of the microflora. An additional key factor in restricting the number of types of micro-organism developing is competition. Competitive pressures are severe and can result in the microflora being dominated by a small number of strains of the same species, although differing affinity for nutrients, especially amino acids, does permit co-existence.

### 2.6.3 Meat and micro-organisms of public health significance

Meat animals and birds are recognized as important sources of zoonotic pathogens (see Chapter 1, pages 32–37). The extent to which meat itself is a source of pathogenic micro-organisms depends on the degree of contamination at slaughter and in subsequent operations and the extent of multiplication, if any, during storage. Many surveys have been published concerning the incidence of the various pathogens in meat. These inevitably show considerable variation, which may be attributed to a number of factors, including differences in methodology. It is not the intention at present to attempt to summarize the different surveys. Under present conditions meat is inevitably contaminated to a greater or lesser extent and it is considered more important to gain an appreciation of the relevance of contamination with the various pathogens to human morbidity. It must be stressed, however, that this is the current situation and that, especially in the US, there are moves towards zero tolerance of pathogens such as *E. coli* O157:H7.

It will be appreciated that meat is usually consumed after cooking and that the major importance, in the case of vegetative pathogens such as *Salmonella*, lies in cross-contamination. A small number of raw meat dishes exist, including steak tartare, which is known to have been associated with *Salmonella* infection, while in Belgium a strong epidemiological link has been established between the widespread consumption of raw pork and *Yersinia enterocolitica* infections.

---

BOX 2.7 **The public buys its opinions as it buys its meat**

Campaigns to improve food safety, both in industry and in the home, have paid considerable attention to the handling of both raw and cooked meat, as opposed to the handling of other foods. It is therefore considered essential for the success of food safety campaigns that people's perception of meat is different from that of other foods. A survey conducted in a UK city (Norwich), involving 94 people of various ages and socio-economic classes, showed that meat is not seen as a different or special food. This finding has serious implications for the planning of food safety campaigns. The survey also showed that people are not an amorphous group at which a food safety message is directed. This too has serious implications concerning the way in which the food safety message should be directed and for the development of policies with relation to food safety. (Maguire, K., 1994, *British Food Journal*, **96(2)**, 11–17.)

---

Contamination with *Salmonella* is common in raw meats of all types. The incidence is generally highest in intensively reared chickens, approaching 100% from some suppliers. Incidence varies according to agricultural practices and hygiene during slaughter and subsequent operations. Chickens are particularly significant in contamination of other foods due to the high incidence of *Salmonella* and the frequent occurrence of large quantities of drip. There is considered to be little doubt that reducing the level of *Salmonella* contamination of meat, especially poultry, would significantly reduce morbidity. The situation is similar with respect to contamination by *Campylobacter* and restriction endonuclease analysis has shown that 61% of human strains could be matched to food animals. Chickens appear to be of major importance in sporadic outbreaks of campylobacteriosis, but red meats are also important. As with poultry, *C. jejuni* predominates in beef and lamb and a high proportion of isolates are those associated with human disease. *Campylobacter hyointestalinis* has also been isolated from beef carcasses, although the significance to humans is not known. The situation with pork is less clear than with beef and lamb. The incidence of contamination is low (*ca.* 10%), probably due to the marked dehydration of pork carcasses during chilling, and most isolates are *C. coli* serovars not usually associated with human disease. Pork is not considered to be an important vehicle of campylobacteriosis in countries such as the UK, although this situation may not be typical elsewhere.

Verocytotoxin-producing *Escherichia coli* (VTEC) is often considered to be associated with beef, but the organism appears to be present at a significant level in most major types of meat at retail level. Beef from cull cows is thought to be infected at a particularly high incidence, reflecting the higher level of infection in older animals. This is considered to be of particular concern since meat from cull cattle is widely used in beefburgers, associated with several outbreaks of VTEC infection. There is considerable geographical variation in incidence of VTEC. In North America, for example, the organism is prevalent in Canada and apparently much more common in northern states of the US than southern. It is thought that differences are at least partly due to variations in reporting.

### 2.6.4 Spoilage of beef, lamb, pork, etc.

In general, spoilage of beef, lamb, pork and other meat derived from mammals follows a similar pattern. Most work has been done with beef and since some variations do exist it is unwise to extrapolate directly to other mammals.

### (a) Storage in air

The shelf life of meat stored in air under refrigerated conditions (ideally <5°C) is limited by microbial growth. Off-odours generally appear when numbers reach *ca.* $10^7/cm^2$ and slime is visible when numbers reach *ca.* $10^8/cm^2$. Spoilage patterns may not be consistent and significantly higher numbers can be present without signs of overt deterioration. The spoilage microflora is dominated by Gram-negative, psychrotrophic, aerobic rod-shaped bacteria. In the past, these have been assigned to a wide range of genera, but it is now recognized that only *Pseudomonas*, *Acinetobacter* and *Psychrobacter* are normally of significance. Of these, *Pseudomonas* is of greatest importance. To some extent this reflects the general property of *Pseudomonas* as a remarkably successful bacterium in high-protein foods stored at low temperatures. More specifically it has been suggested that *Pseudomonas* gains a selective advantage through its ability to metabolize glucose to gluconate and 2-oxo-gluconate via the Entner–Doudoroff pathway. The two acids accumulate extracellularly and are further metabolized by *Pseudomonas* but not by competing micro-organisms after glucose depletion. Evidence, however, appears to be lacking. *Acinetobacter* and *Psychrobacter* are unable to metabolize glucose and presumably obtain energy through oxidation of amino or organic acids.

Other bacteria are present in small numbers and may occasionally form a significant part of the microflora. *Brochothrix thermosphacta* appears to be of more importance on pork and lamb than on beef and may be favoured on fat where the pH value is generally higher. *Brochothrix thermosphacta* is also favoured by temperatures of 5°C or above. Temperatures in excess of 6°C and a pH value in excess of 6.5 appears to open a 'window of opportunity' for *Br. thermosphacta* on some occasions, although it is likely that other favourable factors are involved. On lamb carcasses evidence of a microbial succession has been found in which Gram-negative bacteria, predominantly pseudomonads, were followed by *Br. thermosphacta* and subsequently by coagulase-negative staphylococci. The latter bacteria, predominantly *Staph. xylosus*, appeared to exploit the metabolic activities of its predecessors. Species of both *Micrococcus* and *Staphylococcus* are present on meat stored in air, but their significance is generally considered limited under refrigerated storage.

* Taxonomic studies have identified three major groups of *Pseudomonas* which are of importance in meat spoilage. Two were identified with *Ps. fluorescens* and *Ps. fragi*, while the other group has been assigned to a new species *Ps. lundensis*. Two subgroups of *Ps. fragi* have been described, of which cluster 2 strains are the dominant pseudomonads on meat. *Pseudomonas fragi* cluster 1 strains are second in importance followed by *Ps. lundensis* and *Ps. fluorescens* biotypes I and III. *Pseudomonas lundensis* may, however, be the dominant micro-organism in spoiled meat. (Dainty, R.H. and Mackey, B.M., 1992, *Journal of Applied Bacteriology (supplement)*, **73**, 103–14; Gennari, M. and Dragotto, F., 1992, *Journal of Applied Bacteriology*, **72**, 281–8.)

Psychrotrophic members of the Enterobacteriaceae, including *Serratia liquefaciens*, *Enterobacter agglomerans* and *Hafnia alvei* are also common at low levels, but cannot compete effectively at low temperatures. These organisms become of greater importance during storage at 6–10°C, but *Pseudomonas* usually remains dominant at any feasible commercial storage temperature.

The role of yeasts in spoilage of fresh meat appears limited, although high numbers have been reported on some occasions, notably on lamb. Yeasts may be of greater importance where retail packs of meat have been stored at *ca.* 0°C to obtain an extended storage life. Moulds were of historic importance on carcass meat stored for extended periods at temperatures just above freezing, but have little or no relevance to modern practice.

The spoilage pattern obviously reflects the metabolic activity of the dominant micro-organisms present. Analysis of metabolites present in air-stored meat using gas chromatography, and gas chromatography combined with mass spectrometry, have revealed a complex mixture of compounds (Table 2.10). These include esters, S-containing compounds, ketones, branched chain alcohols, unsaturated hydrocarbons and 3-methylbutanal. The exact composition of the mixture of metabolites varies, but the overall pattern appears consistent.

*Pseudomonas fragi* cluster 2 strains are the major and possibly sole producers of ethyl esters amongst the aerobic microflora and are responsible for the sweet, fruity odours of meat in the early stages of spoilage. *Pseudomonas fragi*, in common with other species of *Pseudomonas*, also produces the sulphur compounds responsible for the putrid odours which follow the fruity odours as spoilage advances. Members of the Enterobacteriaceae isolated from meat also produce sulphur compounds, but may be distinguished from the pseudomonads by production of $H_2S$ but not dimethylsulphide.

Pseudomonads, especially *Ps. lundensis*, are also responsible for production of unsaturated hydrocarbons and are amongst the bacteria which produce ammonia and amines during amino acid metabolism. The prime role of *Pseudomonas*, especially *Ps. fragi* cluster 2, in aerobic spoilage of meat is therefore well established. Amino acids have been identified as the major source of spoilage metabolites, which appear after the exhaustion of glucose. This

**Table 2.10** Metabolites of spoilage micro-organisms in meat stored at low temperatures in air

| | |
|---|---|
| Ethyl acetate | Methanethiol |
|    *n*-propanoate | Dimethylsulphide |
|    isobutanoate | Dimethyldisulphide |
|    2-methylbutanoate | Dimethyltrisulphide |
|    3-methylbutanoate | Methylthioacetate |
|    *n*-hexanoate | |
|    *n*-heptanoate | Ammonia |
|    *n*-octanoate | Putrescine |
|    crotonate | Cadaverine |
|    3-methyl-2-butenoate | Tyramine |
|    tiglate | Spermidine |
| Isopropyl acetate | Diaminopropane |
| Isobutyl acetate | Agmatine |
|    *n*-propanoate | |
|    *n*-hexanoate | 1,4-Heptadiene |
|    tiglate | 1-Undecene |
| Isopentyl acetate | 1,4-Undecadiene |
| 3-Methyl butanol | Acetoin |
| 2-Methyl butanol | Diacetyl |
| | 3-Methyl butanal |

*Note*: Data from Dainty, R.H. and Mackey, B.M. (1992) *Journal of Applied Bacteriology (supplement)*, **73**, 103–14.

occurs when microbial numbers exceed $10^7/cm^2$ in normal meat, but at levels of *ca.* $10^6/cm^2$ in meat of high pH value. This results from the low levels of glucose in high pH meat and it is this factor, rather than more rapid growth, which is primarily responsible for the more rapid onset of spoilage.

Although the role of the Enterobacteriaceae and *Br. thermosphacta* is secondary, they are involved in production of spoilage metabolites. These bacteria may play a role in production of branched chain ester residues and of acetoin and diacetyl produced at the same time as the esters. The appearance of acetoin, diacetyl and 3-methylbutanol at an early stage, when microbial numbers are relatively low, cannot be explained.

### (b) Storage in vacuum

Unless special precautions are taken, large vacuum packs contain *ca.* 1% $O_2$ which, theoretically, would support growth of a rela-

**Table 2.11** Major lactic acid bacteria isolated from meat stored at low temperatures in vacuum packs

| | |
|---|---|
| *Lactobacillus* | *Leuconostoc* |
| *bavaricus* | *carnosum* |
| *curvatus* | *gelidum* |
| *sake* | *mesenteroides* sub sp. *mesenteroides* |
| | |
| *Carnobacterium* | *Lactococcus* |
| *divergens* | *raffinolyticus* |
| *piscicola* | *lactis* |

tively large population of pseudomonads. Continuing respiration by the meat rapidly depletes $O_2$ and increases the $CO_2$ concentration to *ca.* 20%. *Pseudomonas* is usually unable to grow under these conditions, although numbers as high as $10^6/cm^2$ may be present on primal joints. The conditions under which *Pseudomonas* is able to develop in intact vacuum packs are obscure. Under most circumstances conditions in vacuum packs selectively favour lactic acid bacteria, although there may also be significant growth of *Br. thermosphacta*, '*Shewanella putrefaciens*' and the Enterobacteriaceae. More recently, psychrophilic species of *Clostridium* have been recognized as a significant potential problem.

The predominant lactic acid bacteria of vacuum-packed meat are homofermentative species of *Lactobacillus*, *Carnobacterium* and *Leuconostoc*. *Lactococcus* is much less common and other genera have not been reported (Table 2.11).

Lactic acid bacteria are able to grow rapidly at low temperatures and low $O_2$ tensions and are also strongly favoured by their tolerance of $CO_2$. They are able to compete effectively with other micro-organisms capable of rapid growth under these conditions.

* *Carnobacterium* is a relatively new genus which accommodates the so-called 'non-aciduric' strains of *Lactobacillus*, which are unable to grow on acetate agar at pH 5.6. These bacteria have an atypical heterofermentative metabolism, producing L(+)-lactic acid, acetic acid, ethanol, formic acid and $CO_2$. Detection of heterofermentative activity using standard techniques may be difficult due to production of formic acid in addition to $CO_2$. It is salutary that use of acetate agar at pH 5.6 continues in the meat industry, despite the fact that it has been known for many years that this medium will not recover a significant proportion of the lactic acid bacteria present. (Collins, M.D., Farrow, J.A.E., Phillips, B.A. *et al.*, 1987, *International Journal of Systematic Bacteriology*, **37**, 310–6.)

This may be a function of a greater affinity for glucose, although the relative affinity of lactic acid bacteria and *Br. thermosphacta* appears to vary. A number of lactic acid bacteria isolated from meat, including common spoilage species such as *Lactobacillus bavaricus* have been shown to produce bacteriocins. There is no evidence of activity against *Br. thermosphacta* or the Enterobacteriaceae but bacteriocin production may be of importance in determining the dominant strains of lactic acid bacteria. In an extreme case, the lactic acid microflora of vacuum-packed beef was reported to consist almost entirely of a single bacteriocin-producing strain of a *Lactobacillus* species.

Both *Br. thermosphacta* and '*Sh. putrefaciens*' are favoured by high pH values.'*Shewanella putrefaciens*' is unable to grow below pH 6.0 during storage at low temperatures, whereas *Br. thermosphacta* is unable to grow below pH 5.8 on meat packed in film of low $O_2$ permeability. In the case of '*Sh. putrefaciens*' this appears to be a direct effect of pH value, but inhibition of *Br. thermosphacta* is primarily due to the effect of lactic acid at low pH values. Low $O_2$ tension plays a secondary role.

At temperatures below 5°C, Enterobacteriaceae are inhibited in vacuum packs by $CO_2$, low pH value and lactic acid. At higher temperatures and pH values, $CO_2$ is markedly less inhibitory and growth is possible. *Serratia liquefaciens* appears particularly well adapted to these conditions, but other members of the family may also be present, including *Providencia* where pH and storage temperature are markedly higher than the norm.

The predominant spoilage pattern of vacuum-packed fresh meat is souring. This type of spoilage is relatively innocuous compared with aerobic spoilage and is not normally detectable until numbers reach $10^8$/cm$^2$ or even higher. For many years it has been assumed that the odours arise from lactic acid and other end-products of fermentation by the dominant lactic acid bacteria. L- and D-lactic acids together with acetic, isobutanoic and isopentanic acids do accumulate, but the quantities are small in relation to the L-lactic acid produced in normal muscle by post-mortem glycolysis. It has been suggested that methanethiol and dimethylsulphide may be at least partly responsible for acidic odours, although this has not been confirmed. A 'sulphide' type of spoilage has also been described in normal pH meat, in which packaging in a film of high $O_2$ permeability was a contributory factor. The causative organism was *Lb.*

*sake* and H$_2$S production was sufficient to cause greening through formation of sulphmyoglobin. Hydrogen sulphide production, however, only occurs after the depletion of glucose.

A comparison of spoilage by a homofermentative *Lactobacillus* and *Leuconostoc* in inoculated meat showed that with *Lactobacillus*, the concentration of acetate and D-lactate increased during storage, while glucose and L-lactate decreased. Deterioration of flavour coincided with the maximum number (*ca.* 10$^8$/cm$^2$) being reached. In contrast deterioration of flavour was noted before the maximum number was reached with *Leuconostoc*. Ethanol and D-lactate increased, while glucose decreased. Fermentation patterns of homofermentative lactic acid bacteria can be changed by low glucose levels, which may cause a switch to heterofermentative metabolism.

The contribution of sulphur compounds to spoilage is much greater when meat of high pH value is vacuum packed. This results from the presence, in significant numbers, of '*Sh. putrefaciens*' and the Enterobacteriaceae, which produce large quantities of H$_2$S. This results in an extremely unpleasant form of spoilage.

Psychrophilic clostridia, such as *Cl. laramie* (which has a maximum temperature for growth of *ca.* 20°C), produce large quantities of H$_2$ and CO$_2$, resulting in swelling of the pack. Associated off-odours are also present and are attributed to a range of sulphur compounds, butanol, butanoic acid, ethanol, acetic acid and various esters. Greening occurs on prolonged storage and the texture of the meat is soft due to proteolysis. Spoilage by psychrophilic *Clostridium* is reported infrequently and predisposing conditions are not known, since the meat involved has been of normal pH value, with a normal lactic acid microflora and stored at correct temperatures. Psychrophilic *Clostridium* is not recovered using conventional methods for clostridia, however, and spoilage by these bacteria may be more common than is generally recognized.

*(c) Storage in modified atmosphere packs*

The microflora of meat stored in commercially used modified atmosphere packs (20–40% CO$_2$:60–80% O$_2$) is in general similar to that of vacuum packs. At temperatures below 2°C, the microflora is dominated by lactic acid bacteria. *Leuconostoc* is of much greater importance than in vacuum packs. The reason for this is not known, although it has been suggested that *Leuconostoc* is stimulated by oxygen as a result of production of acetate rather than ethanol as a metabolic end-product. This explanation is not entirely convincing. *Brochothrix thermosphacta* is relatively $CO_2$ tolerant and the presence of $O_2$ permits growth of this bacterium at pH values below 5.8 (cf. vacuum packs). Enterobacteriaceae and *Pseudomonas* are more prevalent in modified atmosphere packs than vacuum packs during storage at *ca.* 5°C. Raising the storage temperature permits faster growth by lactic acid bacteria, but Enterobacteriaceae and *Pseudomonas* benefit from the markedly reduced bactericidal effects of $CO_2$ at higher temperatures. These bacteria and *Br. thermosphacta* are also present in greater numbers where prior conditioning of the meat has been in air and are more common on pork than on other meats. The situation is not straightforward, however, and 'hard-to-explain' spoilage by *Pseudomonas* may occur under commercial conditions when prior conditioning has been in vacuum packs. *Aeromonas* can reach significant numbers either during storage at 7°C or above, or during prolonged storage at lower temperatures. Yeasts may be capable of limited multiplication where packaging material of relatively high $O_2$ permeability is used, but they are not a major component of the microflora.

A very long storage life may be obtained by packing meat in an atmosphere of 100% $CO_2$ (Captech process, see page 67). At a temperature of 0–1°C, a storage life of 3 months or more has been claimed for meat of both normal and high pH value. Even longer storage lives (up to 6 months) have been obtained by adding sufficient $CO_2$ to saturate the meat tissues and storing at −1.5°C. Spoilage of such meat has been attributed exclusively to lactic acid bacteria, but it is now recognized that *Br. thermosphacta* can be involved if present in high numbers at the time of

---

* Enterobacteriaceae may show an unusual growth pattern when meat packed in 100% $CO_2$ is stored for extended periods. Even fairly large populations diminish rapidly after packing and become undetectable. This situation persists for a number of weeks, after which the micro-organisms 'reappear'. It has been reported that numbers of Enterobacteriaceae as low as $10^4/cm^2$ can cause spoilage due to putrid odours. Such a pattern may not be unique to $CO_2$-packed meat, or to the Enterobacteriaceae, since a similar phenomenon has been observed with *Pseudomonas* in vacuum packed meat stored for extended periods at low temperatures (Authors' unpublished observation). It may be postulated that the course of spoilage in such situations is dictated by the minority microflora, which appears late in the storage period. (Gill, C.O. and Harrison, J.C.L., 1989, *Meat Science*, 26, 313–24.)

packing, or if high pH meat is packed in film of relatively high permeability.

Meat stored in 100% $CO_2$ is repacked in air for retail display. Lactic acid bacteria continue to dominate the microflora during storage in air and are involved in spoilage. Numbers of *Pseudomonas*, however, increase very rapidly and spoilage may result from metabolism of amino acids following depletion of glucose. Prolonged storage in $CO_2$ compromises subsequent retail display life in air, especially where higher storage temperatures are involved. At a temperature of 8°C, it has been calculated that each 6-week period in a $CO_2$ atmosphere leads to a reduction of 1 day of retail life.

Spoilage of meat packed in modified atmospheres may involve souring similar to that of vacuum-packed meat. Other spoilage patterns have been described, including 'rancid' and 'cheesy' odours. Chemical rancidity does not appear to be primarily involved and it is probable that the metabolic end-products of lactic acid bacteria, and under some circumstances *Br. thermosphacta*, are the major cause. D-lactic, isopentanoic, isobutanoic, acetic and propanoic acids are all significant metabolic end-products. Other compounds possibly involved in production of off-odours, such as acetoin/diacetyl, straight chain aldehydes and unsaturated hydrocarbons, may well be the products of non-microbial processes.

## 2.6.5 Spoilage of poultry and game birds

### (a) Eviscerated chickens, turkeys and ducks

The microflora of eviscerated (oven-ready) poultry stored in air at *ca.* 4°C is similar to that of red meat, being dominated by *Pseudomonas* and to a lesser extent *Acinetobacter* and *Psychrobacter*. '*Shewanella putrefaciens*' may also be present. Growth of microorganisms occurs primarily on cut surfaces and in feather holes. Drying of the skin occurs where birds are sold unwrapped and this is effective in restricting microbial growth. In modern practice, poultry is virtually always retailed in a film overwrap, under which a small elevation of $CO_2$ content may occur, favouring *Br. thermosphacta*. In general, however, *Br. thermosphacta* is relatively uncommon on poultry, although phenotypically similar but distinct Gram-positive bacteria are sometimes present in significant numbers in poultry.

Chickens and turkeys usually have a microflora which is both quantitatively and qualitatively similar. It has been stated that microbial numbers are lower on turkeys, but differences are small and insignificant in a commercial context. Relatively little is known of the microflora of ducks, but it is generally accepted that microbial numbers on these birds are lower by at least one decade than those on chickens or turkeys. There can be considerable variation from one part of the bird to another, reflecting contamination during processing. A relationship also exists between the occurrence of specific types of micro-organism and different parts of the carcass. *Pseudomonas* is dominant in all parts, but both *Acinetobacter* and '*Sh. putrefaciens*' are restricted in breast muscle by the lower pH value (5.7–5.9) compared with leg (6.4–6.7).

The method of chilling can have a significant effect on the $a_w$ level and thus on microbial growth rates. Air-chilled chickens had an $a_w$ level of 0.970 before packing compared with 0.996 for water-chilled. The $a_w$ level of air-chilled, however, increased to 0.990 within two hours of packaging and differences in growth rate are small. The washing effect of water chillers may also reduce the initial microbial load where hygiene management is good. Equally, of course, the initial load may be increased where hygiene management is poor.

In recent years it has become common practice, at times of peak demand, to store poultry, especially turkeys, at $-2°C$. At this temperature the water may freeze, but not the meat, resulting in a significant fall in $a_w$ level. Storage life is markedly extended and yeasts may comprise a significant proportion of the microflora. Species of *Cryptococcus* and *Candida* are the most common yeasts. Poultry does not lend itself to vacuum packing and the use of modified atmospheres at retail level is restricted by poor appearance following exposure to $CO_2$ even when $O_2$ is present. Considerable use is made, however, of bulk storage in $CO_2$ at $-1$ to $-2°C$, before retailing the birds in air (see page 68). This practice is effective in controlling *Pseudomonas* and other Gram-negative bacteria and in extending the storage life. There is little published information concerning the dominant microflora under these conditions, but lactic acid bacteria, especially *Lactobacillus* and *Carnobacterium* are favoured.

Spoilage of uneviscerated poultry stored in air is apparent when microbial numbers reach *ca.* $10^8/cm^2$. Spoiled poultry has a characteristic odour, which is the result of contributions from a wide range of volatile compounds. The most important appear to be $H_2S$, methyl mercaptan, dimethyl sulphide, dimethyl disulphide, methyl acetate, ethyl acetate, heptadiene, methanol and ethanol. No single member of the spoilage microflora produces all of these compounds.

### (b) Uneviscerated (New York dressed) turkeys and ducks

Uneviscerated turkeys and ducks have a longer storage life than eviscerated birds, the skin being dry and unbroken and supporting only very limited microbial growth. New York dressed turkeys were considered acceptable after 18 days storage at 4°C, while eviscerated birds were spoiled after not more than 9 days at this temperature. Spoilage is an internal phenomenon which stems from production of large quantities of $H_2S$ in the intestine and diffusion of the gas into the muscle tissues. Near the surface, where the $O_2$ tension is high, $H_2S$ combines with myoglobin to form green-coloured sulphmyoglobin, a phenomenon known as 'greening' or 'greenstrike'. Green discolouration often first appears around the vent as a consequence of the high level of $H_2S$-producing bacteria in the large intestine. Subsequent growth occurs mainly in the duodenum and other parts of the small intestine, where nutrient levels are higher. Greening is minimized where the birds have been starved prior to slaughter or where a diet leading to a low final pH value in the caecum is fed.

Rate of greening is directly related to growth of $H_2S$-producing bacteria and thus to storage temperature. Most intestinal bacteria are unable to grow at 5°C and, while a $H_2S$-producing microflora does develop, the onset of greening is delayed (typically *ca.* 17 days). At 25°C, significant greening can take place within 1 day.

### (c) Hung game birds

Spoilage of hung game birds (pheasant, partridge, grouse, etc.) follows a similar pattern to that of uneviscerated poultry and largely involves growth of micro-organisms in the intestine. The pheasant, however, appears notably resistant to greening and under many conditions microbial growth in the intestine of these birds is limited. Handling after shooting appears to be a major factor in determining the likelihood of spoilage. It is not uncommon for birds to be kept warm in a large pile for several hours before hanging. Under these conditions rapid onset of greening may be expected.

Significant but localized microbial growth may occur around gunshot wounds, and may lead to off-odours and unpleasant taste. External spoilage may also occur if birds are wet and hanging conditions do not permit adequate drying.

### 2.6.6 Microbiological examination of meat

Under most circumstances microbiological examination of fresh meat (including poultry) is straightforward. When using conventional cultural methods, standard selective media are generally suitable (Table 2.12). It is recommended that an incubation temperature of 20–22°C is used for 'total aerobic' counts, rather than the 30°C still in common use. It has also been suggested that anaerobic counts may be a more accurate means of assessing the microbiological status of meat after prolonged storage in an atmosphere of 100% $CO_2$. Higher recoveries of lactic acid bacteria, especially *Leuconostoc* and *Carnobacterium*, are often obtained by incubation at 30°C, or even 25°C, than at 37°C. At lower temperatures, however, the selectivity of the medium is reduced and the possibility of non-lactic acid bacteria developing is higher. *Carnobacterium* is highly sensitive to acetate used as selective agent and recovery may, in fact, be higher on non-selective media.

**Table 2.12** Media suitable for the isolation of micro-organisms from meat

| | |
|---|---|
| 'Total aerobic' count | Plate count agar |
| | Nutrient agar |
| *Pseudomonas* | Cephaloridin–fucidin–cetrimide agar |
| Lactic acid bacteria | de Man, Rogosa and Sharpe agar |
| *Escherichia coli* | Anderson and Baird–Parker agar |
| *Brochothrix thermophacta* | Gardner's STAA agar |

* A number of game birds appear remarkably resistant to growth of spoilage micro-organisms. In some cases, this has been attributed to the presence of naturally occurring antimicrobial compounds derived from the diet. Rock ptarmigans, for example, feed on crowberries containing benzoic acid and possibly other antimicrobials, while an antimicrobial identified in the muscle tissue of willow ptarmigan was derived from willow and birch buds. Heather, which is eaten by red grouse, may also contain antimicrobials.

Conventional means of recovering clostridia from meat are not suitable for psychrophilic species. The only methods available so far involve use of pre-reduced media, such as blood agar, reinforced clostridial agar and tryptone–soya–yeast extract agar incubated anaerobically at 1–2°C for *ca*. 3 weeks. These methods are not always successful and in some cases detection has involved the inoculation of fresh beef with purge from spoiled packs. General methods for *Escherichia coli* are well established, but most-probable number methods, although sensitive, are time-consuming and expensive, while problems with conventional media arise from the presence of lactose-non-fermenting strains. The direct plating method using tryptone–bile agar is a suitable alternative, but there has been much interest in commercial β-glucuronidase-based media. Of these Petrifilm *E. coli*™ is an effective screening method, while Fluorocult™ is suitable where contamination levels are relatively high.

Recovery of *Salmonella* and *Campylobacter* presents no specific problems. In the case of *Salmonella*, at least two selective plating media should be used. Some authors recommend that, for poultry, one of these should always be deoxycholate citrate medium. Good results have also been reported with more recently developed media including Rambach agar, mannitol–lysine–crystal-violet–brilliant-green agar and diagnostic *Salmonella*-selective semi-solid medium. In all cases it is necessary to be aware of the possibility of atypical lactose-fermenting *Salmonella* of subspecies II to VI. These salmonellas, which are of uncertain pathogenicity to humans, can cause particular problems in examination of imported exotic meat from reptiles such as alligators, which is often heavily contaminated. Isolations have also been made from turkey and sheep and the organisms may be more widespread than generally appreciated.

* Rambach agar is one of a new generation of selective plating media for *Salmonella*. The ability of *Salmonella* to produce acid from propylene glycol is exploited as a diagnostic feature and the medium also contains a chromogenic substrate to detect β-D-galactosidase production by lactose-fermenting Enterobacteriaceae. Colonies of *Salmonella* (except *S. typhi*) are red pigmented, while other Enterobacteriaceae able to develop are blue-green. Lactose-fermenting *Salmonella*, however, also fail to produce acid from propylene glycol and cannot be distinguished from other Enterobacteriaceae. It is therefore necessary to carry out confirmatory testing on all colonies developing on Rambach agar, although this is always good practice on any selective medium. (Kuhn, H., Wonde, B., Rabsch, W. and Reissbrot, R., 1994, *Applied and Environmental Microbiology*, **60**, 749–51.)

Verocytotoxin-producing strains (VTEC, Shiga-like toxin-producing) of *E. coli* continue to present problems of isolation. A common and relatively simple method for serovar O157:H7 is the use of MacConkey–sorbitol agar, this serovar differing from most others in being unable to ferment sorbitol. This property is by no means unique to serovar O157:H7, although specificity is increased by incorporation of a fluorogenic substrate for β-glucuronidase, which is not produced by this serovar. The method is presumptive and confirmation is required using a commercially available latex co-agglutination test. Convenient biochemical markers are not available for non-O157 VTEC serovars and no straightforward method exists for their isolation or detection.

Over the years a number of rapid and instrumental methods have been used for microbiological examination of meat. Simple chemical and physical methods have been proposed which, it was hoped, would provide a more direct assessment of the quality of meat than could be obtained using conventional microbiological methods. Physical methods include determination of extract release volume or water holding capacity (see Chapter 3, pages 152–153), while chemical methods include determination of lactic acid, low molecular weight nitrogenous compounds and amino sugars. All of these methods are characterized by low sensitivity and, in the case of chemical markers of spoilage, extreme variability. Other methods which have been proposed include dye reduction, degree of hydration of the meat and titratable acidity. None of these has been widely used. A further rapid method, the Limulus lysate assay, has been used as a screening test for Gram-negative contamination in restricted circumstances.

Several types of instrumental technique are successfully used in the microbiological analysis of meat. Impediometry probably finds the greatest application and use has been extended to detection of *Salmonella* and *E. coli*. Impediometry, however, is not a true rapid technique. A number of attempts have been made to use ATP measurement as a truly rapid means of determining microbial numbers on meat but, when applied on a routine basis, elimination of somatic ATP continues to cause difficulties, though determination of ATP is successfully used for monitoring hygiene of food contact surfaces. Some use has been made of the direct epifluorescent technique (DEFT) but sample preparation can cause difficulties. Both ATP determination and DEFT are of relatively low sensitivity (*ca.* $10^5/cm^2$).

A number of kits based on either immunological or genetic methods are available for determination of *Salmonella*, *Campylobacter* and *Listeria monocytogenes*. Such kits are expensive and there can be a considerable amount of interference from other micro-organisms present in meat. In the case of *Salmonella*, the cultural Oxoid Salmonella Rapid Test™ may offer a satisfactory compromise between conventional methodology and immunological or genetic methods.

EXERCISE 2.1

You are employed as a technologist by a large retail chain, which has developed a process for production of 'traditional' beef, which is retailed at a premium price. During development both taste panels and expert tasters found the 'traditional' beef to be more tender and of better flavour. Tasting was carried out on beef at the end of the maturation period, before distribution and retail display. Although sales of the 'traditional' beef have met expectations, a significant number of customers have expressed disappointment with the quality and a report by a consumer magazine considered the product to be no more tender than conventionally handled beef, although of rather better flavour.

You have been requested to organize a trial comparing 'traditional' and conventional beef using samples obtained from your retail outlets. Design an outline bearing in mind the extent of natural variation and the vagaries of distribution and retail display. Which of the various taste panel tests would you consider most suitable? To what extent do you feel that taste panels are relevant to an investigation of this type and what alternative approaches are possible?

The 'traditional' beef was marketed, probably fortunately, without specific reference to tenderness or flavour. A number of members of the marketing department had, however, advised use of much more specific claims. Discuss, from both a technical and marketing viewpoint, the advisability of making specific claims for a product such as beef. Balance this against the need to maximize sales and recoup investment as quickly as possible.

EXERCISE 2.2

You are a graduate food scientist, but currently employed as marketing manager by a medium-sized abattoir and meat packing company. Much of the output is supplied as primal joints to major retailers, but attempts are being made to increase sales of branded retail packs, retailed through smaller outlets. At present the packs are overwrapped with an air-permeable film and your Board is unwilling to invest in modified-atmosphere packaging until the market can be proven. The complexities of distribution mean that a relatively long storage life is required and, while there are no major problems with microbial spoilage, there is a high rate of return due to discolouration (metmyoglobin formation). It has been suggested that this problem may be overcome by packing the meat in a pre-formed polystyrene tray with a heat-sealed aluminium foil lid to significantly reduce exposure to light. The lid would be colour-printed with a photograph of the same type and cut of meat, brand logo, etc.

Your comments that the consumer wishes to see meat before purchase are countered by the Managing Director, who points out that major supermarket chains retail fresh liver in this way and that similar packaging is used for cooked cured meats. You are therefore ordered to carry out a feasibility study.

Using your background as a food scientist, determine how successful protection from light is likely to be in preventing discolouration. At the same time assess the likely impact on the consumer of this type of packaging and discuss the various approaches to test marketing.

On a wider scale, do you consider that your company policy of establishing the market with an overwrapped product before switching to MAP is correct? What are the respective merits of continuing to market branded retail meat to small outlets (typically neighbourhood supermarkets) compared with attempting to establish a brand across a much wider spread of outlets, or concentrating on pre-packing for large multiple retailers?

## EXERCISE 2.3

Psychrophilic species of *Clostridium* can cause particular problems due to the inability to detect these organisms by routine microbiological examinations and due to spoilage developing towards the end of the storage period in meat which is otherwise normal. It has been stated that 'the source of the organism is not known but clearly ought to be established so that appropriate preventative measures can be established' (Dainty *et al.*, 1989). Consider the difficulties posed by the lack of an effective isolation method for psychrophilic *Clostridia* and review the various means by which these may be overcome. Draw up a strategy for developing a means of isolating or detecting the organisms, discussing the relative merits (with respect to this specific situation) of cultural, serological and genetic techniques. Assume that subsequent work shows psychrophilic species of *Clostridium* to be widely distributed in soil and water and that elimination of contamination with the organisms is not considered feasible. What alternative methods could be used to minimize problems of spoilage and what further information would you require about the organism to test their suitability? (Dainty, R.H., Edwards, R.A. and Hibbard, C.M., 1989, *Journal of the Science of Food and Agriculture*, **49**, 473–86.)

# 3

# UNCOOKED, COMMINUTED AND RE-FORMED MEAT PRODUCTS

---

## OBJECTIVES

After reading this chapter you should understand
- The nature of the various types of uncooked comminuted and re-formed meat products
- Factors affecting choice of ingredients and formulation
- The basic manufacturing technology
- Variations in processing between the various different types of comminuted product
- Chemical and physical changes during comminution
- The structure of comminuted products
- The role of preservatives in comminuted products
- Comminuted products and foodborne disease
- Spoilage of uncooked comminuted meat products

---

## 3.1 INTRODUCTION

Uncooked comminuted meat products are typified by the ubiquitous burger and the British fresh sausage. Products such as re-formed 'steaks' are also included in this category, although these are not strictly comminutes, being made from flaked rather than comminuted meat.

The development of comminuted meat products was primarily driven by economic factors and the need to utilize lower quality meat, trimmings, etc. In the past, comminuted products, such as sausages, often formed the main source of meat for the poor, who were unable to afford the more 'noble' cuts. Economic factors are still important to many consumers and re-formed steaks, for example, are considerably cheaper than whole cuts and can be more palatable. A further factor, of seemingly ever-increasing importance, is convenience. This is of importance both in home preparation of food and in catering, where burger bars are now a feature of every town.

---

### BOX 3.1  **Flowers of the forest**

The major burger chains, such as MacDonald's and Burger King, operate on a truly global scale and are themselves major beef processors. Much of the beef, especially for the US market, is imported from South America. There has been considerable concern amongst environmentalists that significant ecological damage is being caused by deforestation to produce the cattle grazing required to meet the demand for burgers in the US. This is denied by the burger companies, who claim that the beef is produced under conditions of fully sustainable agriculture. Accusations and counter-accusations have been made and the true situation is difficult to assess. On a wider scale, however, the controversy does seem to typify the conflict between the rich and over-consuming countries of the north, and the poor, producing countries of the south.

---

Comminuted meats often carry a high initial load of micro-organisms and preservatives are sometimes added. Sulphite is widely used in the UK and is also permitted in some other countries. In other cases nitrite is present, either as an ingredient or through use of pre-cured meat (e.g. bacon burgers). In the context of this book, such products are not considered to be cured (see Chapter 4, page 169).

Comminuted meat products have been made for many years and have a well established market position. In common with other food products, considerable product innovation has taken place in recent years. Some innovations involve extension of existing product groups, especially with respect to added-value products, while there is also considerable interest in low-fat products. Advances have been made in technology and, while many are of an evolutionary nature, the processor now has a significantly wider range of options available. Of particular significance was the development of the algin/calcium gelation system, based on the ability of alginates to form instant gels in the presence of calcium salts.

## 3.2 TECHNOLOGY

Although all uncooked comminuted meat products share a common basic technology, there is considerable variation in starting material and processing details. This results in a wide range of products which, to some extent, fill different market niches.

### 3.2.1 Comminution

Comminution plays an obvious role in reducing the size of meat pieces. In addition the process is important in extracting salt-soluble proteins and thus enabling the meat mixture to be bound together. Comminution also reduces the obtrusiveness of fat and connective tissue. Cooking properties are improved by break-up of connective tissue.

A number of different principles may be applied to comminution. The nature of the comminuted meat varies somewhat according to the method used and different methods of comminution may be used for different end-products.

### (a) Mincing and milling

Mincing is a relatively simple process and equipment is available ranging in scale from hand-operated domestic mincers to large industrial plant. The process is widely used at retail level for producing fresh mince as well as comminuted products. The most common type consists of a screw auger operating in a horizontal chamber (Figure 3.1). Meat pieces are fed in from a hopper mounted at one end of the chamber. Mincers cannot usually handle

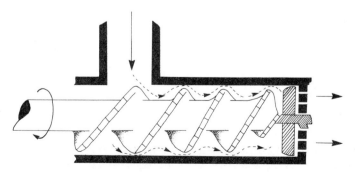

**Figure 3.1**   Schematic diagram of an auger-type mincer.

frozen meat and a temperature rise as great as 10°C may take place. Some types do permit semi-frozen meat to be used and latent heat of melting limits temperature rise. As the pieces pass through the chamber the meat is placed under considerable pressure and also undergoes tearing between the auger and the chamber wall. Final comminution takes place as the meat leaves the chamber. The outer part of the exit consists of a stationery perforated plate, which is mounted adjacent to either a rotating knife or a rotating perforated plate. Bone chips are a potential problem with all methods of comminuting meat, and some mincers are fitted with bone removal systems. These exploit differences in the density of bone and meat to force the bone into separate channels at the exit plate. Systems currently available are of limited efficiency.

Mincers produce a relatively coarse comminute, which is of an irregular nature due to the tearing action. Connective tissue causes problems unless cutting surfaces are properly maintained. This is a skilled operation which can cause some difficulty.

Milling is a similar process to mincing, the main difference being the absence of an auger and the higher speed of operation. The equipment is of vertical configuration, meat being fed to the comminuting knife and plates under its own weight. Material is more finely comminuted than by mincing, but some tearing occurs during passage through the fixed perforated plate. The operation is much faster than mincing and better suited to high throughput operations. Milling also has the advantage that additional ingredients are readily introduced with the meat and fully incorporated during passage through the mill. This has advantages in terms of convenience and technology, the most important being the dispersion of NaCl into lean meat and the consequent working of the meat in the presence of NaCl. This improves extraction of salt soluble proteins, with enhanced binding properties, yield, etc. (see page 167).

### (b) Chopping

The most common chopping equipment is the bowl chopper (Figure 3.2). There are many types but the basic design is similar in all cases. Chopping takes place in a circular, slowly rotating, curved bowl. A set of 3-12 knives rotate at high speed in a vertical plane close to the surface of one side of the bowl. The meat can be chopped very finely, the degree largely depending on residence

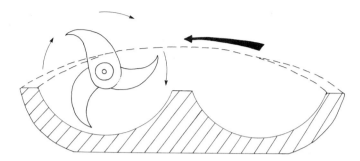

**Figure 3.2** Schematic diagram of a bowl chopper.

time in the bowl. The temperature of the meat is important. It is usual practice to introduce the meat at *ca.* -1°C; the final temperature should not exceed 22°C and, ideally, should be in the range 18–20°C. Lower initial temperatures are likely to lead to damage to knives, while at higher temperatures fat is overchopped and free fat is released.

The main action in bowl choppers is cutting and a high level of skill is required in sharpening and setting the blades. The sides of the knives and movement against the walls of the bowl also have a massaging effect. As with milling, bowl choppers are effective mixers of meat with other ingredients.

### (c) Flaking

Two basic types of equipment are available for flaking meat, block flakers and impeller flakers. Block flakers consist of a guillotine which cuts relatively thick flakes from the end of a frozen block of meat. The equipment is simple in design and operation, but requires the meat to be formed into a block of the correct dimensions and frozen. The meat temperature is important and tempering may be required to ensure an initial temperature of -2 to -4°C. Temperatures outside this range lead to quality faults, due to tearing of the meat at temperatures above -2°C and crumbling at temperatures below -4°C. Very little temperature rise (*ca.* 1°C) should occur during flaking.

The basic design of an impeller flaker is an impeller, consisting of a horizontally mounted shaft bearing a series of blunt blades, or paddles, which rotates at high speed within a static ring of blades.

Meat pieces enter at one end and are thrown against the blades. This results in the meat being sliced, flakes leaving the machine through the spaces between the blades. The width of the space between the blades determines the flake thickness. In general the flakes produced by an impeller flaker are thinner than those produced by a block flaker and more regular. Either frozen or unfrozen meat may be used and pre-forming the meat into blocks is unnecessary. The meat does, however, have to be cut into appropriate sized pieces of 200–500 g depending on machine size. Neither the block flaker, nor the impeller flaker, permit the incorporation of other ingredients.

### 3.2.2 Mechanically recovered meat

Mechanically recovered meat (MRM) is that recovered by machines from bones which have already been hand trimmed. Mechanically recovered meat consists of meat and fat from the bone, periosteum and, in some cases, bone marrow. Depending on the equipment used and yield of MRM extracted, there will also be granules of hard bone. These are normally few and should not exceed 0.05%, but levels as high as 5% have been reported.

Various machines have been developed for recovery of MRM. A widely used type consists of a rotating sieve device, which is fed with ground bone. The soft meat particles are forced through the sieve perforations (*ca*. 0.5 mm diameter) by back pressure created by an annular valve, which controls discharge of the waste bone from the sieve. The valve can be adjusted to vary the back pressure and thus the level of extraction of MRM. An alternative system uses a hydraulic ram to crush the bones at a pressure sufficient to semi-liquefy the residual meat. This enables separation of the meat from the solid bone material by a series of filters and channels.

Mechanically recovered meat is used in a number of comminuted meat products. Levels of use are highest in low quality products, but the amount tends to be limited by the cohesiveness of MRM causing mixing difficulties. Where bone marrow is present, functionality may be improved by the higher pH value, which increases water binding. In many cases, however, this increase is reversed by high levels of copper, iron and magnesium. The presence of high levels of metals, especially iron, may result in rapid development of oxidized flavours from lipids. Mechanically recovered meat

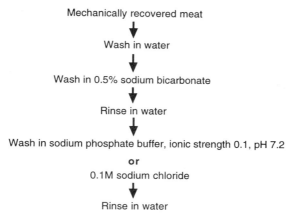

Mechanically recovered meat
↓
Wash in water
↓
Wash in 0.5% sodium bicarbonate
↓
Rinse in water
↓
Wash in sodium phosphate buffer, ionic strength 0.1, pH 7.2
**or**
0.1M sodium chloride
↓
Rinse in water

**Figure 3.3**   A method for production of meat surimi.

is often of poor microbiological status; it may adversely affect the storage stability of products and has been implicated as a source of pathogenic micro-organisms. There has also been concern, especially in the US, over the high level of calcium in some types of MRM.

There is now interest in increasing the functionality of MRM by washing the meat in a process similar to that used for production of surimi from fish. This involves repeated washing of minced meat with aqueous solutions to remove pigments, fat, etc. and produce a crude myosin extract (Figure 3.3). The process was first applied to poultry meat, especially back and neck, which is of poor functionality and, due to the high haem content, a dark colour. Meat surimi is now also prepared from red meats, including beef and mutton. Surimi has improved texture and binding properties and is whiter in appearance.

### 3.2.3 Manufacture of uncooked (fresh) sausages

It is considered that the original economic function of sausages was the utilization, in a manner acceptable to the consumer, of relatively large quantities of fat. This function still exists, but to some extent has been diminished by dietary considerations and the development of low-fat sausages. In addition, in the UK at least, there has been a tendency in recent years to develop premium sausages of high lean meat content, which may contain added-value ingredients.

Uncooked sausages, containing sulphite as a preservative, are a popular product in the UK and sausages of this type are often considered to be synonymous with the 'British fresh sausage'. Similar products are produced in the US and in continental Europe, although sulphite is not in general use outside the UK and a few other countries.

### (a) Meat and fat

Sausages can be made from meat of any species. Pork is most widely used in the UK, but beef is also common. Mutton sausages are most common in the major sheep-producing countries. In recent years poultry, especially turkey, has become a common ingredient as part of a general trend towards wider utilization of poultry meat. Expansion of deer farming means that venison is more widely available for processing and there is also small-scale production of sausages from game birds, especially pheasant.

Lean meat is the most important ingredient of sausages due to its role in water binding, maintaining the fat component of the mixture and in determining product cohesiveness (see pages 149–150). The higher the lean meat content, the higher the quality of the end-product and the fewer the problems during manufacture. In the UK, the minimum total meat content for pork sausage is 65% and for other sausages 50%, of which lean meat must be not less than 50% of the minimum meat content. Similar legislation exists in many other countries. In practice it would be difficult to make a satisfactory sausage containing less than the minimum legal meat content even when extensive use is made of extenders. In the UK, but not in some other countries, rind is legally considered to be meat. Rind contains a variable proportion of fat, which must be considered as part of the fat portion of the recipe.

In practice much of the meat used in sausage manufacture is semi-lean consisting of various trimmings, pork belly, flank, etc. This has consequences during processing, since it is not possible to add the meat and fat at separate stages. Fresh meat is preferred, but good quality frozen meat retains adequate functionality to make a satisfactory product. Frozen meat is best used without defrosting. Pre-rigor meat has superior water binding properties and is used in fresh sausages, especially in the US. Utilization of meat in a pre-rigor state can cause logistic problems, but desirable properties can be maintained temporarily by addition of NaCl to the meat after comminution. This presumably solubilizes protein before a strong bond is formed between actin and myosin. Attempts have been made to freeze salted, pre-rigor meat, but this results in a loss of functional properties. This problem can be overcome by cryostabilization of the meat by the pre-freezing addition of Polydextrose®, a non-sweet dextrose polymer, which maintains the protein in a less denatured form (see Chapter 8, page 371). Salting is not required, which means the meat can be used in low-NaCl products. Mechanically recovered meat may be used, the percentage being dictated by the quality of the sausage being manufactured and the binding properties. Part of the meat and fat content may be derived from reworked mix, broken sausages, etc. Such material may be retorted before re-use, although this adversely affects binding properties.

Fat is sometimes added as a pre-formed emulsion. Such emulsions are a convenient means of utilizing surplus fat, such as trimmings, from other operations. Fat loss during the cooking of sausages made from pre-formed emulsions is also reduced. Emulsions are based on a protein source, usually soya isolate or sodium caseinate. Manufacture involves high-speed chopping of fat with an equal quantity of water and a small quantity of protein. A protein–water matrix is formed, which sets on heating to form a solid structure incorporating the free fat. Soya-based emulsions, which are probably not true emulsions, are of poor keeping quality, although this may be improved by addition of *ca.* 2% NaCl at the end of the manufacturing process. The water in the emulsion is bound and not available for recipe purposes.

Meat and fat play a major role in determining the organoleptic properties of sausages and must be of suitable quality. Taints derived from the meat are almost invariably of microbiological origin. Micro-organisms may also be the cause of taints in fat but, where frozen fat is used, oxidative rancidity resulting from excessively long storage is more usual. The initial load of micro-organisms can be related to storage life and special 'clean' recipes may be used where longer than normal lives are required to meet the demands of special occasions, such as sales promotions.

### (b) Water

Water usually comprises 20–40% of the total recipe and is important in increasing meat binding and providing fluid conditions during fat chopping. It is usual to add up to 50% of the water as ice to minimize temperature rise, and microbial growth, during chopping. Water, used directly or for manufacture of ice, should be of potable quality and satisfactory chemical composition.

## (c) Rusk

Rusk contributes to the mouthfeel and texture of the sausage and absorbs any free moisture present. Rusk is made from a chemically aerated wheat dough, which is baked and ground to specified particle sizes. Most types of rusk are capable of absorbing 3–4 times their weight of water. Rusk is sometimes pre-soaked, but this is not normally recommended.

## (d) Casings

Casings comprise *ca*. 1.0% of the total weight of the sausage and are of two broad types, natural and artificial. Natural casings are made from cleaned intestines of cattle, pigs or sheep. 'Rounds' or 'runners' are made from small intestine and 'middles' and 'bungs' from large intestine. After cleaning, the intestines are packed into solid NaCl and have a virtually indefinite shelf life at *ca*. 5°C. The casings must be desalted by soaking and wound on to a spool before use. Natural casings are often considered to impart traditional virtues to sausages and may give a better chewing sensation, but use is now limited to small-scale operations.

Artificial casings offer considerable advantages in terms of convenience, requiring no special preparation before use. They are also more robust and of consistent diameter. Sausages made with artificial casings are straight rather than curved, which facilitates automated packaging. The most widely used material is regenerated collagen, which is prepared by dissolving hides in acid and extruding the collagen solution into a concentrated solution of ammonium sulphate. This results in the collagen being precipitated as a continuous tube. Various other compounds, including cellulose and its derivatives, may be added and are intended to make the

---

* Some added value ingredients of fresh sausages, such as chestnuts, contain anthocyanin pigments. Under some conditions, these react with metal ions, especially iron and aluminium, to form brightly coloured grey, blue or purple complexes. In such cases, the source of the metals has often been found to be food handling equipment or utensils. Occasionally, however, the problem has been caused by the high iron content of the water, reflecting either the geology of the catchment area or contamination during distribution and storage.

properties of the collagen casing more closely resemble those of natural casing.

In the past cellulose has been used either as an additive to regenerated collagen casings or to make casings for skinless sausages (see page 134). More recently, cellulose casings have been introduced as a direct alternative to collagen. The cellulose is structured and is said to closely resemble natural casing during chewing.

### (e) Sodium chloride and phosphates

Sodium chloride plays an important technological role in solubilizing myofibrillar proteins and increasing water-holding capacity. There is also its role as a flavour enhancer. Sausages usually contain only 0.5–0.75% NaCl, but initially the concentration in the extracted moisture surrounding the meat is much higher, enhancing protein solubilization. Low sodium sausages may be prepared by replacing up to 50% of the NaCl with KCl.

Pyrophosphates and tripolyphosphates are effective in increasing water binding. Sodium acid pyrophosphate, tetra sodium pyrophosphate and sodium tripolyphosphate are most commonly used, although the potassium salts are also permitted. Phosphates also enhance the effect of NaCl on meat proteins and thus contribute to meat binding. In addition, the activity of some antioxidants may be enhanced. Phosphates are common ingredients in sausages, especially those of relatively low meat content.

### (f) Extenders

Soya products (primarily soya concentrate or isolate) are the most common extenders, although use is also made of dried milk powder, sodium caseinate, etc. Chicken skin is used in low-cost poultry products, but the quantity added is limited by the high lipid and collagen content leading to poor quality. Extenders are

---

* During the late 1970s and early 1980s, in the UK, intense price competition occurred between supermarket chains. A number of very low cost products appeared, many of which were of correspondingly low quality. Among these was a pork sausage of low meat content, in which sodium carboxymethylcellulose played a major functional role. The meat was extremely finely chopped and in many ways the structure of the sausage was a carboxymethylcellulose gel, in which the meat was suspended. In a number of cases, the meat content of the sausages was found to be illegally low, and successful action was initiated against some retailers.

used to improve the body and reduce cooking losses of sausages with low meat content. As such, use of extenders is often associated with poor quality products. More recently, however, soya protein, especially textured protein isolate, has been used in low-fat sausages (see pages 142–144). Soya proteins also provide some protection against lipid oxidation.

### (g) Preservatives

Sulphur dioxide, usually added as sodium metabisulphite, is used as a preservative in sausages and similar products in the UK, Eire and some countries of the British Commonwealth. Elsewhere, its use is banned in meat products due to loss of major vitamins. A small number of European uncooked sausages contain nitrite as a preservative, but in other cases and in the US no preservatives are present. Products made without preservatives have a very short shelf life, even under adequate refrigeration, and trials have been made using sodium lactate and other 'more acceptable' preservatives.

A number of phosphates have been shown, by *in vitro* plate assays, to have antimicrobial properties but the extent to which these are significant in comminuted meats is a matter of some debate. It should be recognized, however, that phosphates are not added for their antimicrobial properties and that any activity as a preservative is coincidental.

### (h) Other ingredients

Herbs and spices are usually added and may act both as flavour enhancers and as flavourings in their own right. Herbs used include sage, parsley and mint, while common spices are paprika, coriander and nutmeg. Herbs and spices are usually added as a commercially prepared finely ground mixture, but large, visible pieces of herbs, such as sage and rosemary, are characteristics of some regional types of sausage in the UK. Other flavouring, including nuts, tomato and even celery, is added to some variants.

In some countries, colouring cannot be added to sausages. Colouring is permitted in the UK, although in practice use is restricted. Red 2G is used, which imparts a rather lurid pink colour to the sausage. Colouring is usually incorporated with the mix, but in the case of skinless sausages it may be derived as a surface layer from the removable casing.

## (i) Comminution

The degree of comminution of sausages varies, but in most cases is fairly coarse. A few continental types are effectively emulsion sausages (see Chapter 5, page 245). Bowl choppers are widely used in sausage manufacture and permit incorporation of all ingredients in a single piece of equipment.

Optimal conditions of comminution represent a compromise between a number of conflicting requirements. It is necessary, for example, both to have a high level of meat binding and to retain as much cellular fat as possible. Optimum conditions for meat binding, however, include a long chopping time at low temperature, while those for cellular fat retention include a short chopping time at a relatively high temperature. At the same time, optimum conditions for meat binding include the conflicting conditions of high NaCl content and moderate water content.

Procedures for comminuting the ingredients and preparing the mix vary according to individual factory practice, whether the sausage is made with lean meat, semi-lean meat or a mixture of the two types, and whether a coarse or a fine degree of comminution is required. In Europe pre-salting the meat for 1–2 days at <5°C before adding the water is common practice for some finely chopped products. Alternatively, all the ingredients are added together and chopped, or milled, at high speed. In the latter procedure it is critical to avoid excess heating.

## (j) Filling and linking

Filling is a technically straightforward operation, but good control is essential for end-product quality. Sausage mix is transferred to the hopper of the filling machine and fed to a nozzle by a piston pump. The casing is filled from the nozzle on a continuous basis and linked, either manually or mechanically, to form a string of individual sausages. The operation should be as gentle as possible to avoid 'creaming'. This involves formation of a layer of emulsified fat in a 'cream' of worked meat in the sausage meat adjacent to pipe walls. Resulting sausages have a poor texture and an unattractive pale colour. Creaming is most common with high speed machinery, operation of which involves a high level of mechanical action on the meat. Problems may also occur if there are significant increases in mix viscosity due to delays between making the mix and filling. Avoidance of delays under normal circumstances is a matter of sound plant management, but breakdowns inevitably occur. Attempts to reduce viscosity by stirring, or reworking, usually lead to creaming and are only partially successful.

Fully automated equipment (Chapter 5, pages 246–247) may be used for the manufacture of fresh sausages. At present, however, traditional methods remain completely dominant.

### (k) Packaging

Sophisticated packaging is not used for sausages. The usual pack is an overwrap of air permeable film or, less commonly, the sausages are packed on a polystyrene tray and overwrapped.

Attempts have been made to use vacuum packaging to extend the storage life, but these have been abandoned due to the adverse effect on the appearance of the sausage. Systems have been developed in which film wrapped packs of sausages are assembled in larger packs in a $CO_2$ atmosphere. This bulk pack is used for transportation and pre-display storage. The system is unwieldy and does not appear to have been widely used. Vacuum packaging in chubs is successfully used for sausage meat.

### (l) Special types of uncooked sausages

Skinless sausages are filled into a special cellulose casing, which effectively acts as a mould. The sausages are then heated by passage through hot water, or moist air at 60–70°C. Heating is sufficient to heat-set the proteins at the surface and bind the sausage together. There is also a small degree of cooking of the interior. After cooling the casing is split, peeled away from the sausage and discarded.

---

* A comparison of microwave sausages from different manufacturers suggests that the products had received different degrees of cooking at the factory. In some cases, only warming was required in the domestic microwave oven, but in other cases it appeared necessary to complete the cooking in the home. Products which have the appearance of being fully cooked but which are, in fact, raw or semi-cooked are considered inherently dangerous (cf. burgers, page 157), unless the labelling adequately informs the consumer. It is not sufficient to rely on the cooking instructions being read; the information must be prominently displayed on the front of the pack.

'Microwave' sausages are effectively pre-cooked sausages, intended for reheating in a microwave oven. The sausages are cooked in either steam or hot air and are then charred in a flame to simulate the appearance associated with grilling. Sausages of this type are cooked products and full precautions must be taken to avoid cross-contamination from uncooked material (Chapter 5, page 232).

### 3.2.4 Manufacture of burgers and related products

Burgers are derived from hamburgers, a large beef sausage which is cut into slices before cooking. These were imported into the US with German immigrants and became widely popular. Hamburger sausages are still made, but the term is now applied more commonly to a product of similar organoleptic properties made as flat slices. In the UK the same products are referred to as beefburgers, or burgers, possibly due to the misapprehension that hamburgers are made from ham. Similar products, such as baconburgers and lambburgers are named for their constituent meat.

---

BOX 3.2   *Ich bin ein Berliner*

It was common practice in Germany to name food products, especially sausages, after the town where manufacture commenced. President Kennedy's speech at the Berlin Wall, in which, to demonstrate solidarity with the inhabitants of that beleaguered city, he proclaimed '*Ich bin ein Berliner*', was hailed by western politicians and the press as epoch-making. Rather less impressed were thousands of schoolchildren, who failed to understand why the President of the United States of America should describe himself as a sausage.

---

*(a) Ingredients*

The basic ingredients of burgers are the same as sausages but the meat content is higher: the UK minimum is 80% and products with

---

* The ever-increasing popularity of burgers and a continuing high level of competition between retail outlets means that quality attributes are of major importance in many markets. This has led to research into cattle rearing practices to ensure not only high yield but also optimal quality. The results are not always clear-cut. In the case of feeding regimen, for example, steers fed on a diet of grain and silage produced a frozen burger of superior flavour to forage-fed steers. No difference was detected in fresh burgers.

100% meat content are not uncommon. The fat content is correspondingly lower and while UK regulations permit 40% of the meat content to be fat, this level is rare in practice.

Burger products containing significant quantities of extenders have previously been considered as down-market products. As with sausages, however, plant proteins are seen as a means of lowering fat and cholesterol content, while improving cooking yield. It has also been reported that soya products significantly decrease lipid oxidation and discolouration.

Sulphite is permitted as a preservative in the UK. Elsewhere preservatives are not used, although there has been interest in sodium lactate. Nitrite is permitted in baconburgers, made from pre-cured meat.

The high meat content, especially when beef is used, means that herbs and spices are less important in flavouring of burgers than of sausages. NaCl, however, is always present both for technological and organoleptic reasons. 'Novelty' burgers are also available which have high levels of flavouring, including chilli, cayenne pepper and curry powder. These products may also be made with a high level of colouring. In other types of burger, colouring is not generally used. Burger products are also available in which liquid smoke extracts are used to impart a 'barbecue' flavour.

### (b) Comminution

Burgers are coarsely chopped to produce a fibrous, crumbly texture. Ideally, mechanical action, together with NaCl, should be just sufficient to bind the product before and after cooking. In practice, the need to disrupt connective tissue means that more extensive comminution is necessary. The use of low-quality meat containing large amounts of connective tissue can lead to a burger of an undesirable, highly cohesive texture.

* Connective tissue is a problem in many types of comminuted meat product. Attempts have been made to solve or at least minimize the problem by enzymic degradation. Ante-mortem injection of chicken legs with papain has been used successfully to degrade collagen, but is an expensive procedure and not universally acceptable to consumers. An alternative approach has been devised, the addition of microbial collagenase derived from species of *Clostridium*, but it failed to improve the quality of beef comminutes. The use of *Clostridium* as source may also be considered unacceptable.

Although bowl choppers have been used in burger manufacture, the degree of comminution is too high for a good quality end-product. For this reason, mincing, milling or flaking are more common. A separate mixing stage is required if mincing or flaking is used. Mixing may take place after comminution, but this procedure can lead to overworking. Alternatively comminution can be carried out in two stages, with an intermediate mixing stage.

## (c) Moulding

The simplest method of moulding burgers to shape is the hand press. This method, however, is slow and only suited to small-scale manufacture. Extrusion processes are most widely used for large-scale purposes. Two main types exist: extrusion moulding and extrusion and slicing. In either case mechanical work is involved in pumping, etc. and this should be minimized to avoid over-firm binding. Extrusion moulding involves forcing the mix through an orifice into a moulding chamber. Depending on the design of the machine, considerable shear stress is induced. Meat fibres tend to become aligned along stress lines, along which shrinkage occurs during cooking, resulting in misshapen burgers at point of consumption. Although a cosmetic problem, this is considered unacceptable by retail chains and many burgers are formed into an oval shape, which becomes circular on cooking. Extrusion and slicing is a simpler process in which mix is extruded from a pipe and cut into slices with a knife or guillotine. The process is less well suited to high speed manufacture and can result in poor appearance due to tearing.

## (d) Packaging

Packaging of burgers is simple. Chilled burgers are placed in polystyrene trays, separated by paper to prevent adhesion, and over-wrapped in an air-permeable film.

## (e) Special types of burger

Burgers lend themselves to variation better than uncooked sausages and in recent years a wide range of 'novel' products has been developed. Some are, in fact, evolutionary developments which, to some extent, reflect the higher quality now expected of burger products. The basic cheeseburger, for example, has grated cheese incorporated into the mix. This results in a product of 'chewy' consistency and often a poor flavour balance. A more sophisticated product, which also more closely resembles the cheeseburger served in burger houses, consists of a basic burger topped with mozzarella cheese, or cheese base. The product is coated and breaded, before freezing to maintain integrity.

Some innovation involves presentation rather than technology. An example is the do-it-yourself burger kit, which consists of burgers, packaged together with sesame buns and individual packs of sauce and relishes. Such products provide an easy, albeit expensive, means of preparing a burger house meal in the home and are effectively a counter to the growing market share of burger houses.

---

BOX 3.3 **The great all-day breakfast**

A growing number of products are available which emulate meals produced in fast-food outlets. In addition to 'real' burgers, mixed products are available which imitate the popular 'all-day' breakfast, using a similar packaging and presentation. It is even possible to purchase 'burger accessories', paper napkins, etc. with generic logos vaguely resembling those of burger chains.

---

'Microwave' burgers are usually intended to be fully cooked in a microwave oven. Special packaging is used to ensure full efficiency and even heat distribution. Microwave heating does not induce browning and burgers cooked in this way are anaemic in appearance unless special precautions are taken. One method is to sear the surface with a flame to produce the appearance of grilling. Alternatively proprietary browning mixtures may be included in the formulation.

*(f) Related products*

A basic burger mix lends itself to a number of applications. Meat balls are a similar product formed into a ball. In many cases the meat balls are combined with a sauce as part of a complete dish. Other added value products of this type are also available, extensive use being made of poultry meat. Meat mix, for example, can be formed round a centre of cheese base. In the UK, meat balls have a minimum legal meat content of only 35% (cf. croquettes, etc.)

Various roasting products are also available based on burger-type mix. In these cases the mix is simply extruded on to an aluminium foil tray, or filled into an aluminium foil container. A stuffing may also be present. Such products may be referred to as meat logs, or meat loaf, and have a minimum meat content, in the UK, of 65% (50% with stuffing present). In the original form, however, meat loaf was made from a more finely comminuted mix, reflecting the high proportion of meat trimmings in the raw material. Products of this type may also be referred to as 'roasts', but care must be taken to avoid confusion with re-formed or frozen restructured products.

## 3.2.5 Manufacture of croquettes, rissoles, etc.

Croquettes, rissoles and similar products superficially resemble burgers. The meat content, however, is significantly lower (35% minimum in the UK) and binders such as potato flour or wheat flour maintain integrity rather than meat and fat binding. Products of this nature are usually coated and breaded.

In most cases, any form of comminution can be used, although flaking is preferred for poultry 'fingers' to obtain a better texture. Comminution procedures are less critical than with other comminutes and NaCl, flavouring and other ingredients can be mixed in at any stage. Care must be taken to avoid excessive release of free fat. Coating with batter and a layer of breadcrumbs (rusk) is usually carried out in consecutive operations on the same equipment. Flour, water and baking powder are the basic constituents of batter, which also contains a soluble protein, such as egg or milk powder, to provide body and to aid in adhesion. Rusks are available on a proprietary basis and particles vary in size according to purpose. Rusks may also be coloured and flavoured. Coating is a simple process which involves the product passing on a perforated conveyer through a curtain of batter and a shower of crumbs. Excess coating is removed by shaking or air jets. The batter may be set by passage through a bath of hot oil.

## 3.2.6 Manufacture of re-formed meat products

Manufacture of re-formed meat products exploits the binding properties of meat, in the absence of NaCl at $-2°C$. Re-formed meat products are intended to mimic, as far as possible, the texture, taste and colour of whole meat joints. Re-formed products are particularly popular in mass catering and have the added advantage of a short cooking time, due to breakdown of connective tissue.

### (a) Ingredients

The only major ingredient is meat. Less expensive cuts such as forequarter meat and trimmings from other processes are widely used, together with mechanically recovered meat and rind, sinews, etc.

### (b) Processing

Flaking is highly effective in dealing with the high levels of connective tissue. The meat is flaked at $-2°C$, preferably using an impeller flaker to obtain clean cutting of the connective tissue. The flaked meat is then held for several hours at $-2°C$ and stirred occasionally. Flakes are then filled into moulds, formed and shaped under pressure, the low temperature being maintained throughout the process. Shapes are chosen to approximate to whole joints or steaks. The re-formed meat may then be frozen at *ca.* $-20°C$ or held at *ca.* $-2$ to $-5°C$. In the latter case, the product is usually allowed to defrost during predisplay storage at the retail outlet and displayed with fresh meat products at $2-4°C$. In some cases, fully frozen products may also be defrosted before display. Care must be taken not to confuse re-formed meat products with restructured, which employ a different technology and which are usually fully frozen products.

### 3.2.7 Manufacture of comminuted and restructured products using the algin/calcium gelation system

Considerable effort is being expended in development of unfrozen restructured and comminuted products bonded by the algin/calcium gelation system. The need for NaCl is avoided, permitting manufacture of low-sodium products. At present, however, a number of problems are still to be resolved (see below).

The system is based on the ability of sodium alginate to gel spontaneously in the presence of free calcium ions. The extent of gel formation depends on the availability of $Ca^{2+}$ ions, which is controlled by the solubility of the calcium salt and the pH value of the solution. In meats, calcium carbonate is used as a source of $Ca^{2+}$ ions. This salt is relatively insoluble in cold water and the resultant limited availability of $Ca^{2+}$ ions leads to the presence of unreacted sodium alginate, which has been associated with undesirable flavour and mouthfeel. This problem may be overcome by use of a slow release acid, to change the pH value of the solution and thus control the ionization of $CaCO_3$ and the level of available $Ca^{2+}$ ions. Initial work, using glucono-δ-lactone encapsulated in vegetable oil, was not fully successful, but satisfactory results have been obtained using adipic acid encapsulated in water-soluble maltodextrins.

Some problems have been reported with development of atypical colours during storage but reports tend to be conflicting and, on some occasions, the algin/calcium system appears to sustain quality through minimizing water loss and maintaining desirable colouration. Similarly, there have been conflicting reports concerning the effect of the presence of an algin/calcium gel on the microbial stability of meat products. On most occasions, it appears that microbial growth rate is unaltered or slightly lower than in conventional products, but it has also been reported that growth is faster and that gas production by micro-organisms is enhanced.

Although the algin/calcium gelation system may be applied to both comminuted and unfrozen restructured meat products, the technological objectives are rather different. In the case of comminuted products, it is necessary to form a continuous gel to bind the meat particles into a cohesive structure. This means that the importance of meat binding is much reduced and comminution may be controlled to optimize product quality without the compromises necessary during conventional manufacturing.

The most common method of producing an algin/calcium gel involves mixing the comminuted meat with Na alginate, resting for a short period and then mixing with $CaCO_3$ (internal setting). Alternatively, the $CaCO_3$ can be introduced by passage through a tank containing a solution of the salt.

In restructured meat, the role of the alginate/calcium gel is to 'glue' meat pieces together. This requires a gel which imparts maximum juncture binding strength and minimum thickness. Where meat pieces are relatively small, internal setting (as used for comminuted meat) can be successfully used. Future developments, however, are likely to involve bonding of relatively large muscle blocks, which has greatest scope for new product development. In this case a pre-formed gel is used, which is coated evenly over contact surfaces of the meat pieces before moulding.

Although much attention has been given to the algin/calcium gelation system, other binding systems for low-sodium products are being developed. Possible alternatives are whey proteins or surimi.

### 3.2.8 Manufacture of low-fat comminuted products

Although there is interest in producing low-fat meat products of all types, most effort has been expended on uncooked comminutes, especially of the burger type. This results from the economic importance of comminutes in the US, where concern over dietary fat is greatest. Fat is important in determining the organoleptic properties of meat and, while a variety of strategies may be applied in development of low-fat products, the basic concern is always to reduce fat but retain traditional flavour and texture. Simply reducing the fat content of the recipe leads to a product of poor organoleptic quality.

Two basic means exist of producing low-fat meat products: the use of fat replacers and the physical removal of fat. At present, the most widely used and successful strategies involve fat replacers, although there is increasing interest in methods of physically removing fat.

### (a) Use of fat replacers

The sensory characteristics of juiciness and mouthfeel are associated with the fat content of comminuted meat products. To a large extent these may be retained in low-fat products by use of binders. Binders have, of course, been added to meat products for many years both for technological reasons and to reduce cost. In formulating low-fat products it is essential not to confuse objectives; economy should not be the main consideration.

Binders are available with a number of different properties, but all those used in low-fat products increase water binding. Carrageenan is currently very widely used in the US. Three basic types are available: iota-, kappa- and lambda-carrageenan, of which iota- and kappa-carrageenan are gelling agents and lambda-carrageenan a non-gelling thickener. Iota-carrageenan will gel with calcium ions (cf. algin/calcium gelation system) to form a synersis-free, clear plastic gel with good resetting properties after shear and it is recommended for use in low-fat products. Iota-carrageenan has very good water retention properties and some types enhance processing through cold solubility and freeze–thaw characteristics. The presence of NaCl in solution inhibits swelling of the carrageenan but this difficulty may be overcome by the use of NaCl encapsulated with partially hydrogenated vegetable oil (encapsulated salt). Extraction of proteins is much reduced, but meat binding is of minimal importance in products made with carrageenan.

Oat bran and oat fibre can be used in low-fat comminuted meats to increase water retention and enhance texture. Oat bran also improves mouthfeel and a product, Leanmaker™, is available which consists of a blend of specially processed oat bran, flavours and seasoning. This product has been successfully used in manufacture of pork patties, in a process in which 90% lean pork is mixed with Leanmaker™. The mixture is chopped, ground, extruded into casings, chilled overnight and then sliced.

Soya is a traditional binder that has been used in comminuted meat products for many years. Soya protein is available in several forms but hydrated, textured protein isolate is probably most effective in low-fat products. This material is familiar and can be handled in a similar way to the meat; use is straightforward and the end-product is nutritionally complete.

Modified food starches are used in low-fat products to maintain juiciness and tenderness. Starches are cheap and they use a familiar technology; the end-products (burgers and sausages) are highly acceptable to consumers. Various types of starch have been proposed; Slenderlean™ is based on modified tapioca starch and flavouring and can be used in manufacture of products with less than 10% fat.

Maltodextrins of low dextrose equivalent (<20) are permitted as binders in the US and are also used as low-cost, technically simple fat replacers. An oat-based maltodextrin (oatrim) is produced by α-amylase and sets as a fat-like gel. This product is marketed as Leanesse™ and has been used in several applications, including a chub-packed comminuted beef product.

Plant oils may also be used as fat replacers. Partially hydrogenated corn oils or palm oils are particularly effective in replacing beef fat.

Soya oil emulsion is also effective at levels of up to 25%, especially when used in conjunction with isolated soya protein.

A number of functional blends (defined as ingredient blends formulated to meet specific technological aims) have been developed as fat replacers. One type, Prime-O-Lean™, is used as a direct replacement for animal fat at levels up to 25% and comprises water, partially hydrogenated canola (rape) oil, hydrolysed beef plasma, tapioca flour, sodium alginate and NaCl.

The technology of low-fat meat products is still fairly new and continual development of alternative products is likely. Wild rice, for example, has been proposed as a binder and is also claimed to have beneficial effects on flavour. It also seems likely that there will be further developments in the alternative technology: the physical removal of fat.

### (b) Physical removal of fat

At present there are two main approaches to physical removal of fat: the use of supercritical $CO_2$ to reduce cholesterol levels and separation processes to produce fat-reduced meat. Supercritical $CO_2$ has been succesfully used on a pilot scale, although cholesterol reduction is not universally considered to be desirable. Separation processes producing fat-reduced beef have been developed on a commercial scale in the US but fat-reduced beef is not currently permitted in comminuted products. A rather different concept involves the use of microwave cooking pads, which absorb fat lost during microwave cooking, minimize contact between the meat and released fat and consequently allow a higher level of 'cooking out'. The principle is the same as grilling comminutes on a griddle, rather than pan frying, and is only applicable to products which can be adequately cooked by domestic microwave ovens. Most products of this nature are made from non-woven polypropylene fibres; wood fibres are also used and can be formed into a cook-in tray.

### 3.2.9 Quality assurance and control

Traditionally, considerable reliance is placed on selection of raw materials (especially meat) and end-product testing. Visual inspection may be supported by laboratory tests, especially for water-holding capacity. Comminution and other processes are controlled to predetermined standards for each formulation and there is only rarely any attempt to optimize processing according to the characteristics of the meat. This situation is likely to change with the development of in-line control devices which have the potential for computer-based feedback control. In this area, optical sensor technology is currently dominant. Optical sensors are available, for example, that measure lipid content, pH-related aspects of functionality and collagen content. The ideal is the development of optical sensors for monitoring the main functional properties of the comminute mix. This would permit feedback control of composition and processing, coupled with minimum-cost formulation. In most plants, however, control of processing operations still relies heavily on operator control and supervision.

Some meat products manufacturers have on-site abattoirs and are able to exert control over the meat throughout the production cycle. Formal systems are required to ensure that refrigeration is adequate at all stages of processing. This is particularly important with storage of bones before mechanical recovery of meat and of off-cuts. In some cases, meat is brought in from outside sources and this requires control to be extended to procurement. The weakness of the control procedures instituted for incoming meat was illustrated some years ago by the findings that burgers and other comminuted products were adulterated either with meat of other species, including horse and kangaroo, or with knacker

---

BOX 3.4  **Rich man, poor man, knacker man . . .**

Considerable publicity was given to the adulteration of burgers which involved, amongst others, a large multi-national chain. The extent of the problem was never fully known, but it seems likely that deception was widespread within the meat industry. Equally, the manufacturing companies were not necessarily innocent victims and in at least one case executive management were fully aware that pork was being substituted for higher priced beef. The use of knacker meat, which involves risk to public health, was probably also far more common than appreciated. The extent to which manufacturers were aware of the dubious origins of brought-in meat is not known; the only certainty is that someone, somewhere, became very rich.

meat (meat classified as unfit for human consumption). Meat species may be determined by relatively simple serological tests (see page 153), but it is not possible to detect knacker meat unless gross defects are present. The only real protection lies in obtaining meat only from suppliers of known probity and integrity.

Close control of meat quality, whether brought in from an outside supplier or butchered on site, also reduces the reliance on laboratory examinations for ensuring microbiological quality. Microbiological problems can also stem from delays in refrigerating the products after manufacturing. This stage is a critical control point with respect to prevention of premature spoilage and should be under the supervision of a responsible person.

End-product testing should be restricted to verification that earlier control procedures have been effective. Chemical analysis is required for meat content and, where appropriate, sulphite or nitrite to ensure legal requirements are met. Additional analyses to ensure compilation with specifications may be made for fat, protein and NaCl content. Cooking is widely used to assess organoleptic qualities and also to ensure that cooking losses are within acceptable limits. The importance of rheological properties is now recognized and instrumental measurement is now being introduced. Rapid microbiological testing (see pages 163–164) has been used as part of a positive-release system. This is not considered necessary if control procedures for raw materials and handling during processing are adequate. Microbiological analysis can, however, be of value in detection of both long and short term trends indicative of falling or rising control standards.

## 3.3 CHEMISTRY

### 3.3.1 Nutritional properties of comminuted and re-formed meat products

Nutritional properties of the meat component are unaffected by comminution and other technological processes. In products made using sulphites as preservatives, however, there is loss of the B-complex vitamins thiamine, riboflavin and $B_{12}$. Meat is an important source of these vitamins and their loss, especially that of thiamine, is of considerable dietary significance.

### 3.3.2 Colour of comminuted meat products

Brown or grey discolouration of comminuted products, such as burgers, is a common problem. The most common cause is formation of metmyoglobin, which is a consequence of oxygen-dependent meat enzymes and aerobic micro-organisms successfully competing with meat pigments for oxygen. The mechanism of metmyoglobin formation is common to all fresh meat (see Chapter 2, pages 93–97 for discussion), but micro-organisms are particularly significant in comminuted products, due to the large surface area and the presence of high numbers of micro-organisms. In deeper products, such as meat rolls or roasts, reduced myoglobin may be present in the interior and distinct zones of colouration may be observed, reflecting the extent to which oxygen is able to diffuse from the surface. Formation of metmyoglobin can be variable and, occasionally, discoloured areas are present adjacent to, and fully demarcated from, areas where colouration is bright pink. In some cases, this appears to be caused by conversion of metmyoglobin to myoglobin derivatives by spoilage micro-organisms, including *Kurthia* and *Lactobacillus*. The pink colour resembles that of fresh meat but is usually distinguishable by a fiery appearance.

Oxidation of meat pigments can result from use of low-quality fat containing high levels of peroxides or, in burgers, from $H_2O_2$ production by lactic acid bacteria. Discolouration can also occur where onions are incorporated in the mix, due to high concentrations of pyruvic acid formed during conversion of precursors to the lachrymator, thiopropan-*S*-oxide.

Sausages are subject to a number of specific problems. 'Pressure marks' are the result of oxygen deficiency where packed sausages are in close contact. Pigment is initially converted to reduced myoglobin and subsequently, as some diffusion of oxygen occurs, to metmyoglobin. 'White spot' appears to be an oxidative defect, which involves formation of circular grey or white areas that increase in size with continuing storage. There is an association with low $SO_2$ levels and use of fats with a high peroxide content. The use of pre-soaked rusk also appears to be a contributory factor. 'Blue spot' or 'purple spot' can be a common cause of complaint. In most cases, the cause is an optical effect associated with the presence of rind pieces directly below the sausage skin. Similar optical effects have been responsible for other coloured spots.

Formation of blue spots has also been attributed to interaction of rusk constituents with residues of iodine-based sanitizers.

### 3.3.3 Flavour of comminuted and re-formed meat products

Flavour is derived primarily from the meat (including fat) and any herbs, spices or added flavourings present. The use of poor quality fat can introduce off-flavours and there may be rapid (within 1 hour at ambient temperature) development of off-flavours as a result of oxidation of unsaturated lipids. This phenomenon is usually associated with cooked meats ('warmed-over flavour'), but can occur in comminutes as a result of rupture of the muscle membrane system and exposure of labile components to oxygen. Warmed-over flavour is discussed fully in Chapter 5, pages 260–266.

Sodium chloride enhances flavour but is not usually present at sufficiently high concentration to impart a definite saltiness. Substitution of KCl for part of the NaCl can result in an unpleasant metallic taste. Sodium chloride, but not KCl, acts as a pro-oxidant in all types of lipid oxidation.

Additional ingredients, such as binders, should be neutral in effect on flavour. Additives can introduce undesirable flavours if used at high levels. Some types of phosphate, for example, impart a bitter or metallic taste at concentrations in excess of 0.3 to 0.5%. Some persons appear to be much more sensitive to this than others. Atypical flavours have also been associated with soya flours, 'beany' tastes being detectable when some types are used at concentrations above 5%. Problems of this nature have been minimized by modern milling and processing techniques for soya.

### 3.3.4 Physical and chemical nature of comminuted and re-formed meat products

The functional properties of the myofibrillar proteins of meat are of major significance in conventionally made comminuted and restructured meat products. Three types of interaction are involved: protein:protein (meat binding), protein:water (water holding, or binding) and protein:fat (fat binding). The relative importance of these interactions varies according to the nature of the product.

## (a) Sausages and burgers

During comminution there is considerable physical disruption of muscle tissue through damage to sarcolemma, endomysium and integrity of muscle fibres. In the presence of NaCl, at ionic strengths of 0.6 or higher, comminution causes swelling and depolymerization of the myosin and extraction of myofibrils from the muscle fibres. Proteins within swollen fibres and solubilized monomeric myosin have much greater water-holding capacity than highly ordered and aggregated native myosin. This results from the exposure of binding sites to solvent. Myosin contains a total of 38% polar amino acids and has a high content of aspartic acid and glutamic acid residues, each of which can bind 6–7 molecules of water.

Water binding by myosin is further enhanced by the effect of NaCl in increasing the net negative charge and in breaking ionic bonds. These factors lead to molecular swelling and enhanced water uptake.

Muscle fibre swelling and myosin solubilization act to increase the viscosity of the protein matrix in the exudate. The viscosity may also be increased by low temperature gelation of the protein matrix. Exudate provides sufficient binding for the product to maintain a structure in the uncooked state and gels to form a solid structure on cooking (Chapter 5, pages 270–271). Myofibrillar proteins are fibrous in nature and high levels of solubilization into exudate mean a decrease in fibrous texture in the cooked product. Exudation can be controlled by varying the degree of comminution, or other mechanical working, and the quantity of NaCl added. It is also affected by type of meat, chicken and pork producing exudate most easily and mutton and beef least easily. There is also wide variation between meat of the same species and even the same cut. White muscle myosin and actomyosin from chicken and white muscle myosin from beef have a higher water-holding capacity and solubility than the equivalent red muscle. This is attributed to the presence of different myosin and actomyosin isoforms in different muscle fibres. Myosin is very readily extracted from meat processed pre-rigor, due to the relatively high pH value and levels of adenosine triphosphate. Limited proteolysis of the myosin heavy chain occurs during ageing and has an adverse effect on functional properties.

Myofibrillar proteins also play a major role in fat stabilization. The extent to which free fat is released during comminution depends largely on the length and temperature of the process. In the case of coarse-cut burgers and sausages, fat cells are not disrupted to a significant degree and fat cell membranes remain intact to stabilize the fat. The other extreme is represented by finely comminuted 'emulsion' sausages, in which a high proportion of the fat is liberated during comminution and which require a high level of fat binding for stability. Two mechanisms have been proposed for fat binding. The first, the physical entrapment theory, suggests that coalescence of free fat globules, in uncooked material, is prevented by the viscosity of the exudate. In contrast the second theory, the emulsion theory, suggests that myofibrillar proteins form a stabilizing film or membrane around the fat globules. These theories are discussed in greater detail with respect to finely comminuted cooked sausages in Chapter 5, pages 272–275.

Phosphates (Figure 3.4) enhance meat hydration, although the effect is not fully understood. Several possible mechanisms for phosphate activity have been proposed. These include the influence of changes of pH value, effects of ionic strength and specific interactions of phosphate anions with divalent cations and myofibrillar proteins. Complexation of calcium and consequent loosening of the tissue structure is widely thought to be a major function of phosphates. Swelling of the muscle system and increased water uptake are also probable as a consequence of the binding of phosphates to the protein and simultaneous cleavage of actomyosin and myosin cross-linkages. In addition the increase in ionic strength may lead to reduced interaction between proteins and the formation of a colloidal solution of myofibrillar proteins.

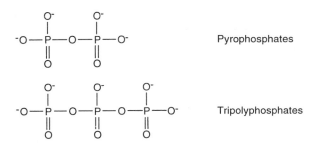

**Figure 3.4** Structure of polyphosphates.

## (b) Re-formed meat products

The mechanism of binding without NaCl at temperatures of *ca.* $-2°C$ is not understood. Some extraction of proteins may occur, but this seems unlikely unless free water is available. Free water, however, may be present as a consequence of heat generated during working or through high pressure.

## (c) The algin/calcium gelation system

Algins are extracted from brown algae, the principal source being *Macrocystis pyrifera*, the giant kelp. The molecule is a copolymer of homogeneous sections of D-mannopyranosyluronic acid (M) units and L-gulopyranosyluronic acid (G) units linked by sections of alternating M-G (Figure 3.5). The ratio of M:G varies according to source and influences the properties of alginates. Alginates with predominant M-blocks give weak elastic gels, while those with predominant G-blocks give stronger and more thermostable gels. Calcium (or other alkaline earth cations) is required for gelation of unesterified alginates. A gel structure described as resembling an egg-box has been proposed to account for the binding of $Ca^{2+}$ in G-blocks between two parallel polymer chains. It seems probable that ionic bonds may be formed between the carboxyl groups and $Ca^{2+}$ and hydrogen bonds between the carboxyl groups.

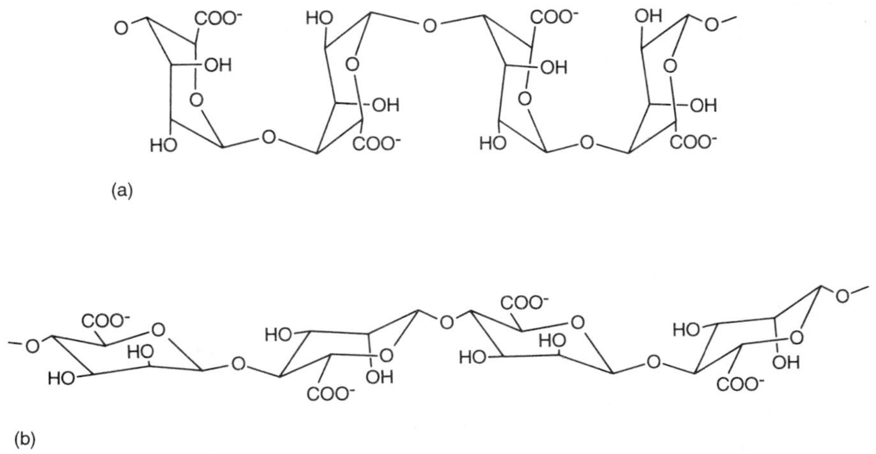

**Figure 3.5** Structure of alginate. (a) Poly (L-guluronic acid) segment; (b) poly (D-mannuronic acid) segment.

**Table 3.1** Chemical examination of comminutes

**Meat composition**

| | |
|---|---|
| Moisture | Vacuum oven or approved instrumental method, e.g. thermogravimetric analysis |
| Fat | Babcock (or other wet extraction method) or approved instrumental method, e.g. near infrared transmittance |
| Protein | Kjeldahl or approved instrumental method, e.g. Elementar Macro-N |
| Ash | Wet ashing or approved instrumental method, e.g. thermogravimetric analysis |

**Other ingredients**

| | |
|---|---|
| NaCl | Silver nitrate titration or approved instrumental method, e.g. specific ion analysis |
| Nitrate/ nitrite | Colorimetric or approved instrumental method, e.g. specific ion analysis |
| $SO_2$ | Any approved method |
| Other additives | Any approved method |

**Overall composition**

| | |
|---|---|
| Meat content | Any approved method |

### 3.3.5 Chemical analysis

Determination of meat content, fat and protein, as well as NaCl, sulphite and nitrite, if applicable, may be made using standard techniques (Table 3.1). In many cases instrumentation and automation is possible and near infra-red spectroscopy has found a wide range of applications.

A number of methods exist for determination of water-holding capacity. Measurement of drip loss from meat held under standard

* The classic Kjeldahl technique remains the reference method for protein determinations and is used to standardize rapid methods, such as near infra-red reflectance spectroscopy. Such methods are more rapid than classic Kjeldahl, but less accurate. Classic Kjeldahl does, however, have a number of disadvantages, including slowness, lack of automation and use of hazardous chemicals. A recent instrument, the Foss Elementar Macro-N, overcomes these disadvantages while maintaining a very high level of accuracy. The instrument is based on the Dumas method, in which nitrogen is released by pyrolysis and combustion of the test sample. Released nitrogen is swept out by $CO_2$ gas and measured in the mixture of gases using a thermal conductivity analyser. The concentration of nitrogen is converted to protein by application of a conversion factor. (Wilson, P.R., 1994, *Food Technology International Europe 1994*, 181–3.)

conditions for a standard time period is the simplest, but alternative methods involve expelling water by physical pressure (Grau–Hamm press method) of centrifugation.

Meat species can be determined by serological methods. A tube precipitation method is technically simple and highly sensitive, but immunoelectrophoresis may be preferred. Kits based on ELISA are commercially available and are often highly suitable for use in laboratories where technical expertise is limited. A possible limitation with all serological techniques is that antisera to exotic animals may not be readily available. If adulteration with animals such as kangaroo is suspected, it may be necessary to use alternative methods such as isoelectric focusing.

Texture and cohesiveness of comminuted products is of considerable importance in determining acceptability. Sensory evaluation is widely used for monitoring these factors but can give variable results, unless very closely controlled. A very good objective means of measurement is the Instron tester fitted with a penetrometer head, but this system is very expensive. Simple alternative methods have been devised, including determination of force required for a polished steel ball to penetrate a standard sliced meat sample.

## 3.4 MICROBIOLOGY

Uncooked comminuted meat products generally have a higher incidence of contamination with pathogenic micro-organisms and are of poorer overall microbiological quality than whole meat. This stems from three main factors:

1. The use of lower quality ingredients, which have been subject to high levels of handling and, possibly, temperature abuse. The use of mechanically recovered meat has been of particular concern.
2. The thorough mixing of a wide range of ingredients leads to spread of contamination throughout the product.
3. Cellular constituents are released during comminution and subsequent operations, providing a readily available source of nutrients. At the same time, the surface area available for microbial growth is greatly increased and the organisms, originally on the surface, are distributed throughout the meat.

Results of microbiological sampling shows that wide variations can occur between manufacturers, with respect to both numbers present and the incidence of pathogens. In many cases, these reflect abattoir practice and handling of ingredients, rather than handling of the end-product. High numbers of spoilage organisms are often due to poor refrigeration, or excessive storage, of ingredients, such as meat trimmings. In the case of contamination by pathogens, however, the major cause of variation can result from differences at the producing farm, or cross-infection at the lairage.

### 3.4.1 Uncooked comminuted meat products and foodborne disease

In overall terms uncooked comminuted meat products have a good safety record although the incidence of pathogens may be high. Products of this type are usually cooked at, or shortly before, point of consumption and foodborne disease can result only from conscious raw consumption, undercooking, or contamination of the cooked product. In addition uncooked comminutes can act as a source of contamination of precooked products.

Conscious raw consumption is rare and usually involves young children. Consumption of raw sausage is, however, a feature of parts of the English midlands and north-west and foodborne disease has been attributed to this practice. It should also be noted that consumption of raw minced meat is relatively widespread in continental Europe, especially Belgium, France and Holland.

A risk of undercooking exists with any meat product. In the past, sausages, burgers, etc. were considered less likely to cause problems than large joints of meat or whole turkeys, but in recent years infection with both *Salmonella* and *Escherichia coli* O157:H7 has occurred due to undercooking. In each case, poor practice was involved (see below).

* Consumption of uncooked foods, including meat, is a risk factor for illness amongst food handlers and should be strongly discouraged. Sporadic salmonellosis occurred over a 7 month period amongst members of a raw-sausage eating cult in an English meat products factory. Subsequently a single outbreak involving 13 workers in the cooked meat part of the plant was attributed to a single meal prepared by a canteen worker, who had eaten raw sausage meat. (Varnam, A.H. and Evans, M.G., 1991, *Foodborne Pathogens: An Illustrated Text*, Wolfe Publishing Ltd, London.)

The risk of contamination of sausages, burgers, etc. after cooking is generally the same as for other meat products. It has been claimed that dehydration of the surface during grilling reduces growth of contaminants, but this claim is unsubstantiated. The role of uncooked comminutes as sources of contamination is difficult to evaluate in relation to other meat products. The level of contamination may be higher but the quantity of drip lower than with whole meat joints and steaks. Particular note should be taken of the continuing practice in some small shops of using strings of sausages as decorative garlands around cooked meat products.

Meat is of overwhelming importance as a source of pathogens in uncooked comminutes and the important pathogens are those of animal origin. Much attention has been given to herbs and spices as a potential source of pathogenic species of *Bacillus* and *Clostridium*. Under most circumstances, endospores of these bacteria are unable to outgrow and the risk is considered to be much exaggerated.

## (a) Salmonella

*Salmonella* is inevitably present in raw meat and poultry and thus in uncooked meat products. Pork usually has a higher incidence of contamination than other red meats and surveys of pork sausages and sausage meat have shown levels of contamination as high as 70%, with a norm of *ca*. 20–25%. There is considerable variation between surveys, some of which is attributable to differences in sampling procedures and isolation methods. Some caution is also required with older surveys, the findings of which may not be relevant to modern practice. Contamination rates with *Salmonella* inevitably rise following infection at farm level from, for example, contaminated feed. A survey of two UK plants, however, attributed the markedly higher incidence of *Salmonella* at one site to overcrowding in the lairage. The rate of contamination in comminuted poultry products is not known, but is assumed to be very high.

The number of reported cases of *Salmonella* infection associated with comminuted meat products is small. Sausages have been more commonly implicated than burgers, possibly reflecting the wider use of pork. Cases reported have all involved undercooking or conscious raw consumption. In some recent cases, undercooking has involved use of barbecues. Barbecue grilling can be a very uncertain type of cooking and food is consumed which would be rejected under normal circumstances. Until recently, barbecues have been a purely domestic social gathering, but much larger-scale events are now organized by public houses, clubs, etc., with a far larger number of persons at risk.

## (b) *Verocytotoxin-producing* Escherichia coli

Verocytotoxin-producing *E. coli* (VTEC) has become recognized as being of considerable importance. The organism has been the subject of much recent research, but despite this relatively little is known of the ecology of the organism and its occurrence in the food processing chain. A number of foods have been implicated as vectors of the organism, but consumption of comminuted beef products, especially burgers, has emerged as a significant risk factor. Surveys of the incidence of VTEC in beef products inevitably show variation, but work in the UK, published during 1993, showed 17% of beef products to be contaminated with VTEC. This incidence exceeded that of *Salmonella*, *Campylobacter* and *Yersinia* combined (13.6%). Serogroup O157, most commonly associated with human disease, was not detected, but the dominant serovars present were associated with disease. This survey broadly correlated with surveys carried out earlier in other countries, although the incidence of VTEC was rather higher. It has also been suggested that the incidence of VTEC is particularly high amongst older dairy cattle. Cull dairy cattle are widely used to provide meat for processing and may be responsible for the high level of contamination in burgers.

All reported outbreaks of VTEC infection associated with comminuted meats appear to involve undercooking and there is no suggestion that the organism is of enhanced vegetative heat resistance. In a number of cases it has been possible to trace the source of infection to individual retail outlets within a large chain and it appears that undercooking is due to failings at local level, rather than to inherently unsafe procedures. Children, who are major users of

---

* Some care is required over terminology. Verocytotoxin-producing *E. coli* is the term applied to all strains capable of producing that toxin. These are a diverse set of organisms, which usually have no other distinguishing features. *Escherichia coli* O157 is the most common serogroup involved in human disease and has a characteristic phenotype. Within *E. coli* O157, serovars H7 and H⁻ are the most common human pathogens. The term 'enterohaemorrhagic *E. coli*' is sometimes used synonymously with VTEC. This is potentially misleading, however, since not all VTEC strains cause bloody diarrhoea.

burger houses, are susceptible both to infection with VTEC and to haemolytic uraemic syndrome. Infections with VTEC can be life-threatening and a number of deaths have occurred amongst children and the elderly. There is, however, no enhanced risk through use of burger houses in general, or through consumption of comminuted beef products.

---

BOX 3.5 **Have a nice day, Sir**

For many years, burgers were considered to present a low risk of undercooking due to their flat profile. More recently, however, undercooking at burger houses has been linked with a number of cases of food poisoning involving comminuted meat products. Various reasons have been suggested, including inadequate facilities and poor staff training. In some burger houses, however, the cult of service is taken to the extreme. Food must be available very quickly and staff, with no effective job security, face dismissal if a customer complaint is received. There is therefore considerable pressure, especially when faced with an unexpected influx of customers, to serve the food irrespective of the adequacy, or otherwise, of cooking.

---

Infection with VTEC has also resulted from undercooking of burger-type products during preparation of dinners for school-children. The underlying cause was a misunderstanding of the nature of the product, partly as a consequence of poor labelling. The burgers, intended for microwave cooking, had been flame-charred on the surface, to simulate the appearance of a grilled product. During preparation of the dinner, the burgers were mistakenly thought to be ready cooked and were only warmed before service.

*(c) Campylobacter*

*Campylobacter* is present at a high level in poultry and at a lower but still significant level in beef and lamb. *Campylobacter* enteritis has been attributed to consumption of undercooked beefburgers. In countries such as the UK, pork is not considered to be of importance as a vehicle of *Campylobacter* (see Chapter 2, page 103). This is not the case universally, and comminuted pork products have been implicated as cause of *Campylobacter* enteritis in Holland and the former Yugoslavia.

### (d) Listeria monocytogenes

*Listeria monocytogenes* has been detected in comminuted beef products at an incidence of as high as 70% and in fresh pork sausage at an incidence of 43%. The extent to which these figures are representative is not known. An epidemiological link has been made between undercooked burgers and listeriosis, but no direct causal link established.

### (e) Yersinia enterocolitica

The situation with respect to *Y. enterocolitica* in comminuted meat products remains enigmatic. The organism is known to be present in pigs, but while isolation may be consistently made from fresh pork tongue, isolation from other fresh pork products is sporadic. No direct evidence exists for uncooked comminuted products being a cause of infection with *Y. enterocolitica*, although the high incidence of sporadic yersiniosis in Belgium has been attributed to the national habit of consuming raw minced pork.

*Yersinia pseudotuberculosis* is often considered to be an animal pathogen, although human infection is well known. Contact with animals is recognized as the major risk factor, but water and foods have been implicated as vehicles in Japan, where the incidence of human infection is high. A wide range of foods has been involved, including comminuted meat products, although definite causal links have not been established. Pigs are a reservoir of *Y. pseudotuberculosis*, although the serogroup normally isolated (III) is rare in human infections.

---

* Symptoms of illness caused by the three species of *Yersinia* recognized as pathogens, *Y. enterocolitica*, *Y. pestis* (the causative organism of plague) and *Y. pseudotuberculosis*, vary widely. The three organisms, however, share a distinct pattern of pathogenicity: 'Although these three micro-organisms infect the host via different routes and cause diseases of a very different severity, they share a marked tropism for lymphoid tissue and a remarkable ability to resist the primary immune system of the host. Moreover, they share the common feature of invasion via mucosal interaction be it in the gut (*Y. enterocolitica*, *Y. pseudotuberculosis*), or in the lung (pneumonic plague).' (Cornelis, G. *et al.* 1987. *Reviews of Infectious Disease*, 9, 64–87.)

## 3.4.2 Spoilage of uncooked comminuted meat products

The large microbial population associated with uncooked comminuted meat products and the potential for rapid growth means that storage lives are short. In many countries, refrigeration is the only significant growth-limiting factor. For this reason distribution of burgers in the US and elsewhere is largely in the frozen form. In other cases, 'fresh' products are distributed at 0 to $-1°C$ to obtain a longer life. It is notable, however, that comminutes can remain stable at temperatures of $4°C$, or below, for relatively long periods despite high numbers of micro-organisms. Problems of quality loss and overt spoilage tend to arise during poorly controlled retail display, when both microbial growth rate and metabolic activity increase markedly. Temperature differences of as little as $1-2°C$ can be critical and display lives are often very short at temperatures in excess of $5-6°C$.

Particular attention has been paid to mechanically recovered meat (MRM) as a source of spoilage micro-organisms. In some cases, high initial microbial numbers in sausages, etc. have been attributed to the use of heavily contaminated MRM, even at relatively low levels. It has also been suggested that microbial growth rates are significantly higher in products containing MRM, although this does not appear to have been investigated systematically. Mechanically recovered meat is itself a good medium for microbial growth because of its finely divided state and large surface area.

In surveys, MRM has usually been found to have higher counts than carcase meat, although differences are rarely dramatic. Problems with MRM arise primarily from poor handling of bones before processing, contamination during meat recovery and failure to refrigerate meat immediately after recovery. Control is through good manufacturing practice, although it is recognized that meat recovery equipment can pose special problems. These arise from opportunities for cross-contamination, the high operating temperature of some equipment and difficulties in cleaning. These latter factors should be taken into consideration when purchasing equipment.

Metabisulphite, as a source of $SO_2$, is used as a preservative in the UK and a few other countries. Metabisulphite has a selective effect, being markedly more effective against Gram-negative bacteria, especially *Pseudomonas* and other genera associated with fresh meat spoilage. Inhibition of fermentative Gram-negative bacteria, such as the Enterobacteriaceae, is less marked and it has been suggested that adaptation to the preservative occurs amongst members of this family.

The antimicrobial effect of metabisulphite is exerted through the undissociated $SO_2$ molecule. The degree of dissociation is dependent on the pH value and is reduced under acidic conditions. For this reason, metabisulphite and other sulphites are of relatively low activity at the pH values common in meat. Activity is further reduced in the presence of yeasts as a result of acetaldehyde production and consequent binding of sulphite. Despite these shortcomings, sulphite is sufficiently powerful to act as an elective agent and to determine the usual spoilage microflora.

Nitrite is added as a preservative to a small number of European fresh sausages and is also present in comminuted products containing cured meats as an ingredient. The relatively low levels of NaCl mean that nitrite is less effective than in cured meats and, in the absence of nitrate, rapid depletion can occur. Under many circumstances, however, nitrite is effective in extending the storage life and also acts as an elective agent.

Despite the basic similarities of uncooked comminuted meat products, significant differences occur in the spoilage microflora. In a number of cases it is not possible to attribute these differences to presence or absence of preservatives and so other, less obvious factors must be involved. Various suggestions have been made, including differences in non-meat ingredients, degrees of comminution and physical form, but none are entirely convincing.

### (a) Fresh sausages

The 'British fresh sausage' (containing metabisulphite) has been extensively studied by Professor R.G. Board and colleagues at the University of Bath, England, and it is probable that more is known of the microbiology of this product than any other comminute. Several different spoilage patterns are possible, depending on initial microbial load, level of preservatives and storage temperature. Comparison of sausages from a large number of manufacturers also suggests that factory-specific microflora may develop in some cases.

Under most circumstances the spoilage microflora of the British fresh sausage is dominated by *Brochothrix thermosphacta*. This organism, derived from fresh meat, is relatively resistant to sulphite and able to grow at 4°C. *Brochothrix thermsophacta* is of moderate spoilage potential, sour and/or cheesy taste and odour appearing when $10^8$ to $10^9$ cfu/g are present. *Brochothrix thermosphacta* has also been associated with discolouration of the sausage meat.

Yeasts are not usually of significance where initial numbers of *B. thermosphacta* are high. Yeasts, some of which are highly resistant to sulphite, are able to attain numbers of $10^5$ to $10^6$ cfu/g if initial bacterial numbers are low. Yeasts are also favoured by use of very low storage temperatures (*ca.* 1°C). Sulphite binding by yeasts enhances growth of bacteria and a succession may occur, with the proportion of yeasts declining as the total microbial numbers increase. *Brochothrix thermosphacta* usually follows yeasts, but on some occasions growth of members of the Enterobacteriaceae occurs. This may either follow growth of *B. thermosphacta*, or the two types of micro-organism may develop more or less concurrently. The most common form of yeast spoilage is slime production on the exterior of the sausage. This may progress no further than a slight stickiness, even when cell numbers are high. On other occasions, a thick white or yellow slime develops. Yeast growth also occurs within the sausage and results in fruity off-flavours. Visible colony formation can occur between the casing of the sausage and the meat filling, although this form of spoilage is rare. A wide range of yeasts have been isolated from sausage, but *Candida* and *Debaryomyces hansenii* are most common in sulphited sausages. There is some evidence that the yeast microflora can be factory-specific.

Although growth of members of the Enterobacteriaceae is restricted by metabisulphite, bacteria of this family can be involved in spoilage when initial bacterial numbers are high and storage temperatures are above 8°C. *Enterobacter*, *Citrobacter* and *Klebsiella* are amongst the most common members of the Enterobacteriaceae isolated from sausages. *Proteus* may also be present in high numbers, but only in sausages from a small number of factories.

Gram-positive rod-shaped bacteria, other than *B. thermosphacta*, are dominant in the microflora of sausage on some occasions. *Bacillus* and rather ill-defined 'coryneform' bacteria are most common, *Lactobacillus* being very rare in fresh sausages. The conditions favouring this type of microflora are obscure, but high storage temperature and low levels of sulphite appear to be involved.

Mould growth on the outer surface of sausages can occur but usually follows bacterial spoilage. Moulds, like yeasts, are favoured by low storage temperatures and limited competition from bacteria. Attempts to extend the storage life of sausages for special occasions, such as sales promotions, by use of a specially 'cleaned up' recipe and storage at *ca.* 0°C can be frustrated by mould growth. A large number of genera are potentially involved but, in practice, *Penicillium*, *Rhizopus* and *Cladosporium* are the most common.

Relatively little work has been published concerning nitrite-containing fresh sausage but nitrite also acts as an elective agent. *Brochothrix thermosphacta* is of importance in sausages of this type, but develops more slowly than in sulphited sausages. Members of the Enterobacteriaceae are more commonly involved in spoilage, even at low temperatures and spoilage by yeasts is also more common.

The microflora of sausages made without preservatives more closely resembles that of fresh meat. The usual microflora, however, tends to be dominated by the Enterobacteriaceae, even when storage temperatures are as low as 4–5°C. Species of *Pseudomonas* and other Gram-negative rod-shaped bacteria become of greater importance at lower temperatures. Spoilage by yeasts and moulds is not normally of importance.

### (b) Burgers and related products

Despite the similarity between fresh sausages and burgers, there are often significant differences in the spoilage microflora. The major difference is that lactic acid bacteria, especially *Lactobacillus* and *Carnobacterium*, which are only rarely of importance in sausages, are often the dominant component of the microflora in burgers. Spoilage is by souring and, where heterofermentative species are present, by gas formation. Both *Lactobacillus* and *Carnobacterium* are also associated with discolouration and formation of grey-green pigments. The mechanism of discolouration is not clear, but may be associated with $H_2O_2$ production by the bacteria and low levels of catalase activity in poor quality beef used in burger production.

On some occasions, lactic acid bacteria fail to multiply significantly, even when initially present in relatively large numbers. In these circumstances spoilage is usually caused by *B. thermosphacta*, although when sulphite levels are low and storage temperatures are high members of the Enterobacteriaceae may be involved.

As with fresh sausages, yeasts and moulds are usually involved in spoilage only when initial numbers of micro-organisms are low, high levels of sulphite are present and storage is at low temperatures. There appears to be a correlation with meat species, in that yeasts are more prevalent on products made with lamb than with other meats.

### (c) Re-formed meat products

Microbial growth obviously does not occur during storage at *ca.* −20°C and is slow at −2 to −5°C. Yeasts are favoured in the latter case, although mould spoilage may occur during prolonged storage at these temperatures. Yeasts usually remain dominant after transfer to higher temperatures and are usually responsible for ultimate spoilage. Significant growth of Gram-negative psychrotrophic bacteria and *Br. thermosphacta* may also occur.

### 3.4.3 Microbiological analysis

The initial microbiological status of meat used in comminute manufacture is of critical importance in determining the numbers of micro-organisms initially present in the comminuted product and its subsequent storage life. For this reason, microbiological examination of the meat before processing is common. In most cases, reliance is placed on 'total' viable counts at 22°C, although examination may also be made for known spoilage bacteria, such as *B. thermosphacta* and members of the Enterobacteriaceae (see Chapter 2, page 114). The value of selective counts at this stage is dubious although enumeration of the Enterobacteriaceae, or a 'coliform' count, can give an indication of general conditions of hygiene in the abattoir, especially with meat obtained from an outside source.

A major problem with colony counts in any situation involving perishable products is the length of time needed to obtain results. There has, therefore, been considerable interest in application of rapid and instrumental methods. Two main approaches have been taken: luminometric determination of adenosine triphosphate (ATP) and impediometry (see Chapter 2, page 116). Determination of ATP has not been widely adopted, due to problems of interference from somatic ATP, but a number of plants have invested in high capacity impedance monitoring equipment. Impedance monitoring is not a true rapid method, although unsatisfactory batches can realistically be detected before use. In practice, the value of impedance monitoring equipment appears to be variable from one factory to another, although this may reflect the technical competence of the user rather than the technique itself. In any case, it is unwise to place too much reliance on microbiological analysis; good practice with respect to meat handling offers the best protection against disaster.

Microbiological analysis of end-products is also common practice. Modern distribution systems often mean that products are stored for only a very short period and positive-release systems are neither possible nor necessary. On most occasions results are obtained on a retrospective basis, even when impedance monitoring systems are employed. In itself, such information can be valuable in detecting trends and providing early warning of problems. However, this requires that results should be fully analysed; all too often potentially valuable information is ignored, filed and forgotten.

Examination of comminutes for specific pathogens requires special consideration. Some manufacturers continue to examine both raw materials and end-products for *Salmonella*, although the value of this must be doubted. The emergence of VTEC as a significant problem in beefburgers has led to the development of methods for detecting the organism in raw material. Methods involving cultural techniques tend to be cumbersome and difficult to apply on a routine basis. Gene probe techniques are, potentially, suitable for screening purposes, but many detect all verocytotoxin-1- and verocytotoxin-2-producing strains. The human disease significance of VTEC strains is unknown in many cases and the value of detection is dubious unless the organism can be characterized further. This would require the elucidation of virulence factors other than verocytotoxin production and the development of simple assays to distinguish pathogenic from non-pathogenic strains. A DNA probe has been prepared from a segment of the *E. coli* O157:H7 60 Megadalton plasmid. This probe hybridizes with 99% of O157:H7 strains, 77% of O26:H11 strains and 81% of all other VTEC strains involved in human disease.

## EXERCISE 3.1

You are employed as technical manager of a company operating a regional chain of burger houses. Most outlets are in small towns and your company has been relatively successful in meeting competition from the much larger multi-national chains. In recent months, however, an erosion of market share has been attributed to ill-defined problems of poor eating quality with the biggest selling line (90% meat beefburger). These problems have not been detected by in-house taste panels and the chairman of the company has now decided that there is no actual quality problem, but that the burgers are 'different' to those of your competitors. You have been asked to carry out a full investigation to determine the nature and extent of any quality problems. Discuss the various means by which the quality 'problem' (as perceived by the consumer) can be identified. What are the relative merits of a consumer panel and an in-house taste panel in operations such as a burger house chain?

To some extent the situation is exacerbated by lack of information, arising from the absence of a formal means of collating and analysing customer complaints. Describe the design of a new complaints system, paying attention to the gathering of data from the complainant, the use of computerized systems in trend analysis and a means of ensuring that, where required, corrective action is taken as early as possible.

EXERCISE 3.2

Obesity is a major health problem in a number of developed countries and is associated with a high fat diet and a generally sedentary lifestyle. You are a scientific adviser to a health pressure group, which has recently supported a publicity campaign aimed at persuading the population to adopt a lower fat diet and a less sedentary lifestyle. Although apparently successful, close analysis showed that the campaign made little impact on the prime target groups, the obese and those at high risk of becoming obese. You are asked to prepare a discussion document reviewing the fundamental approaches to health education. The members of your group are split between a totally objective factual approach and a 'shock–horror' campaign. You are therefore asked to pay special attention to these two approaches and to assess the relative likelihood of success.

EXERCISE 3.3

In the US, it has been stated that *E. coli* O157:H7 and other human pathogenic VTEC strains present the greatest risk of any current 'emerging pathogen'. Discuss the factors which must be taken into account in determining risk in this context. To what extent are mathematical models likely to be of use in determining risk associated with different foodborne hazards?

Discuss the use of surveys of the incidence of the organism in foods and the environment and epidemiological data in determining risk.

# 4

# CURED MEATS

---

## OBJECTIVES

After reading this chapter you should understand
- The nature of cured meats and the curing process
- The function of curing agents and other ingredients
- The different methods of manufacture of cured meats
- The smoking of cured meats
- Quality assurance and control
- The chemistry of meat curing
- The effect of NaCl on meat structure
- The risk posed by nitrosamine formation
- The antimicrobial function of curing agents and other ingredients
- Public health hazards associated with cured meats
- The microbial spoilage of cured meats

---

## 4.1 INTRODUCTION

The curing of meat, like other methods of preserving meat at ambient temperatures, developed from drying. Subsequently salt (NaCl) was used to assist the drying process and it is assumed that nitrate, the original curing ingredient, was derived adventitiously from crude salt. It is not known where the deliberate use of salt containing nitrate first began, or at what time. Various suggestions have been made, including China and the area between the Tigris and Euphrates rivers, many years before the birth of Christ. Alternatively, in Europe, the use of nitrate may have commenced in the 16th century, or later, when the compound had become readily available and was widely used for medicinal purposes.

Curing originally involved packing the meat in a dry mixture of NaCl and curing agents, a process which is still in use today. The

BOX 4.1 **The live savings bank**

During the hungry years of the 19th century bacon was of very considerable importance in the diet of agricultural workers. In many cases bacon, eaten once a week or less frequently, was the only meat available, while bacon fat was widely used to add some flavour to meatless and insipid dishes. Better off families were able to keep a pig, the 'live savings bank', to provide both meat for curing and some income. All too often, however, the bulk of the proceeds were accounted for before slaughter by debts to the feed merchant.

traditional dry cure is not well suited to large-scale production and, especially in the UK, tank cures were developed using concentrated brines. This development permitted the curing of pork on a very large scale. The south-western county of Wiltshire was a major bacon-curing area and the tank cure is often referred to generically as the 'Wiltshire cure', although there are a number of variants. Large quantities of 'Wiltshire' bacon are made today in a number of countries, although there have been a number of technological changes. Newer methods of curing have also been developed, including the relatively mild 'sweetcure'. In developed countries, the general availability of refrigeration means that the importance of curing as a means of preservation is reduced and this is reflected in a lowering of the levels of NaCl and curing agents in accordance with modern tastes and health concerns.

In Europe and the US, pork is the most widely cured meat, although smaller quantities of beef are also cured. Beef curing is of greater importance in exporting areas, such as South America and Australia. Mutton is also cured in exporting countries and in parts of Africa, where the consumption of pig meat is proscribed. This chapter is primarily concerned with curing of pork, although much of the material is equally applicable to other meat.

## 4.2 TECHNOLOGY

### 4.2.1 Definition of cured meats

Although products such as bacon are distinctive and easily recognized, the definition of cured meats can present some difficulties.

The problem lies in the fact that NaCl and curing agents are used as ingredients in a wide range of meat products, which may not be generally considered to be cured. A number of definitions have been proposed, usually based on the addition of NaCl and the formation of nitroso derivatives of myoglobin, but none are considered fully satisfactory. In the present work, cured meat products are defined as those with intact muscle structures which have been subjected to a discrete process, intended to ensure the distribution of NaCl and curing agents throughout the product, with the intention of producing an end-product of typical colour and organoleptic character. Nitrite-containing products excluded from this definition are discussed in Chapters 3 (raw comminutes), 5 (cooked comminutes) and 7 (fermented sausages).

### 4.2.2 The function of curing ingredients

#### (a) Sodium chloride

Sodium chloride is the major curing ingredient by weight, although levels in the final products vary widely from less than 2% in some mild cured bacon to over 6% in some traditional types of ham. The traditional function of NaCl is to act as a preservative and to impart the characteristic salty character. Sodium chloride also increases the water-holding capacity of proteins and thus enhances water uptake.

For a number of years there has been a trend to lower NaCl contents in cured meats. This reflects consumer taste and, more recently, concern over the adverse effects of NaCl on health. Bacon has been made using 'salt replacers', which consist either of a salt such as potassium chloride alone, or a combination of potassium chloride with other salts, such as potassium sulphate and potassium glutamate. Salt replacers are markedly less inhibitory to micro-organisms than NaCl and this may have implications for both the

---

* It is a common misconception to attribute the inhibitory effect of NaCl on micro-organisms solely to the lowering of the $a_w$ level. There are in fact two inhibitory processes, the lowering of the $a_w$ level and the specific inhibitory effect of the $Na^+$ ion. The two processes are interrelated, but at NaCl concentrations found in foods, the specific effect of the $Na^+$ ion is usually of greater importance. With some exceptions, micro-organisms which are sensitive to reduced $a_w$ levels are also sensitive to inhibition by $Na^+$ ions. In contrast to the situation with pH value, there is only very limited adaptation to growth in elevated NaCl concentrations and very little variation within a population of cells.

safety and stability of cured meats. Potassium chloride has been associated with adverse effects on health when consumed at high levels and the benefits of salt substitutes must be doubted when viewed in the overall context of food safety.

### (b) Nitrate and nitrite

Nitrite, as either the potassium or the sodium salt, is the active curing agent, responsible for inhibitory effects on micro-organisms and providing a source of nitric oxide for formation of the characteristic cured meat pigment. Nitrite is also thought to have a beneficial effect on flavour. In contrast nitrate, which may also be added as either the potassium or the sodium salt, is generally accepted to serve only as a reservoir of nitrite. This itself is valuable in stabilizing the cover brines used in the Wiltshire bacon process, reducing the incidence of spoilage of some bacon cuts and, under some circumstances, reducing the risk of toxigenesis by *Clostridium botulinum*. Specific concern over the risk of *N*-nitrosamine formation (see pages 203–204) and general concern over the levels of preservatives in foods have led to a general reduction in nitrite levels and nitrate (as a reservoir of nitrite) is now often omitted from cures.

### (c) Curing adjuncts

Curing adjuncts are used for a number of technological reasons. Curing adjuncts are most common in cures of the sweetcure type and are only rarely used in the Wiltshire and other more traditional cures.

*Polyphosphates* are used primarily as a means of improving water retention. This permits greater weight gain during curing and minimizes moisture loss during cooking. Anti microbial properties have also been claimed for some polyphosphates. The situation, however, is by no means clear and there is variation between different types of polyphosphate and resulting from interactions with other factors, such as pH value.

---

* Ascorbate should always be used in brines and not ascorbic acid. At the low pH value of ascorbic acid, formation of nitric oxide is extremely rapid, and the compound may be released from the brine as a gas. In air, nitric oxide gas combines with oxygen to form the highly toxic, brown-coloured nitrogen dioxide.

*Ascorbate and erythorbate (isoascorbate)* are permitted additives to cured meats in the UK and the USA respectively. The major technological function is to improve colour formation and retention, although nitrosamine formation is also reduced. Inhibition of *Clostridium botulinum* has been claimed, although evidence is contradictory.

*Sugar* is added to some curing brines, especially but not only those of the eponymous sweetcure type. Sugar is reputed to provide a characteristic flavour and, especially in the case of reducing sugars, to improve colour. Sucrose is most commonly used.

### 4.2.3 The Wiltshire cure and technologically similar cures for bacon and ham

Wiltshire bacon is a generic term for traditional tank-cured bacon. Wiltshire bacon itself is virtually unique to the UK, although many countries (notably Denmark, Holland and Eire) manufacture for this market. Similar processes are also used elsewhere, especially for ham curing. Details of the Wiltshire process can vary considerably, but the basic process is the same in all cases (Figure 4.1). Some

---

BOX 4.2 **The Campaign for Real Bacon**

For some years it appeared that the popularity of Wiltshire bacon was in irreversible decline and that the process itself was in danger of falling into disuse. This stemmed from competition from bacon made by more modern cures, such as the sweetcure, which not only reflected current taste for more bland foods but, in contrast to Wiltshire bacon, was of highly consistent quality. More recently, the quality and consistency of UK-produced Wiltshire bacon has been improved by various initiatives, notable amongst which is the *Specification and Code of Practice for the Production of Tank Cured Wiltshire Bacon* (British Meat Manufacturers' Association, London). Wiltshire bacon is now viewed as a 'real' product and compared favourably by many consumers with the sweetcure type. The situation is similar to that with beer, where the poor and inconsistent quality of many traditional ales led to the development of bland but consistent keg beers, followed by a revival of 'real' ale.

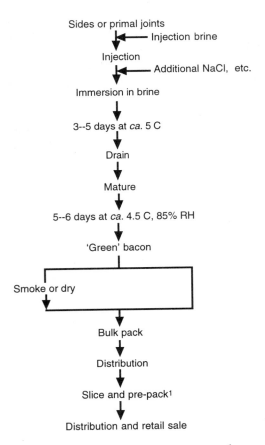

**Figure 4.1** The Wiltshire method of bacon curing. [1] Small quantities of bacon are sliced at point of sale.

modern variants have been developed (see pages 179–180), but are relatively little used.

### (a) Selection and handling of meat before curing

Most Wiltshire bacon curers procure and slaughter their own pigs, slaughtering procedures being the same as those for fresh pork (see Chapter 2, pages 51–52). The condition of the meat is important. Pale, soft, exudative meat, which often results from stress is undesirable for curing since although curing ingredients penetrate easily, the water-holding capacity is significantly reduced and the final product is of poor quality. Dark, firm, dry meat of high ultimate pH value (>6.2) is also undesirable since the fibres of the meat are enlarged and penetration of curing agents is restricted. This, together with the high pH value, can mean that the end-product is prone to rapid microbial spoilage. Dark, firm, dry meat is a consequence of the exhaustion of glycogen and can be minimized by resting and feeding pigs before slaughter. Feeding of sugar has also been used to reduce the incidence of high pH value meat, it being common practice in Northern Ireland to feed 1 kg of sugar some 16 hours before slaughter. A process involving the feeding of liquid sugar has been developed, but adversely affects the quality of the liver, an important by-product.

In conventional practice the carcass is cooled before butchery. Cooling is usually planned to reduce the temperature of the carcass to 4–5°C by the morning after slaughter. Meat butchered before the onset of rigor mortis has also been used in Wiltshire cures, but this is more common in the sweetcure process (see pages 180– 182).

In classic practice, the Wiltshire process involves curing an entire side of bacon – a trimmed half carcass. Butchery involves removing the head and then trimming by removal of feet, cervical vertebrae, sternum, scapula, pubic and caudal bones. The rib bones, jowl and belly are trimmed and the psoas (fillet) removed. The hind legs may be removed and cured separately as ham. In modern practice, it is increasingly common to break down the carcass into primal joints, usually shoulders, middles and hind legs, and curing each type separately. Primal joints are usually but not invariably deboned before curing. Primal joints are more convenient to handle and permit better control of curing. There are also economic advantages in that a single carcass can be split between fresh and cured meat to provide optimum return.

### (b) Injection brine

The first stage of curing proper is injection with a curing brine. This is intended to ensure that curing ingredients are present at an even distribution throughout the interior of the meat. Immersion brines can be used for injection, but this is considered undesirable and a freshly made brine is usual. The composition is similar to that of immersion brines (Table 4.1), although the concentration of curing ingredients is usually rather lower.

**Table 4.1** Composition of injection and immersion brines in Wiltshire bacon curing (%, w/v)

|         | Injection brine | Immersion brine |
| ------- | --------------- | --------------- |
| NaCl    | 18–22           | 24–25           |
| $NaNO_3$ | 0.1–0.2         | 0.3–0.5         |
| $NaNO_2$ | 0.07–0.1        | 0.1–0.15        |

*Note*: Concentrations of nitrate and nitrite are slightly higher when the potassium salt is used.

Whole sides were originally injected manually, using a process of 'stitching' or 'pumping' to a predetermined pattern to ensure even distribution of brine. A single hand-held needle connected to a piping system was used, the brine being injected at a pressure of not more than 60 $lb/in^2$. Sides were injected to a weight of 8–10%, most being added to the fore and hind leg, less in the loin and none in the belly. Special care was required to ensure adequate levels of brine in the vicinity of bones. Manual injection is effective when carried out by a skilled operator, but variation is inevitable and multi-needle injectors are now used in the great majority of large-scale operations.

Multi-needle injectors consist of a conveyer that feeds sides or primal joints to an injection head with two to four rows of needles, which make *ca.* 400 individual brine injections per side. Some older types of machine require supplementary hand injections to be made, but this is not usually considered necessary with more modern equipment. The most effective equipment holds the meat in a clamp and injects the rind and meat side simultaneously. This ensures good brine distribution, especially in deboned meat.

High numbers of micro-organisms are considered undesirable in injection brines, although there is little evidence of a direct connection with spoilage. Making and maintaining an injection brine containing low numbers involves using good quality water and ingredients and a high standard of equipment hygiene. Particular problems can arise with multi-needle injectors which recirculate surplus brine from the pumping operation. This is filtered to remove meat debris, but heavy contamination with micro-organisms may result. Recirculation is not universal practice, but the loss of brine is expensive and can cause effluent disposal problems.

Various methods have been used to reduce the microbial load in recirculated brines including filtration, centrifugation and ultraviolet irradiation, the latter process being highly effective on a commercial scale.

Most Wiltshire hams are injected using multi-needle equipment, although the brine may be of rather lower NaCl content. Polyphosphates may be present to minimize loss during cooking. Some hams are produced using arterial injection, a process developed in the USA rather than Wiltshire that involves injection of brine into the open ends of arteries and requires considerable skill. Its use is now limited to high quality products manufactured on a small scale.

### (c) Immersion in brine

Immersion in brine is intended to cure the shallow, unpumped tissues such as belly, rind and exposed meat surfaces. The immersion brine also prevents the growth of micro-organisms during curing and provides reducing conditions, which favour the formation of nitrosylmyoglobin and minimize formation of metmyoglobin. Following injection, sides or primal joints are stacked, meat upwards, in large tanks made of concrete, rendered or glazed brick, stainless steel or fibreglass. Traditionally, NaCl, or a mixture of NaCl and $NaNO_3$, was placed in the shoulder cavity (pocket) remaining after removal of the scapula. This procedure is considered unneccesary when multi-needle injection is used, although the practice persists, especially during the summer months. Sodium chloride is also distributed evenly between the layers of sides or, more economically, at contact points. Immersion brine is then run into the tank and should completely cover the sides, wooden battens being used to hold the upper layer below the surface. Immersion lasts for 3–5 days and the temperature should be maintained at 4–5°C.

Although the immersion brine is of similar composition to the injection brine, there are important differences in that the immersion brine is used on a continuous basis and contains a large population of micro-organisms. The bacteria of brines are responsible for reduction of nitrate to nitrite (and subsequently the reduction of nitrite), but their role in flavour is obscure. It is recognized, however, that bacon made by modern curing techniques, without immersion in a Wiltshire-type brine, is often considered tasteless.

A recent investigation into the possible role of three brine isolates, *Halomonas elongata*, *Micrococcus roseus* and *Vibrio* sp. has shown that the *Vibrio* may play a role through production of free fatty acids, while the aminopeptidase activity of *M. roseus* may produce amino acids which serve as flavour precursors.

Despite a number of studies, the microflora of Wiltshire curing brines remains generally poorly characterized. Total numbers, as determined by direct microscopic count, are typically in the order of $10^8$/ml, although there is wide variation from factory to factory. Brines with bacterial numbers markedly in excess of $10^8$/ml tend to be unstable due to rapid reduction initially of nitrate and then nitrite and direct microscopic counts are successfully used as a means of controlling brines (see page 220). The types of bacteria present may be divided broadly into three categories:

1. Bacteria derived from the meat. These include pathogens, such as *Salmonella*, and survive in the brines for varying periods. Some types are of index and indicator significance (Table 4.2).
2. Halophilic bacteria of relatively low NaCl tolerance and relatively high biochemical activity. These bacteria are predominantly species of *Vibrio* and are unable to grow in the bulk of the brine. Growth is possible in the vicinity of meat, where the brine tends to be diluted, and the bacteria are also able to grow rapidly in bacon and spoil it. Large numbers in brine are predictive of bacon spoilage.
3. Halophilic bacteria capable of growth in very high NaCl concentrations. These bacteria, which form the great majority of those present, are also psychrotrophic and of relatively low biochemical activity. Special cultivation techniques are

---

* The recognition of the marine bacterium *Vibrio parahaemolyticus* as a significant cause of food poisoning coincided with a general reduction of the NaCl content of curing brines in continental Europe. These factors led to speculation that *V. parahaemolyticus* might become established in brines with obvious public health implications. Fears, however, were totally unfounded. *Vibrio parahaemolyticus* is unable to grow at temperatures below 5°C, or at the NaCl concentrations even of weaker brines. Bacteria identified with *V. vulnificus* have, however, been isolated from Wiltshire-type brines in Denmark. *Vibrio vulnificus* causes severe and often fatal intestinal and extra-intestinal symptoms in susceptible persons and its presence in brines must be regarded as highly undesirable. (Anderson, H.J. and Hinrichsen, L.L., 1991, *Proceedings of the 37th International Congress of Meat Science and Technology*, volume 2, Kulmbach, Republic of Germany, pp. 528–533.)

**Table 4.2** Index and indicator significance of meat-derived bacteria in Wiltshire bacon curing brines

| Bacteria | Significance |
| --- | --- |
| *Escherichia coli* | Faecal pollution |
| *Pseudomonas* | Quality of pork being cured |

required for cultivation and only a small proportion can be recovered. Although slow growing, these bacteria are largely responsible for reduction of nitrate and nitrite. Recognized genera, such as *Halomonas*, have been isolated but many of these bacteria are very poorly characterized.

The use of Wiltshire brines for indefinite periods necessitates careful management, supported by chemical and microbiological analysis (see pages 198, 220–221). During the cure, the NaCl content of the brine falls by *ca.* 2.5% and the volume increases by *ca.* 7% as a result of leakage of injection brine from the meat. There may also be changes in the level of $NO_2$ and, if present, $NO_3$, but these changes are usually small in a well managed brine containing nitrate. At the end of the curing cycle, following removal of the sides by flotation, the brine is pumped from the tank and in many cases it is filtered to remove particulate matter. The brine is then reconstituted to the correct level of NaCl, $NO_2$ and $NO_3$. Brine should be maintained at 4°C between curing cycles. Tanks should be cleaned to remove particulate matter, accumulation of which can lead to instability.

It may be necessary to discard unstable brines, but this is expensive and can cause effluent disposal problems. Dilution with freshly prepared brine or acidification with HCl can be used to stabilize brines, but these procedures are not always effective. The most efficient method is filtration and at one time, in Northern Ireland, a trailer-mounted mobile filtration unit was available to curers to 'save' unstable brines. It should be appreciated, however, that unless underlying problems are solved, brines will soon return to an unstable condition.

Nitrate-free brines are used in some circumstances in response to concern over high levels of nitrite in bacon. Such brines can produce bacon of satisfactory quality, but tend to be less stable than conventional brines and require a higher level of monitoring and control. Where used in long-established curing cellars, however, substantial quantities of nitrate may enter the brine from the environment.

### (d) Maturation

Maturation involves drainage and drying of the bacon and equalization of the concentration of curing ingredients through the meat. It does not involve chemical changes to the meat and should not be confused with maturation of dry-cured hams (see pages 207–210). After removal from the brine, the new bacon is stacked rind side up on pallets. Pallets should be constructed of stainless steel or anodised aluminium, but wooden pallets are still in use and often serve as a source of microbial contamination. Where bacon is cured as whole sides, particular care must be taken in stacking to prevent distortion. Conditions in the maturation room are important, the temperature should be controlled at 4–5°C and the relative humidity at 82.5–85%. Satisfactory control of relative humidity can be obtained by keeping the floor wet with brine. Maturation for 6–7 days is considered ideal, but 4–5 days is now more common and periods as short as 2 days have been used. In the latter case, the bacon is of poor quality. Extensive microbial growth can occur during maturation and may adversely affect the quality and stability of the final product.

### (e) Post-maturation processing

The majority of Wiltshire bacon is sold without further processing (green) but a proportion is smoked (see pages 194–195) and a small quantity is dried without smoking. Although some Wiltshire curers possess butchery and packing facilities, most bacon is sold from the factory on a wholesale basis. Sides are traditionally wrapped in muslin stockinette, protected by sacking, but it is now more usual to transport and store bacon as vacuum packed,

---

* Under conditions of high temperature and humidity, bacterial growth is sufficiently rapid for microcolonies to merge and form visible slime on the maturing sides. A particularly spectacular form of spoilage involved production of large quantities of purple slime by *Janthinobacterium lividum*. The bacon had been cured in a highly unstable cover brine and was of very low NaCl and $NO_2$ content. *Serratia marcescens*, by repute, is also able to grow on maturing bacon, producing large quantities of pink pigment.

deboned primal joints. Most Wiltshire bacon is butchered and, where appropriate, sliced and prepacked at a central operation before transport to retail outlets (see pages 187–188), although a certain amount is still sliced at point of sale.

### (f) Variations on the Wiltshire process

In addition to minor differences in curing between factories, a number of modifications have been made to the basic Wiltshire process. In most cases, these have been concerned with reducing cost, especially with respect to shortening the length of the cure. Few have met with any long-term success and in some cases it may be argued that the bacon produced should not be described as 'Wiltshire'.

Tumbling of deboned joints has been used as a means of enhancing distribution of curing ingredients before immersion (see pages 183–184). This is then followed by 1–2 days immersion in a conventional Wiltshire brine and 2–3 days maturation. The bacon produced is of acceptable taste, but colour can be poor.

The slice cure involves a more radical change in curing technology. The meat is pre-sliced and immersed in a tank of brine for *ca*. 1 minute. After immersion the bacon is drained and packed directly

---

**BOX 4.3   Impediments to great enterprises**

The full reasons behind the demise of the slice cure process are shrouded in commercial secrecy. The packing difficulties, however, illustrate the risks which can be involved when introducing even a well researched innovation. In many cases, innovation involves a central 'big idea' and much of the development effort involves the practicalities of exploiting that idea. Under these circumstances peripheral problems tend to be overlooked, or their importance underestimated. The situation is complicated by the interface with the consumer. In the case of slice cure bacon, the consumer was, apparently, perfectly happy with the quality of the bacon itself which, no doubt, had been a major concern of the process developers. The neatness of the pack, however, a peripheral issue from a technical viewpoint, was a major reason for rejection by the consumer.

into retail vacuum packs. Colour formation takes place in the reducing conditions of the vacuum pack. The curing process can be highly automated and composition can be closely controlled. The slice cure has been used on a large commercial scale but was abandoned, apparently because of difficulties in packing the slices sufficiently neatly.

### 4.2.3 Sweetcure bacon and ham

Sweetcure is a generic term for a group of curing methods used to produce a bacon which is usually of lower NaCl content and significantly blander flavour than the Wiltshire cured equivalent. In most cases, sugar is added to the brine, but a similar process is used to produce 'mild' bacon without the addition of sugar. The technology is often proprietary and the bacon may be known by a number of names, including 'Canadian-style' and 'tendersweet'. The process is widely used in conjunction with re-forming to produce a variety of 'ham' and other 'joints' (roasts) as well as 'steaks'.

The basic sweetcure process (Figure 4.2) is superficially similar to the Wiltshire process but differs in a number of ways. There may

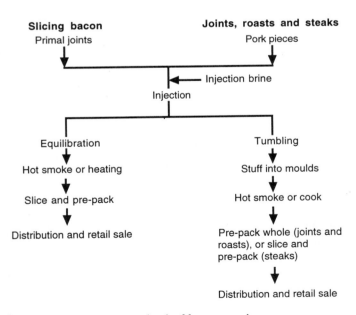

**Figure 4.2** The sweetcure method of bacon curing.

also be significant differences between the curing of meat for slicing and for re-formed joints.

### (a) Slicing bacon

Selection of pork is based on similar criteria to those for Wiltshire bacon. The meat is now almost invariably cured as deboned primal joints and injection by multi-needle injector is universal practice. The basic injection brine consists of NaCl (usually 17–22%) and NaNO$_2$ (*ca.* 0.1%). Sugar is present in many brines, usually at a concentration of *ca.* 2%. Sucrose is most commonly used in the UK but glucose derived from hydrolysis of corn syrup is in widespread use in North America. Brines have also been made containing honey, treacle or molasses. Polyphosphates are usually present to increase brine uptake and thus enhance curing gain. Binding agents such as milk powder have also been incorporated into the brine, although this practice is fairly rare.

Following injection, the meat may be tumbled (see below), although this stage is often omitted. Bacon may be sliced and packaged immediately after injection, the final stages of curing taking place in the pack. Such bacon, however, is generally considered to be of poor quality and an equilibration stage is usual. Equilibration may involve immersion in a cover brine, or holding in a plastic film bag ('bag' or 'brineless' cure). In some cases, both procedures are used.

Cover brines used in the sweetcure differ from those of the Wiltshire cure in being freshly prepared for each curing cycle. The characteristic microflora of the Wiltshire brine is thus absent. The composition is similar to that of the injection brine, although the NaCl content may be slightly higher and polyphosphates are often omitted. Equilibration in brine usually lasts 1–3 days, although in a

---

* The absence of nitrate from the brines used in the sweetcure means that no reservoir of nitrite exists in the bacon. This, together with the generally lower NaCl content, means that sweetcure bacon is of lower microbial stability than sweetcure and of shorter storage life. Attempts have been made to increase the storage life by incorporation in the immersion brine. This is only effective if a nitrate-reducing microflora is established in the factory. At the same time, the presence of nitrate can encourage the growth of spoilage bacteria including members of the Enterobacteriaceae, such as species of *Proteus*. The addition of nitrate to immersion brines of the sweetcure type is not, therefore, usually recommended.

variant procedure equilibration in brine lasts no more than a few hours before the bacon is transferred to a bag.

Equilibration can take place in the absence of immersion, simply by placing the bacon in a plastic film shroud. It is now more common to vacuum pack the meat as primal joints. Equilibration is considered to take place more rapidly in a vacuum pack, although in either case a minimum period of 1–2 days is usual.

There is no significant wholesale trade for sweetcure bacon, which is usually smoked and packed at the curing plant. In many processes, smoking (see pages 194–196) is carried out at a relatively high temperature, protein denaturation being necessary to produce a firmer texture. In some cases the bacon is heated in the absence of smoking. The temperature of smoking is sufficiently high to reduce the microbial load markedly and tendersweet bacon often has very low counts at time of packing. Smoking is not, however, intended as a means of processing for safety.

*(b) Joints, roasts and steaks*

Although sweetcure hams can be produced in the same way as slicing bacon, it is now common practice to cure the meat as relatively small deboned pieces and to re-form the 'joints' after curing. There is usually no equilibration stage, tumbling being of considerable importance in ensuring even distribution of brine ingredients,

---

**BOX 4.4  Ham and sherry?**

The practice of describing re-formed products with the name of the joint from which the meat was cut is controversial, and is not permitted in some countries. Such descriptions are, however, permitted in the US and UK, sweetcure 'joints' of ham, collar and shoulder all being readily available. Joints re-formed from meat taken from several different parts of the pig must not be labelled with a descriptor such as ham and so generalized names, for example picnic steaks, are applied. Some members of the UK industry are opposed to the use of 'ham', etc. for re-formed products, arguing that the practice, while legal, is misleading and, in the long term, potentially damaging.

and heating after curing is an essential part of the re-forming process.

Similar criteria to those for Wiltshire bacon are applied to selection of meat. Some curers, especially in the US, prefer to use pre-rigor meat. This has advantages in reducing the length of the process, possibly improving colour formation and in minimizing plant refrigeration load. Electrical stimulation may also be applied to carcasses immediately after slaughter, despite the fact that electrical stimulation of pork is generally considered to be associated with pale, soft, exudative meat. Electrical stimulation reduces tumbling times through cellular disruption, which improves distribution of curing ingredients, increases salt soluble protein extraction and, up to 180 volts, improves binding of the meat pieces.

The injection brine is of similar composition to that used for slicing bacon. Polyphosphates are added and are important not only in reducing cooking losses but also in enhancing the effect of NaCl in extracting proteins and binding the meat pieces. Binding agents are used more commonly than in bacon for slicing and ascorbate may also be present to aid colour formation, especially where the end-product is to be fully heat processed. Spices are occasionally added to the brine to improve flavour.

Tumbling is an important part of the production of re-formed meat of any type. Tumbling is a process which utilizes the kinetic energy of falling meat pieces to fracture the myofibrillar membrane structure. This results in the extraction of salt-soluble, heat-coagulable proteins, with increase in cohesiveness of meat pieces and in yield and tenderness. In the case of cured meats, penetration and distribution of cure ingredients is enhanced. Tumblers are horizontally or eccentrically mounted rotating drums. As the drum rotates, meat pieces are carried towards the top before falling back to the base and impacting against drum walls or against other pieces of meat. Drums either are specially shaped or contain baffles to assist movement of the meat pieces. Some tumblers are hermetically sealed and equipped with vacuum pumps. Tumbling under vacuum is considered to improve uptake and distribution of curing ingredients but the capital cost of equipment is significantly higher.

Tumbling is a technologically simple operation but careful selection of the optimum length is necessary. Sufficient mechanical

**Table 4.3** Commonly used tumbling processes in cured meat production

---

1. Tumble continuously for 2–3 hours.
2. Tumble for 0.5–1 hour; rest for 16–24 hours; tumble for 0.5–1 hour.
3. Tumble for 5–20 minutes in each hour for 18–24 hours; rest during interim periods.

---

action is necessary to obtain maximum yield and adhesion of individual pieces but extended tumbling results in an excessive loss of the fibrous structure of the meat, leading to poor texture and perhaps reduced adhesion between pieces. There is considerable variation in operating conditions (Table 4.3) which in many cases are established on a purely empirical basis. In general, it is agreed that intermittent tumbling, in which the meat is tumbled for a period in each hour, is the most effective. Intermittent tumbling, however, lengthens the overall process time and is not suitable for use with pre-rigor meat.

Although tumbling is most common, a second mechanical process, massaging, is used by some curers. Massaging involves gently rubbing the meat surface against the massager wall or against the surface of another piece of meat. Massagers usually consist of circular tanks containing paddles rotating at low speed on a horizontal axis.

Emulsion coating has been developed as an alternative process to tumbling or massaging. It involves coating the individual pieces of meat with an emulsion prepared by chopping lean ham and fat in curing brine and passing the mixture through an emulsifier. Emulsion coating is claimed to improve yield after cooking, moisture retention, juiciness, tenderness and organoleptic quality. The process is particularly suited to production of low NaCl hams (see pages 192–193).

Re-forming usually involves stuffing meat pieces into a casing and shaping in a mould. Where the meat is to be smoked, it is necessary to use a casing of sufficient permeability to enable smoke to penetrate. Hams for full heat processing (pasteurization) may be filled direct into metal moulds or into large cans for slicing at a remote location. Heating is necessary to denature the extracted myofibrillar proteins and bind the meat pieces by formation of a solid gel. Heating is often concurrent with smoking, but smoking may also be applied separately at lower temperature.

The extent of heating depends on the nature of the end-product and, more particularly, the extent of heating to be applied in the home. Where the product is intended for cooking by the consumer, the heat treatment applied is that sufficient to denature protein (centre temperature of 60–62°C) and it is not intended to process for safety. There is, however, a very significant reduction in the number of micro-organisms present. In other cases, the heat treatment is designed to develop the full 'cooked' flavour and to destroy vegetative pathogens (see Chapter 6, page 299).

Following heating, the product is cooled and removed from the casing. Whole joints, or roasts, are vacuum packed, while steaks are prepared by thickly slicing suitably shaped joints.

### 4.2.5 Dry curing of bacon and hams

Dry curing is the oldest curing method but to a large extent it has been replaced by brine cures. In the UK, dry curing of bacon on a large scale was supplanted by the Wiltshire cure, but dry curing persisted in farms for many years and some large-scale Wiltshire curers continued to dry-cure bacon to special order at least until the 1960s. In recent years, dry curing of bacon has undergone something of a revival with the increase in interest in traditional food processes and development of a technically advanced dry-curing process. Dry curing remains in use in continental Europe, especially for curing very high-fat, heavily smoked products such as katenspeck.

---

* The substantial decrease in microbial numbers occurring when sweetcure bacon is heated to coagulate proteins could be expected to produce conditions favourable to *Staphylococcus aureus*. This organism is able to grow rapidly on cured meats, provided that a competitive microflora is absent. The risk must, therefore, exist that *Staph. aureus* gaining access to sweetcure meat could grow and elaborate enterotoxin if refrigeration were inadequate at any stage. The enterotoxin would then be expected to survive domestic cooking. For this reason, a high standard of hygiene should be applied during handling of heated sweetcure meats, even when the products are intended for cooking before consumption, and strong precautions taken against contamination from raw material. In practice, however, *Staph. aureus* does not appear to present a risk even in situations where precautions are judged inadequate.

The basic dry-cure process involves rubbing the meat with a mixture of dry NaCl and KNO$_3$. The meat is then placed in wooden troughs and surrounded by a similar mixture. At intervals, depending on individual practice, the meat is turned and re-rubbed and the curing ingredients are replaced. Considerable care is required to ensure adequate distribution of curing ingredients and the bacon is often very salty. Excessively high local levels of nitrite can arise as a consequence of reduction of poorly distributed nitrate, leading to a bright green discolouration ('nitrite burn'). The high NaCl content, however, means that the bacon is highly stable.

An advanced dry-curing method, developed in both the UK and Denmark, involves the use of a cloud of finely divided NaCl. The process is not a true dry cure, since the meat is pumped with injection brine and immersed in a curing brine, albeit for a short period, before dry salting. There is a marked increase in shelf life, without excessive saltiness.

Ham can be dry cured using a similar procedure to that used for bacon. Alternatively the joints may be stacked and loosely covered with curing ingredients, or curing may involve rubbing with a mixture of NaCl and KNO$_3$ on several consecutive occasions, the ham being hung in air in the interim. As with bacon, dry curing of ham may be combined with brine curing, production involving brine injection before dry curing, or rubbing with curing ingredients before immersion in a brine.

In some countries dry-cured ham, in common with other types, is usually eaten after cooking, but a wide variety of hams are produced for eating raw. Two extremes are represented by the Italian Parma ham, which is cured using NaCl only and air-dried at low temperatures, and the German Black Forest ham, which is cured using a mixture of NaCl and KNO$_3$, flavoured with juniper berries and smoked for an extended period until the ham is extensively dried and blackened. A common feature of many dry-cured raw eating hams is the extended maturation (ripening) period, during which significant biochemical changes occur. These involve extensive proteolysis and lipolysis and the overall effect of ripening is similar to that of dry fermented sausages. There is also limited acid production. To some extent hams of this type may be regarded as fermented meat products, although there are significant differences with fermented sausages, resulting from the intact muscle structure (Table 4.4).

**Table 4.4** Differences between dry-cured ripened hams and fermented sausages

1. Hams are made from intact muscle, rather then emulsion.
2. Diffusion of curing ingredients is very slow in hams.
3. At the same NaCl concentration, the initial $a_w$ level of hams is significantly higher than that of sausage.
4. Initial pH value of hams lower than that of sausage, but final pH value usually higher.

The technology of dry-cured, fully ripened ham manufacture varies, but Spanish hams may be considered typical. Considerable care is required in selection of meat, which should have a pH value of less than 5.8 to minimize microbial growth before penetration of curing ingredients. In some cases, the diet of the pig is considered to be of major importance in determining the characteristics of the ham. Some types of Spanish ham, for example, are made from pork from pigs fed exclusively on acorns. This apparently imparts a delicate flavour to the ham which is remarkable enough to justify a selling price in excess of £60 per pound in a London department store!

The proper processing of dry-cured hams involves three stages: salting, post-salting and maturation. The salting and post-salting stages are considered to be equivalent to stabilization. Procedures at the salting stage vary but usually involve either repeated rubbing of the meat with NaCl, or a mixture of NaCl and $KNO_3$, or rubbing the meat before packing in boxes with a mixture of NaCl and

---

### BOX 4.5 **The Long March**

China has a long tradition of meat curing and produces several types of ham, which are notable delicacies both in that country and in export markets. The best known is probably Jinhua ham, manufacture of which is a complex process involving salting, washing and sun-drying meat from carefully selected pigs. The ham is then tenderized by beating and manipulation before maturation at 25–37°C for *ca.* 6 months, then finally cleaned and polished. Jinhua ham is sometimes called 'tribute ham' from the practice of giving the product as tribute to feudal emperors. It was also a favourite of Mao-Tse Tung and helped to sustain the future Chairman and his supporters during the hard days of the Long March.

KNO$_3$ and turning at regular intervals. The temperature must be controlled at this stage to <3.3°C to minimize bacterial growth.

Penetration of salt through the interior tissues occurs during the post-salting stage. Hams must also be maintained at <3.5°C during this stage to prevent growth of *Clostridium*. This is a critical stage both in prevention of spoilage and in safety since *Cl. botulinum* is amongst the species of clostridia able to grow. The length of post-salting depends on the methods employed at the salting stage, but is usually in the range 80–100 days. During this period NaCl and other curing ingredients penetrate the tissues with a corresponding reduction in $a_w$ level to *ca* 0.95. At the same time the nitrite level increases due to microbial reduction of nitrate. Significant physico-chemical and ultrastructural changes also occur during post-salting (see pages 207–210). Diffusion is not usually completed during this stage, however, and the product is not in an equilibrium state.

Maturation of hams commences directly after the post-salting stage. Maturation is usually carried out at temperatures in the range 15–20°C. Temperature control is often not available and much higher temperatures may be used. Maturation is carried out for extended periods of 150–300 days or even longer. A considerable degree of drying occurs during maturation, accompanied by extensive proteolysis and lipolysis. Depending on variety, hams may be smoked either at the beginning of or during maturation. Smoking at high temperatures modifies the course of maturation by partial inactivation of enzymes.

In the US, country hams are produced using a similar technology to their European counterparts, although maturation is often for a significantly shorter period. New technology is being introduced in

---

* Control of the post-salting stage of dry cured ham production tends to be of an empirical nature and relatively few studies have been undertaken of the diffusion process. It has been shown, however, that diffusion of NaCl is a uni-directional process from the dry NaCl covering the surface, which is regarded as a semi-infinite medium, through the upper surface of the ham (the meat) into the muscle which, initially, is effectively NaCl free. Fick's second law, which relates time, concentration and concentration gradient, may be applied to this situation and used as the basis of a model for predicting rate of diffusion. Using this method, it has been shown that the effective diffusion coefficient of chloride in semimembranosus muscle is $0.225 \times 10^{-9}$ m$^2$/s with a standard deviation of $0.0191 \times 10^{-9}$ m$^2$/s. (Palmia, F. and Bolla, E., 1991, *Proceedings of the 37th International Congress of Meat Science and Technology*, volume 2, Kulmbach, Republic of Germany, pp. 918–921.)

attempts to shorten the curing process and develop new products. This frequently involves tumbling of the hams in a mixture of dry-curing ingredients. Curing of deboned hams is also used to obtain a high cure rate and a product that is convenient for slicing and pre-packing. Problems arise due to excess moisture loss and internal mould growth and an alternative method involves fabricating a boneless joint from a bone-in. Hams are boned, rehydrated and tumbled before cooking. An emulsion coating process has also been developed, using a slurry of ground dry-cured ham mixed with water under vacuum to increase the extraction of salt-soluble proteins. Products of this type are distinct from traditional dry-cured hams and represent a new category.

### 4.2.6 Microwave bacon

'Microwave bacon' is a recent development that does not appear to have achieved significant market success. Technology is proprietary but is based on a frying process (see Chapter 5, page 228) of sufficient length to develop the characteristic flavour and texture of cooked bacon, but not to cook the meat fully. The bacon is packed in a special pack designed to enhance microwave heating. Additional procedures may be necessary, including the use of binders to minimize moisture loss during further heating and 'texturizing' through rollers to prevent excessive distortion of the rashers.

### 4.2.7 Curing of other meats

Many types of meat other than pork are cured, although in some cases (such as Chinese 'silk bound rabbit') production and consumption are on a local basis. In recent years there has been an increased interest in curing beef, mutton and goat. This results partly from efforts in major producing countries, such as Australia, to offset the loss of traditional markets and partly from attempts to improve meat utilization in developing countries. A range of cured poultry products have been developed in Europe and the US primarily as a means of extending product range away from low-profit basic processing to high-profit manufacture of added-value products.

### (a) Beef

Salt beef is a traditional product but is not widely consumed except in Jewish communities, where hot salt beef is popular as a light meal. Methods of manufacture of salt beef are similar to those for bacon, although individual joints, such as brisket, are cured. Dry curing has been largely replaced by use of brines. In small-scale operations, joints are rubbed with NaCl or a mixture of NaCl and $KNO_3$ before immersion for 4–5 days in a brine of similar composition to Wiltshire brines, but usually made without added nitrite. In the past, small-scale curing was practised by butchers and even in some multiple outlets, primarily as a means of 'saving' fresh beef at the end of its storage life. The resulting salt beef was often of highly variable quality. This practice is now rare and the vast majority of salt beef is made on a centralized basis. Joints are injected with brine before a 2–3-day immersion in a brine containing added nitrite. In some cases the immersion brine is freshly made for each curing cycle. A curing method analogous to sweet-cure has also been developed, in which joints are injected with brine and allowed a short equilibration period before vacuum packaging. Equilibration may involve tumbling and/or immersion. Smoking is not applied to salt beef.

Until recently, salt beef has been produced and retailed as whole joints. There is increasing interest, especially in the US, in the manufacture of a sliced product for the breakfast market. This is sometimes referred to as 'beef bacon', although this term is also used loosely in other contexts. Manufacturing procedures vary and some are proprietary but, in general, the process is analogous to manufacture of sweetcure bacon. Beef, with visible fat removed, is a common raw material, processing being controlled so that brine uptake is fairly low. A mild thermal processing is used to 'set' the bacon and this may be combined with smoking. It is intended to establish beef bacon as a high quality product in its own right, rather than as a cheap substitute for 'real' (pork) bacon. At the same time, the retail price of beef bacon must be highly competitive, at least initially. It has been demonstrated that an acceptable product can be made from good (2nd USDA grade) carcasses after either blade tenderization or enzyme treatment.

### (b) Tongue

Although most tongue is a cooked product (Chapter 6, page 302), small quantities of tongue are produced for home cooking. There is no immersion stage, the tongue being injected with brine and vacuum packed immediately. Substantial quantities of uncooked tongue are frozen immediately after curing and exported from producing countries. These are defrosted and cooked in consumer countries.

### (c) Poultry

Uncooked cured poultry products are uncommon. Cured turkey 'rashers', however, are marketed as a direct alternative to bacon, exploiting the low fat content as a selling point. The technology is essentially the same as sweetcure bacon, the 'rashers' being sliced from re-formed blocks. Polyphosphates and binding agents are required to minimize water loss during frying and flavouring is used to enhance the similarity with bacon.

### 4.2.8 Low NaCl and low nitrite ('nitrite-free') bacon

To a large extent, the driving force behind the development of low NaCl bacon is the commercial wish to exploit specific fears over the adverse effect of NaCl and more general concern over additives in foods. In contrast, development of low nitrite bacon was driven by much more clearly defined toxicological concern over the possible formation of carcinogenic *N*-nitrosamines. A further contrast lies in the fact that during work on low nitrite bacon, the need to ensure continuing product safety, particularly with respect to *Clostridium botulinum*, was fully recognized. Scant attention appears to have been paid to safety during development of low NaCl bacon.

### (a) Low NaCl bacon

Virtually all low NaCl bacon is of the sweetcure type and may be produced simply by reducing the NaCl content of the brines. This is not feasible with Wiltshire and related curing methods, due to instability of the immersion brine at low NaCl concentrations. In production of low NaCl bacon immersion in a brine is often omitted, but an equilibration period is recommended to reduce variability in composition. Reducing the NaCl content of the brine is likely to reduce pick-up and polyphosphates may be added to increase weight gain.

The NaCl content of bacon can be reduced to *ca.* 1.5% while maintaining sufficient 'saltiness' to be acceptable to most consumers. Such a concentration of NaCl is also at the extreme lower end of the quantity required for inhibitory effects to be demonstrated.

Reducing the NaCl concentration further results in a product which no longer resembles bacon and which is microbiologically unstable. Attempts have been made to resolve this problem by the total or partial replacement of NaCl with 'salt replacers'. These products avoid use of the $Na^+$ ion by substituting KCl, or a mixture of KCl with other potassium or ammonium salts (see pages 169–170). Flavour enhancers may also be present. The salt replacer is used to make up the brine, the technology otherwise being identical to that of conventional sweetcure bacon.

### (b) Low nitrite bacon

A number of approaches have been taken to production of low nitrite bacon. In all cases, however, it is necessary to take account of the fact that risk from nitrosamines arises in two ways: preformed nitrosamines in the product at time of consumption (produced during either manufacture or cooking) and nitrosamines formed from product-derived nitrite (residual nitrite) after ingestion. Factors affecting nitrosamine formation in the bacon and residual nitrite levels are summarized in Table 4.5.

Attempts to make a true low nitrite (nitrite-free) bacon have met with variable but usually limited success. Solving the problem of low nitrite bacon requires use of a system which both ensures the microbiological safety of the product and produces a product of typical colour. In addition, nitrite minimizes lipid oxidation and

**Table 4.5** Factors affecting risk from nitrosamines in cured meats

---

**Nitrosamines present at point of consumption**
    Input nitrite concentration
    Method of cooking
    Residual nitrite concentration at time of cooking
    Cooking time × temperature
    Presence of ascorbate (or any other antioxidant)
    Fat content

**Residual nitrite concentration**
    Input nitrite concentration
    Manufacturing process, particularly extent of any heating
    Storage time and temperature
    Product pH value
    Presence of ascorbate (or any other antioxidant)

---

this function must also be fulfilled in low nitrite products. Various alternatives have been proposed as preservatives, including nisin, sorbic acid and sulphur dioxide, but none can be truly described as satisfactory. Extended use of $SO_2$ is unlikely to be favoured in view of the known adverse effects. Sorbic acid and sorbates were considered to be of considerable promise in control of *Cl. botulinum* in cured meat, provided that nitrite was present at a minimum level of 40 mg/l. Some consumers, however, suffered allergic reactions to sorbates in bacon and 'generally recognized as safe' (GRAS) status has been refused in the US.

Pigment production in cured meats is generally considered to occur at lower nitrite levels than those required for inhibition of micro-organisms. It is, therefore, possible to produce a bacon of very low nitrite content but which is of typical colour. The pigmentation, however, tends to be very unstable under these conditions and this method is not considered to be suitable for commercial use. Various pigments have been proposed for addition to bacon, including both artificial and natural colours. Some have been successful under laboratory conditions, but none have been used commercially and it appears that scaling up presents major problems. An alternative is to generate a cured meat pigment without the use of nitrite in the meat. A curing system using direct addition of nitric oxide has been developed, for example, but suffers from problems of control when used in large-scale production.

An alternative procedure that has received considerable attention in recent years involves addition of the pre-formed cooked cured meat pigment dinitrosylferrohaemochrome, which is highly susceptible to light-mediated oxidation and cannot be used in cured meats in its native form due to colour fading. This problem can be overcome by encapsulating the pigment in a coating consisting of β-cyclodextrin and modified starch. In whole-muscle products there are also problems with pigment distribution; these require close control of pigment particle size, the use of special injection

* Dinitrosylferrohaemochrome is prepared chemically from haemoglobin. Iron (III) porphyrin haemin is initially prepared from beef red blood cells. A buffered solution of haemin is then reacted with nitric oxide in the presence of a reducing agent and the pigment recovered as fine, red-black crystals. (Killday, K.B. *et al.*, 1988, *Journal of Agricultural and Food Chemistry*, **36**, 909–914; Shahidi, F. *et al.*, 1985, *Journal of Food Science*, **50**, 272–281.)

equipment and tumbling both before and after injection. In the case of cooked cured meat, the cooking procedure must produce a high rate of moisture retention to assist even distribution of pigment.

The use of dinitrosylferrohaemochrome permits low-nitrite cured products to be made that closely resemble their conventional counterparts. It is still necessary to select a suitable preservative; also use of dinitrosylferrohaemochrome is expensive and likely to add considerably to the cost of the product.

It is currently considered that, under most circumstances, nitrite levels in cured meats are such as to ensure the stability of the meat and control of *Cl. botulinum*, while not presenting a significant risk from nitrosamines. In some cases reducing agents, such as ascorbate or erythorbate, or anti-oxidants, such as tocopherols, are added to minimize nitrosamine formation at any given nitrite level.

### 4.2.9 Ancillary operations

#### (a) Smoking and drying

The original purpose of smoking was to increase the storage life as a consequence of surface drying (and the resulting increase in concentration of curing ingredients) and deposition of antimicrobial compounds. In many cases, the temperature during smoking was also sufficiently high to reduce the surface load of micro-organisms significantly. The situation today is different in that many smoking processes have little effect on the microbiological stability of the bacon and smoking is applied primarily for its effects on organoleptic properties. The temperatures used may, however, be sufficient for a significant reduction in the number of micro-organisms.

In the case of Wiltshire and related types of bacon, the sides or primal joints are smoked when 'green'. Traditional smoking involves the generation of smoke from hardwoods, primarily oak, beech and hickory. Sides of bacon were originally suspended in brick kilns and exposed to smoke from smouldering fires of wood chips or sawdust in the base. Kilns of this type were notoriously difficult to control and have been almost entirely replaced by a more modern type in which the smoke is generated externally and blown into a metal chamber containing the bacon. Humidity and temperature can be closely controlled and consistency of smoking is high, though in some cases it has been found necessary to supplement the smoke from the external generator with sawdust fires in the base of the kiln. Consistency of smoking can also be improved by electrostatic precipitation of smoke particles on to the bacon. Concern has been expressed over the presence of harmful substances, including the carcinogen benzpyrene, in smoke and some generators are equipped with a purification stage. Purification involves the removal of undesirable compounds either by water sprays or by precipitation. It has been claimed, however, that the character of the smoked product is changed by removal of flavour-active constituents.

The more precise control offered by modern kilns means that humidity may be set to minimize moisture loss. Within limits, any temperature regimen may be used, but two are most common. The first, 'cold smoking' (low temperature/long time), involves smoking at 32–38°C for 15–18 hours. The second, 'hot smoking' (high temperature/short time), uses a temperature of *ca.* 60°C for 2–4 hours.

Sweetcure bacon is smoked directly after curing. A hot smoke is almost invariably used and may involve temperatures as high as 65°C. The density of smoke is usually much less than for Wiltshire bacon.

Conventional smoking is time-consuming and expensive. For a number of years attempts have been made to use liquid extracts and essences of smoke constituents as a substitute for smoke generation. Such 'liquid smokes' are claimed to be more convenient to use and more economical. It is possible to fractionate extracts to remove components with undesirable organoleptic properties, or those associated with adverse health effects. Liquid smokes are available for use at different temperatures. In use the extracts are usually atomized and blown into a modified kiln along with heated air. Deposition on to meat can be uneven and electrostatic precipitation is sometimes used to overcome this problem.

Small quantities of Wiltshire bacon are dried without smoking, and retailed as 'pale dried'. Drying involves holding bacon in a chamber at 30–33°C for *ca.* 18 hours and results in a weight loss of 6–8%. In some circumstances the bacon is smoked after drying, but this is likely to lead to unacceptable weight loss.

## (b) Packaging

Virtually all sweetcure bacon, ham, etc. is retailed in pre-packaged form. In the UK and US, pre-packing is also very common for Wiltshire and related types of bacon. A small quantity is sliced at point of sale. Pre-packaging is used to some extent for raw hams, such as Parma ham, but is relatively uncommon in continental Europe. Slicing of raw hams at point of sale is also common in the UK.

Whole bacon joints, as well as reformed 'hams' etc., are individually vacuum packed in films of low oxygen permeability. The irregular shape of some Wiltshire joints and the high entrained air content can cause problems and the use of special high performance vacuum packaging equipment may be necessary. Alternatively the use of getterers to reduce the oxygen content of the pack has been recommended, although little advantage may be gained in commercial use.

The majority of both Wiltshire and sweetcure bacon is retailed as rashers and must obviously be sliced before packaging. Slicing is a technologically simple operation, which may be highly automated. The neat appearance of rashers in the pack is apparently an important aspect of consumer acceptability of bacon, however, and the slicing operation must, therefore, be carefully controlled. Problems can arise due to variation in the cross-section of bacon joints, which causes particular difficulties with high speed equipment. These can be overcome by use of bacon presses, which compress the chilled joint (*ca.* $-2°C$) into a standard shape. Problems may also be caused by a lack of rigidity in the meat being sliced, which leads to slices of uneven shape and thickness. To a large extent problems of this nature are overcome by slicing at $-1$ to $-2°C$, at which temperature the fat is hard. In some cases, however, the fat is particularly soft, usually as a consequence of feeding practices at the rearing farm. In these circumstances, slicing at low temperatures offers only a partial solution and longer term action must be taken at the farm.

In the past, slicing machines were notorious as vehicles of cross-contamination, the problem being greatest at retail level.

Small quantities of sliced cured meats are pre-packaged in air permeable film. Storage life is short, however, and vacuum or modified atmosphere packaging are now almost universal for cured meats pre-packed on a centralized basis.

Vacuum packaging of uncooked cured meats employs the same technology as is used for other products. Problems can occur with bacon, due to fat contaminating the film and preventing a good heat seal being made. This leads to a high incidence of 'leakers' of reduced storage life and quality. This problem can be overcome by heat sealing over the entire area of the pack which is unoccupied by meat. Packs of this type, however, are very difficult to open and tend to be disliked by consumers.

## 4.2.10 Added-value cured meat products

In recent years a number of added-value bacon products have been developed. These are technologically straightforward products, usually based on a re-formed sweetcure steak, cut to the desired shape and size. Cured poultry meat is used in some cases. The simplest products are merely enrobed in breadcrumbs, a variant product consisting of a steak formed from minced bacon and mashed potato, moulded into a waffle shape and enrobed in bread-crumbs. A combination of ham or bacon steaks and melting cheese is also common. An example is a triangular steak, topped with a layer of Mozzarella cheese and encased in batter, designed for eating after grilling or baking. A number of variants are produced in which additional ingredients are introduced with the topping. These include products in which the topping consists of grated cheese with onions and peppers, to approximate to a pizza, and grated cheese with pineapple. Pineapple is a well established culinary companion of gammon steaks and attempts have been made to market gammon with a pineapple slice. Quite apart from

---

* In most cases, added-value cured meat products are based on raw or minimally cooked material and require full baking, frying or grilling in the home. A smaller number of products, however, are based on partially cooked meat and are designed for finishing in a microwave oven. The use of domestic microwave ovens for reheating prepared dishes is considered to be of dubious safety by some microbiologists, although only a small number of systematic investigations have been made. It has been shown, however, that foods containing high levels of ionized molecules, including NaCl, $NaNO_2$ and $KNO_3$, heat less readily in microwave ovens than foods containing low levels. This is believed to be due to decreased microwave penetration. These findings are of obvious importance when formulating foods intended for microwave heating which are based on cured meats, or which contain cured meats as ingredients. (Dealler, S.F. *et al.*, 1991, *International Journal of Food Science and Technology*, 27, 153–7.)

the pointless nature of the product, these have been unsuccessful due to rapid deterioration of the pineapple. Hybrid products are available, consisting of gammon steak packaged together with a small pot of fruit conserve.

Cured meats may also be minced and used as an ingredient in raw comminutes (e.g. baconburgers, Chapter 3, page 122), while precured ham is used as an ingredient in some cooked comminutes (e.g. ham loaf).

### 4.2.11 Quality assurance and control

In the UK, the need for a high level of quality management was not recognized until relatively recently. This resulted in wide variations in quality, which put the industry at a considerable disadvantage with respect to other producing nations, especially Denmark. After intensive efforts this is now being remedied, although systematic control remains virtually non-existent in the case of some small independent producers.

Quality assurance must begin with the pig, which must be of the optimum configuration with respect to lean meat:fat ratio. Post-mortem glycolysis should proceed normally and the $pH_{ult}$ should be neither abnormally low nor high.

In both the Wiltshire and the sweetcure process it is necessary to ensure that brines are of the correct composition; make-up should be the specific responsibility of a responsible and adequately trained person. Maintaining the strength of a Wiltshire cover brine requires laboratory analysis for $NO_2$ and $NO_3$ content. Sodium chloride content can be monitored by hydrometer (salinometer), but occasional laboratory analysis is recommended. There is disagreement over the value of routine analysis of Wiltshire and related immersion brines. In some cases, it is considered that chemical analysis alone is sufficient to ensure effective management of the brines, while in other cases microbiological analysis is applied not only to ensure brine stability but also to assess slaughterhouse hygiene, 'back flow' contamination from cured meats to brines and the quality of the pork being cured. It is currently considered that an evaluation of the numbers of bacteria in the brine is essential for good control. It must, however, be appreciated that examinations must be made at a relatively high frequency, to permit the normal microbiology of the individual brine to be understood.

It is also necessary to ensure that the correct quantity of brine is injected into the meat and that distribution is even. This is also a matter of economic importance since, although the constituents of curing brine are relatively cheap, the quantities used are considerable and loss due to misuse is expensive. The use of multi-needle injectors has much improved distribution of brine, although care must be taken to ensure correct operation, especially with respect to blocked or broken needles. Where older methods of injection such as artery stitching are used, operators must have a high level of experience, but the equipping of pumping mechanisms with a simple monitoring device, such as a brine flow meter, is effective in controlling usage. In the Wiltshire curing process, addition of solid curing ingredients to sides before immersion, or placing curing ingredients in the shoulder cavity, are manual processes which are not easy to control formally and they require an experienced operator and/or close supervision.

Control of temperature during the curing process is of key importance, especially with the longer Wiltshire type cure. Temperature should be monitored on a continual basis and, in the case of Wiltshire bacon, through the maturation period. Relative humidity is also of major importance and should be controlled and monitored.

Temperature and duration of smoking must be monitored and kilns should be fitted with appropriate recording equipment. Meat should be visually inspected after smoking to ensure even appearance.

Chemical analysis of the end product for NaCl, $NO_2$ and $NO_3$ is required. The need for microbiological analysis is not universally accepted and counts tend to have a very poor predictive value. Microbiological analysis is probably best restricted to specific purposes, such as monitoring for 'mis-cures' (see page 221).

## 4.3 CHEMISTRY

### 4.3.1 Nutritional properties of cured meats

There is no evidence to suggest that curing has any significant effect on the nutritional properties of meat. Use of ascorbate obviously increases the vitamin C content of meat, although this is not normally considered to be of dietary significance. Some loss of amino acids can occur by reaction with nitrite, but levels of nitrite in cured meats are usually too low for this to be of any dietary importance.

### 4.3.2 The chemistry of meat curing

*(a) Reaction of nitrite with reducing agents*

Sodium nitrite is a strong oxidant and reacts with endogenous or added reductants, such as ascorbic acid, to produce nitric oxide (NO). Although nitrite is capable of reacting with a wide range of reducing compounds, reaction with ascorbate (or erythorbate), if present in the cure, is quantitatively the most important. Nitric oxide is a gaseous molecule with an odd number of electrons (Figure 4.3). The molecule is very reactive towards radicals and oxygen and is the key to the important reactions of curing.

Nitrous acid ($HNO_2$), the conjugate acid of nitrite, has a $pK_a$ of 3.22 and only a small proportion of that added during curing is converted to the more reactive free acid. Nitrous acid is unstable in solution and decomposes in a reversible reaction:

$$3HNO_2 \rightarrow H^+ + NO_3^- + 2NO + H_2O$$

In solution nitrous acid can behave as both a reductant or an oxidant, but in meat systems the role is primarily as an oxidant, with oxidizing capacity increased at low pH values. Part of the nitrite is, however, oxidized to nitrate during both the curing process and subsequent storage:

$$NO_2^- + 2OH \rightarrow NO_3^+ + H_2O + 2e^-$$

In reaction with ascorbic acid, nitrous acid is reduced by one equivalent to yield the free-radical product NO, while ascorbic acid

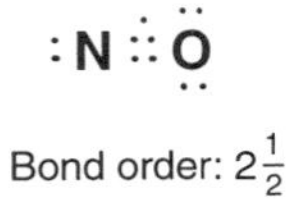

$$:N \stackrel{..}{::} \stackrel{..}{O}$$

Bond order: $2\frac{1}{2}$

**Figure 4.3** Nitric oxide. Nitric oxide is a so-called 'odd molecule' with a three-electron bond. The unpaired electron hides behind the nitrogen atom and the oxygen atom, in effect making nitric oxide a slow-reacting free radical. (Skibsted, L.H., 1992, in *The Chemistry of Muscle-based Foods*, (eds D.E. Johnston, M.K. Knight and D.A. Ledward), pp. 266–86, Royal Society of Chemistry, London.)

is simultaneously oxidized by two equivalents to yield dehydroascorbic acid:

$$\text{asc. acid} + 2HNO_2 \rightarrow \text{dehydroasc. acid} + 2NO + 2H_2O$$

Ascorbic acid is a bifunctional acid ($pK_{a,1}$ = 4.04; $pK_{a,2}$ = 11.34) and because of the different reactivities of the acid and base forms, the reaction with nitrous oxide is highly dependent on pH value. The first stage in the oxidation is formation of $H_2O_3$, which reacts with either ascorbate or ascorbic acid.

With $NO_2$ in excess, dehydroascorbic acid is further oxidized via ring opening to diketogulonic acid and other intermediates to yield polymeric brown substances. Up to seven intermediates are involved, although only diketogulonic acid has been positively identified. At least one of the intermediates is a highly effective nitrosating agent, capable of transferring the NO radical. Nitrosating ability is gradually lost during ageing of reaction mixtures while $NO_2$ is regenerated. Oxygen is not necessary for $NO_2$ regeneration and ascorbic acid is not regenerated simultaneously. 2,3-Dinitrosoascorbic acid is believed to act as a key nitrosating agent, in the nitrosation reaction sequence which generates $N_2O_3$ and dehydroascorbic acid, or transfers NO to other substrates.

In the presence of high chloride concentrations (derived from NaCl in the cure), nitrous acid may be transformed into nitrosyl chloride:

$$HNO_2 + H^+ + Cl^- \rightarrow NOCl + H_2O$$

Nitrosyl chloride is less reactive than $N_2O_3$ and more reactive than $NO^+$. Studies based on model systems suggested that the rate of nitroso (cured meat) pigment formation increased with increasing concentration of $Cl^-$ ions, but effects in meat are limited. Nitrosyl chloride may, however, be involved in generation of other nitrosating agents.

## (b) Nitrosation and transnitrosation

The fate of nitrite in cured meat has been widely studied in order to understand the formation of toxic compounds and to minimize their production. In model systems, all nitrite (labelled with the stable $^{15}N$ isotope) could be recovered and accounted for as nitrate, nitrosylmyoglobin, gaseous nitrogen compounds and residual nitrite.

In non-haem proteins, nitrite reacts with tryptophyl residues to form nitroso derivatives. These derivatives are capable of transferring NO to metmyoglobin (transnitrosation), in the presence of ascorbate, to yield nitrosylmyoglobin. This is highly relevant to oxidative stability in that the protein fraction of cured meats provides a reservoir of nitroso groups. These are available through entropy-driven reactions to regenerate NO-based antioxidants.

A role has also been discussed for cytochrome *c*. During respiration, cytochrome *c* in the electron transport chain alternates between Fe(II) and Fe(III). Cytochrome *c* in the Fe(III) state reacts readily with NO to form the diamagnetic ferrocytochrome *c* nitrosyl compound I:

$$\text{Cyt } c \text{ (Fe}^{III}) + \text{NO} \rightarrow \text{Cyt } c \text{ (Fe}^{II}\text{-NO}^+) \text{ (compound I)}$$

Compound I appears to play a central role in nitrite 'metabolism' in meat. At the pH value of meat, compound I is formed in the presence of ascorbate, the reaction probably being mediated by an unknown nitrosating agent. Ascorbate is capable of reducing compound I to the less stable ferrocytochrome *c* nitrosyl compound II:

$$\text{Cyt } c \text{ (Fe}^{II}\text{-NO}^+) + e^- \rightarrow \text{Cyt } c \text{ (Fe}^{II}\text{-NO) (compound II)}$$

It has been suggested that compound II is involved in transnitrosation or formation of gaseous N compounds through subsequent oxidation.

Hydroxide ions release $NO^+$ from compound II in alkaline solution, effectively regenerating $NO_2$:

$$\text{Cyt } c \text{ (Fe}^{II}\text{-NO}^+) + \text{OH}^- \rightarrow \text{Cyt } c \text{ (Fe}^{II}) + \text{HNO}_2$$

It has been suggested that, at the pH value of meat, other nucleophiles, such as chloride and thiocyanate, may be involved in similar reactions to yield NOCl and NOSCN respectively as part of further transnitrosation. It is also thought possible that compound I is able to transfer NO to metmyoglobin in the presence of a reductant.

### (c) Formation of cured meat pigments

Cured meat pigments have been the subject of studies for many years. Nitrosylmyoglobin is the normal pigment of uncooked cured meats and nitrosylhaemochromogen is formed on cooking (Chapter 6, page 304). Pathways for formation of this, and the other cured meat pigments have been described.

The initial stage in the pathway is the oxidation, by nitrite, of myoglobin to metmyoglobin and the simultaneous reduction of nitrite to nitric oxide (NO). This stage involves the same reaction as that proposed in earlier models. Nitrite is then thought to combine with metmyoglobin to form an unobserved intermediate, nitrosylmetmyoglobin, which undergoes rapid autoreduction to a nitrosylmyoglobin radical cation. A possible but less likely alternative pathway at this stage involves a simultaneous NO coordination and autoreduction to the nitrosylmyoglobin radical cation. (Considerable care is required with terminology, the nitrosylmyoglobin radical cation corresponds to the 'nitrosylmetmyoglobin' described in earlier pathways.) The final stage in unheated cured meat is further reduction of the nitrosylmyoglobin radical cation to nitrosylmyoglobin. This is brought about either by reducing systems within the protein itself, or by migration of the charge to an adjacent histidine residue. The latter mechanism would also explain the incorporation of a second mole of nitrite during heating.

Although nitrosylmyoglobin is the usual pigment of uncooked cured meats, a further pigment, nitrihaemin, may be formed under exceptional circumstances. These normally involve excessive levels of nitrite, especially at low pH value. Nitrihaemin is a green-brown pigment, responsible for discolouration often referred to as 'nitrite burn'.

Nitrosylmyoglobin is unstable in air and discolouration can be rapid. Ascorbate improves the stability of uncooked vacuum-packed bacon, but can accelerate discolouration during storage in air. This is a consequence of the role of nitrosylmyoglobin as an antioxidant and as a buffer for nitric oxide radicals, capable of breaking chain reactions involved in the propagation of lipid oxidation. Light catalyses the dissociation of NO from cured meat pigments and thus leads to rapid discolouration when oxygen is present:

$$MbFe^{II}NO \xrightarrow{hv} MbFe^{II} + NO$$

### (d) Formation of nitrosamines

Nitrites are able to react with both secondary and tertiary amines. The reaction can occur either during cooking or, probably, during digestion at the low pH value of the stomach. Reactive substrates are amino acids, which may be either free or protein-bound.

Proline, tryptophane, tyrosine, cysteine, arginine and histidine are most commonly implicated, many of the resulting nitrosamines or nitrosamides having been identified as potent carcinogens.

The active nitrosating agent under most circumstances is $N_2O_3$, the general type of reaction being:

(i) $R_2NH + N_2O_3 \rightarrow R_2N^\cdot NO + HNO_2$

(ii) $R_3N + N_2O_3 \rightarrow R_2N^\cdot NO + R$

At the pH value of meat, reaction between $N_2O_3$ and ascorbate is far faster than that between $N_2O_3$ and secondary amines. This competition accounts for the inhibition of nitrosamine formation when ascorbate is in excess. Nitrosamine formation follows third order kinetics, the pH optimum being at 3.4, the $pK_a$ of nitrous acid:

$$v = k_n[HNO_2]^2[amine]$$

Improved analytical methods led to the detection of the volatile nitrosamine $N$-nitrosothiazolidine in fried bacon and subsequently in other cured meats. It is now known that a wide range of thiazolidine-based compounds may be found in cured meats, including $N$-nitrosothiazolidine-4-carboxylic acid, $N$-nitroso-2-methylthiazolidine-4-carboxylic acid, 2-(hydroxymethyl)-$N$-nitrosothiazolidine and 2-(hydroxymethyl)-$N$-nitrosothiazolidine-4-carboxylic acid. The last of these appears to be the most common volatile nitrosamine and during cooking, or high temperature smoking, it is converted by heat-induced decarboxylation to 2-(hydroxymethyl)-$N$-nitrosothiazolidine. The potential carcinogenicity of this and other thiazolidine-based nitrosamines has not been fully evaluated.

In the US, there has been considerable concern over the possibility that elasticated rubber netting, used for hanging and shaping hams, may be a significant source of nitrosamines. Both $N$-nitrosodiethylamine and $N$-nitrosodibutylamine have been detected in hams prepared using elasticated netting, the rubber component apparently being a source of nitrosable amines. The problem appeared to have been solved by modifications to the netting manufacturing technique. Subsequently, however, the problem re-occurred due to the breakdown of rubber threads in a new type of netting to form nitrosamine precursors.

### 4.3.3 Effect of NaCl on the structure of meat

Water uptake on curing appears to be due to lateral expansion of the myofibrils, which is accompanied by protein solubilization. During curing, an area of very high NaCl concentration will initially be present at the sites of brine injection. The NaCl concentration at these sites will gradually diminish, while at sites remote from the point of injection the NaCl concentration will gradually increase. Water uptake by meat is at a maximum at NaCl concentrations of *ca*. 1 M and it appears likely that swelling and protein solubilization first occur at sites some distance from the point of injection. It has been postulated that 'tiger striping', a common fault with Wiltshire-type bacon, is the consequence of differences in rates of diffusion of NaCl in different parts of the meat with resultant differences in swelling. Many theories concerning the mechanisms of swelling have been based on an increase in negative charge on the filament due to binding of Cl⁻ ions, which in turn increases long-range electrostatic repulsive forces and, therefore, swelling. At high NaCl concentrations, however, the charge on the filaments is effectively screened and this outweighs the effect of binding of Cl⁻ ions. An alternative hypothesis has been developed, in which swelling induced by NaCl is entropically rather than electrostatically driven. The pro ss begins with the depolymerization of myosin in the thick fil    the presence of moderately high NaCl concentra-tions. J             he myosin tails are flexibly attached to the heads,                fibrils, movement would be severely restri                 hin filaments and the bound myc                 move to a state of highest ent                 r lattice swelling, gi                  m of movement.

adjuncts and they tion. This has been from their presence, of the effect. In the eakens the association synergistically with Cl⁻ ts. It would therefore be erization, swelling of myo-ntrations in the presence of effect of pyrophosphate in py dissociat              e expected to lower the total amount of swe

### 4.3.4 Flavour of cured meat

The flavour of cured meat is partly derived from the same components that are responsible for flavour in uncured meat (see Chapter 1, pages 30–32). After cooking, however, cured meats have a distinctive flavour, which has three basic components. Saltiness, derived from NaCl, is important in itself and in enhancing other flavours. Only cations cause salty tastes, anions being inhibitory. The $Na^+$ ion produces only salty tastes, while the $Cl^-$ ion appears least inhibitory of the anions and has no apparent taste of its own. Potassium and other cations (with the exception of lithium) produce both salty and bitter tastes. Bitterness is disliked by consumers and this limits the acceptability of salt substitutes. Flavour enhancers are often used in salt substitutes to strengthen other flavour components and mask the bitterness.

A minimum level of *ca.* 1.5% NaCl is required to produce the saltiness usually associated with cured meats, but levels above 3.5–4.0% are often considered too salty. Some types of dry-cured product, however, contain in excess of 6% NaCl and this is acceptable in these limited situations. Depending on the nature of the product, the saltiness may be partially masked by a heavy concentration of smoke constituents or other flavours. In the case of sweetcure bacon, the effect of added sugar is primarily to reduce the harshness which can be associated with saltiness. There is generally little direct effect on flavour, unless the sugar source is honey, treacle or some other compound with a distinctive flavour.

Heated fat makes a significant contribution to the flavour of fried bacon and other cured meats. There tend to be strong flavours characteristic of the species. Lactones, especially $\gamma$-C5, C9 and C12, are present in fairly large quantities in pork, but while these may contribute to the 'sweet' flavour of pork, the compounds responsible for the 'piggy' flavour of pork fat remain undefined.

Nitrite has not been associated with any specific flavour compounds. There is, however, a significant indirect beneficial effect on flavour, which is thought to result from the antioxidative activity of nitrosylmyoglobin and S-nitrosocysteine, which is formed during the curing process (Figure 4.4).

Smoke constituents modify the basic flavour of cured meats and impart a distinctive smoked characteristic. A large number of con-

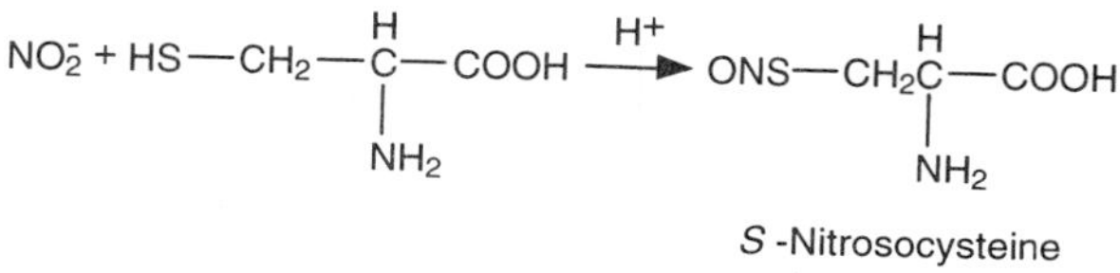

$$NO_2^- + HS-CH_2-\overset{\overset{H}{|}}{\underset{\underset{NH_2}{|}}{C}}-COOH \xrightarrow{H^+} ONS-CH_2\overset{\overset{H}{|}}{\underset{\underset{NH_2}{|}}{C}}-COOH$$

*S*-Nitrosocysteine

**Figure 4.4** Formation of *S*-nitrosocysteine during curing.

stituents are present on meat after smoking, many of which can act as flavour constituents, either singly or in combination. A wide range of chemical groups are present in wood smoke and many members are poorly defined. Among the most important constituents are: phenolic compounds, cresols, aldehydes (including acetaldehyde), ketones, aliphatic acids from formic through caproic, primary and secondary alcohols, catechol, methyl catechol, pyrogallol and the methyl ester of pyrogallol. Smoke constituents also contribute to flavour by interaction with meat components. Many of the smoke constituents, for example, are highly effective as antioxidants.

The situation with dry-cured ripened hams is different in that the product undergoes extensive biochemical change during manufacture (see below). The high concentration of free amino acids, for example, is thought to enhance the natural 'meaty' taste of this type of ham.

## 4.3.4 Biochemical processes during the manufacture of dry-cured hams

Biochemical changes in dry-cured hams are often considered to occur synonymously with the ripening or maturation stage of manufacture. This is not strictly true since, in many cases, change is initiated immediately after salting of the green ham and it is preferable to discuss changes in the context of the entire process. In recent years, there has been considerable interest in the biochemistry of dry ham processing and major advances have been made in understanding the processes involved. The number of types of ham which have been studied in detail is relatively small and differences in manufacture may mean differences in the biochemical processes occurring. It seems likely that these would involve extent of change rather than fundamental nature.

## (a) Proteolysis

Processing of dry-cured ham is characterized by intense proteolysis. This has been attributed, to a large extent at least, to the activities of the muscle proteinases, calpains and cathepsins. Calpains are active in green hams and may be detected after salting, but the enzymes are unstable and activity cannot be detected after this stage. In contrast, cathepsins B, H and L remain active during many months of ripening, although activity is reduced by increasing NaCl concentration. Cathepsins B, H and L retain 10–15% of activity 15 months after initiating the curing process. Cathepsin D is less stable, activity persisting for no more than 5–10 months. Although cathepsin H remains active, the enzyme does not appear to degrade myofibrillar proteins and in the later stages of manufacture cathepsins B and L are responsible for proteolytic activity. Stability is greater than would be expected and it has been postulated that NaCl, while reducing the rate of proteolysis, plays a role in stabilizing the enzyme in an active configuration.

Muscle proteinase activity leads to a progressive disappearance of myosin heavy chain, myosin light chains 1 and 2 and troponins C and I. This process involves degradation of the protein molecules and is distinct from physico-chemical loss of myofibrillar ultrastructure (see Chapter 2, pages 88–91). Protein degradation is accompanied by an increase in breakdown products. The ultimate breakdown products are free amino acids, the concentration of which increases very markedly (*ca.* 7.5×) during ham manufacture. Free amino acids are the product of amino peptidase activity and production continues throughout the processing. The activity of aminopeptidases is reduced as the NaCl concentration approaches 1.25 M. Nitrite also has an inhibitory effect in some cases. In other situations nitrite and, if present, glucose counterbalance the inhibitory effects of NaCl. This is particularly significant in the case of leucyl aminopeptidase, which is activated by nitrite and glucose. Leucyl aminopeptidase retains 39% of its activity from salting to the end of maturation and is the most important aminopeptidase. Other aminopeptidases are involved towards the end of maturation, but are of limited importance.

The overall content of free amino acids increases markedly during maturation. Increases in levels of individual amino acids vary and decreases of some, such as glutamine and taurine, have been observed. There is variation in this respect between different studies, which may reflect different curing technology and differences in conditions during ripening. In any single situation, differences between individual amino acids can be explained by further reactions between amino acids and carbonyl compounds derived from fat autoxidation. Reactions involve Maillard condensation, initiation of which depends on conditions within the ham. A strong Maillard reaction, for example, appears to take place in Iberian hams, which have a pH value of *ca.* 6.0 and an $a_w$ level of *ca.* 0.85, during maturation at 15–25°C. Conditions during ripening of other hams are less conducive to Maillard condensation.

In addition to Maillard condensation reactions, it seems likely that amino acids undergo oxidative deamination–decarboxylation via Strecker degradation. This is the probable origin of a number of types of carbonyl compound, such as the branched chain aldehydes 2-methylbutanal and 3-methylbutanal, phenylacetaldehyde and α-dicarbonyls (butan-2,3-dione, pentan-2,3-dione).

The role of micro-organisms in protein metabolism during maturation of hams has been a matter of discussion in the past. It is now recognized that micro-organisms play no significant role in internal protein breakdown. Endocellular proteinases of moulds and yeasts may play a minor role on the surface but this is of limited significance. It is possible, however, that aminopeptidases from *Pediococcus* may work in concert with muscle aminopeptidases.

## (b) Lipolysis and autoxidation of fat

In meat, lipases (acid, basic and neutral) are present both within lysosomes in the muscle tissue and (neutral and basic) in the adipose tissue. Lysosomal lipases are considerably more stable, basic lipase retaining almost all activity over a 7-month production process and the other lipases retaining 40–50% of activity. Lipases in adipose tissue are markedly less stable, activity virtually disappearing within 5 months. As with proteolysis, the role of micro-organisms in lipolysis is, at most, very limited.

---

* As expected, proteolysis proceeds most rapidly when ripening is at higher temperatures and when NaCl content is lower. Type of pig also has an effect, levels of muscle proteinases being lower in heavy pigs (Large White) than in light pigs. This reflects the lower rate of protein turnover in heavy pigs. Despite advantages in terms of utilization, heavy pigs are not generally used for dry cured ham manufacture, and their introduction would require a modification to the process.

Fat autoxidation is very rapid in the early stages of ham production and is the origin of straight chain aldehydes. The rate of autoxidation in the later stages can be slowed very appreciably by the accumulation of Maillard products with powerful antioxidative properties. Where a strong Maillard condensation occurs the level of volatile carbonyl compounds is likely to fall.

## 4.4 MICROBIOLOGY

### 4.4.1 Cured meats as an environment for micro-organisms

The intrinsic microbiological stability of cured meats varies considerably according to the curing technology used and thus the levels of NaCl and other curing ingredients. At one extreme are traditional products, usually dry cured, including some types of bacon, Iberian hams, etc., which are shelf stable for many months at ambient temperatures. At the other extreme are products of the sweetcure type, which often contain very low levels of curing agents and which are dependent on extrinsic factors, such as refrigeration and vacuum or gas packaging for stability. Most Wiltshire-type bacon is also now dependent on extrinsic factors for stability.

### 4.4.2 Cured meats and food poisoning

The safety record of commercially produced cured meats is extremely good. As far as is known, there are no authenticated cases of food poisoning resulting from consumption of Wiltshire or sweetcure bacon, salt beef or similar products. *Staphylococcus aureus* can be isolated from uncooked cured meats, however, and was thought to present a hazard in vacuum-packed products where the spoilage microflora was suppressed. Various inoculation experiments showed that *Staph. aureus* can grow and, under some circumstances, elaborate toxin at storage temperatures in excess of 20°C. In practice, however, there is no evidence that growth can occur. In Wiltshire bacon, this may be attributed to inhibition by lactic acid bacteria, but this explanation is less convincing with sweetcure bacon where numbers of lactic acid bacteria are often very small in the initial stages of storage. In any case, storage temperature should be strictly controlled during manufacture and distribution.

The presence of other pathogens may be demonstrated on bacon. *Salmonella* is present at a low incidence and, according to serovar, persists but does not grow. Clostridia are very widely distributed and can readily be isolated from meat. Both *Clostridium botulinum* and *Cl. perfringens* have been isolated from vacuum-packed bacon. The reported incidence of both organisms varies widely, although this may be partly due to differences in methodology. In one survey a single lot of bacon was notable for an incidence of *Cl. botulinum* of 73%, but in most cases the incidence was less than 10%. *Clostridium botulinum* has a higher tolerance of NaCl and curing agents than *Cl. perfringens*, but neither of these organisms, nor *Salmonella*, is considered to constitute a risk under foreseeable conditions.

Dry-cured hams are recognized as constituting a higher risk of *Cl. botulinum* than brine-injected cured meat due to the very slow diffusion of curing ingredients to deep muscle tissue and, in some areas, manufacture under technologically unsophisticated conditions, where temperature control may be inadequate. A further consideration is that dry-cured ham may be eaten raw and there is no cooking during which toxin is inactivated. Despite this, there are no known cases of botulism resulting from consumption of commercially produced dry-cured ham. The situation with home-produced ham is different and botulism is relatively common in countries where home curing is widely practised. Home curing of ham in Poland is largely responsible for the high prevailing incidence of botulism, while 12 of 13 outbreaks reported in Portugal

---

BOX 4.6 **The hungry years**

Home curing, like other methods of home preservation, tends to be significantly more common in less affluent societies. This reflects the fact that economic hardship increases dependence on home produced food and thus the use of non-industrialized curing methods. In many cases such methods, unless adequately controlled, can involve a high risk of botulism. This is illustrated by the situation in rural areas of the US during the great depression of the 1930s. Poverty led to a shift away from commercially processed foods and towards the home canning and preservation of vegetables by methods that were inherently unsafe. String beans were most often involved and it was noticeable in accounts of outbreaks that beans were often consumed, presumably through economic necessity, even when signs of spoilage were obvious.

during the 1980s were due to home cured ham. The Portuguese outbreaks all involved type B toxin and were notable for the fact that the nature of the symptoms was identical in each case, but the severity varied greatly. No fewer than 20 of 50 affected individuals remained ambulant and were treated on an out-patient basis.

### 4.4.3 Spoilage of cured meats

*(a) Uncooked sides and joints*

The majority of the large quantity of information available on spoilage of uncooked sides and joints concerns Wiltshire and related types of bacon. Spoilage of other types follows similar general patterns, although some problems are specific to the Wiltshire type of cure.

Spoilage of traditional Wiltshire sides, or joints, can commence immediately after removal from the immersion brine. Prevention of bacterial slime formation at this stage (see page 178) is seen as being an essential part of good manufacturing practice. At this stage, however, the surface of all bacon will carry a bacterial load, the number of bacteria present, and in some cases the type, being predictive of keeping quality. The number of bacteria present on green bacon depends on several factors, all of which are under the control of the curer. Surveys in Northern Ireland have shown that many factories are capable of consistently producing bacon with total colony counts below $10^5$ cfu/cm$^2$ and, while some variation appears inevitable, this figure is considered an attainable target under good manufacturing conditions.

A side or joint of Wiltshire bacon is a heterogeneous environment, consisting of many niches, and a wide range of types may be present and ultimately cause spoilage. Species of *Micrococcus* are usually dominant; the ultimate source is the pig skin, although the bacteria can multiply in many parts of the curing environment. Most strains are both proteolytic and lipolytic and the spoilage pattern tends to differ between meat and rind. Micrococci may also cause slime formation. Most strains are capable of reducing both nitrate and nitrite and growth can lead to an initial increase in nitrite levels followed by depletion.

Species of halophilic *Vibrio* are important in the spoilage of bacon sides. These are often identified with *V. costicola*, a species of

*Vibrio* which differs in a number of ways from other members of the genus. *Vibrio costicola* is not itself a homogeneous species and bacon isolates have been divided into three groups, *V. costicola*, *V. costicola* subsp. *liquefaciens* and an unidentified group, which is biochemically inactive and differs from other groups in being present on the live pig. It is considered, however, that the situation is more complex than that represented by three groups and that a range of vibrios, related to but distinct from *V. costicola*, may be present on bacon. The dominance of any particular strain appears to depend on complex and little understood ecological factors.

*Vibrio costicola* and other vibrios are known to cause spoilage of bacon. Spoilage pattern appears to vary according to the dominant type present, some but not all being proteolytic. The organisms, however, are most commonly associated with slime formation and stickiness, together with taints especially in the shoulder and femur. Vibrios are able to reduce both nitrate and nitrite and thus play a role in depletion of curing agents.

In addition to their spoilage role, the presence of large numbers of *Vibrio* spp. on bacon is considered to indicate that growth has occurred during production as a consequence of poor temperature control, or high humidity during maturation. There also appears to be an independent relationship between the presence of *V. costicola sensu stricto* in bacon and low nitrite levels at the end of curing.

A variety of other Gram-negative bacteria may be present, including *Acinetobacter*, *Alcaligenes*, *Janthinobacter* and members of the Enterobacteriaceae including *Enterobacter*, *Providencia* and *Serratia*. These genera have all been isolated from spoiling or spoiled bacon, but a causal role is difficult to prove, although *Serratia* is alleged to have caused red pigmentation of bacon after prolonged post-maturation storage. A number of the genera present are capable of reducing nitrate and nitrite and may be involved in depletion of these salts. *Alcaligenes*, most strains of which can reduce nitrite but not nitrate, may be of particular significance in reducing the inhibitory properties of spoiling bacon.

Other Gram-positive bacteria include small numbers of lactic acid bacteria and members of a group loosely described as *Arthrobacter–Corynebacterium*. Lactic acid bacteria are only rarely involved in spoilage of unpackaged bacon, while *Arthrobacter–Corynebacterium* species appear to develop as a consequence of spoilage rather than as a cause. Pigmented species have, however, been alleged to cause coloured spots on the rind of bacon.

Yeasts are more common on bacon sides after smoking, but *Torulopsis* and *Candida* have both been associated with spoilage and other species have been isolated. Under most circumstances, moulds are rare on green bacon, but spoilage due to *Penicillium* and *Cladosporium* has been observed in cases where bacon backs were stored for a prolonged period at *ca*. 0°C.

Smoking has a profound effect on the spoilage microflora of Wiltshire bacon. Depending on the temperature of the smoke, the numbers of bacteria are reduced by up to $10^4$ cfu/cm$^2$. Gram-negative bacteria tend to be more heat sensitive than Gram-positive and the proportion of *Micrococcus* and, to a lesser extent, lactic acid bacteria is increased. The dry surface of the smoked bacon also favours *Micrococcus* over other bacteria but growth of *Micrococcus* is restricted and slimes or other overt spoilage are rare. Yeasts and moulds are also favoured and spoilage by these micro-organisms is common. Visible mould growth is usually the most common spoilage pattern, the growth being initially present as small spots, spread being restricted by the reduced $a_w$ level and the mycostatic effect of smoke constituents. Many genera of mould have been implicated in spoilage of bacon, including *Alternaria*, *Aspergillus*, *Fusarium*, *Monilia*, *Oidium*, *Penicillium* and *Rhizopus*. Of these *Aspergillus* and *Penicillium* are the most common, although other genera may be dominant in specific environments. In the past, superficial mould growth on smoked bacon has been of little concern to the industry, remedial action involving wiping the surface of the bacon. Concern over mycotoxins means that this practice is now considered undesirable. The use of potassium sorbate sprays has been suggested as a means of controlling mould growth but is not permitted in many countries, including the US and UK. Application of the principles of good practice, including adequate control of temperature and humidity and avoidance of excessive storage periods, should minimize problems with mould growth.

Although spoilage of bacon usually involves surface growth, internal spoilage can cause severe problems. The incidence has been reduced, largely due to improvements in refrigeration and improved process control, but not eliminated. Pocket taint refers to spoilage in the region of the shoulder from which the scapula is removed, special precautions often being taken during curing to minimize this problem (see page 175). Bacterial slimes may develop in this area and cause taint, especially if the membrane surrounding the scapula has been broken or otherwise damaged and the meat exposed during butchery. *Micrococcus* and *Vibrio* are most commonly involved, although spoilage of this type due to *Alcaligenes* and *Providencia* has also been reported.

Bone taint is spoilage within the deep meat and bone marrows of cured meats. An underlying cause is poor distribution of curing ingredients, which has been much reduced by the application of improved injection techniques. On some occasions it has been possible to trace recent incidents of bone taint to malfunction of multi-needle injectors. The incidence is higher when meat is of high pH value and during summer months when the load on refrigeration systems is higher. Bone taint can affect any joint, but is most common in shoulders and gammon. Under commercial conditions, occurrence of bone taint may follow distinct patterns, which can be very difficult to interpret.

It is now accepted that bacteria may enter the internal tissues and bone marrows of newly slaughtered pigs by the process of agonal invasion, the alternative theory of bacterial migration from the gut having been virtually discarded. Agonal invasion is the rule rather than the exception and, in practice, all bone marrows may be contaminated. In some cases bone taint is restricted to the bone marrow and the meat may be used after removal of the offending bone. In more severe cases, the meat surrounding the bone is also heavily contaminated and the product must be rejected. In the case of the femur, and probably other bones, it appears that micro-organisms grow in the marrow, which has a pH value of 6.8–7.2 and low levels of curing ingredients, and spread via the blood vessels to surrounding tissues. Microbial numbers in material from the blood vessel of badly affected bones exceed $10^8$ cfu/ml and infection may be demonstrated by the exudation of a foamy liquid when the bone is gently heated.

Bone taint may be either a sour or a putrefactive sulphide-type odour, the latter being highly distinctive. A wide range of bacteria have been implicated in the past but halophilic *Vibrio* species and *Providencia* are now most common.

A third type of taint, rib taint, is now largely of historic interest. The causative organism, *V. costicola* (*V. costicolus*), is of course now recognized as being of importance in other types of spoilage.

### (b) Pre-packed uncooked cured meat

As with sides and joints, most of the published information concerning spoilage of pre-packed uncooked cured meat refers to Wiltshire-type bacon. The vast majority of the product is vacuum or modified atmosphere packed and discussion will be restricted to meat in these two types of packaging.

Vacuum and modified atmosphere packaging are not catastrophic events and there is no immediate change to the composition of the microflora. Conditions in both types of pack, however, favour the development of lactic acid bacteria. Discussions in the past concerning lactic acid bacteria in vacuum-packed bacon have often used the terminology of Orla-Jensen and referred to the dominant type as being 'atypical streptobacteria'. Such strains of *Lactobacillus* are common in meat and have subsequently been studied in greater detail. A division is possible on the basis of acid tolerance, aciduric strains being identified with *Lb. sake* and *Lb. bavaricus*, while non-aciduric strains were assigned to a 'new' genus, *Carnobacterium*. In the case of Wiltshire bacon, however, a number of lactobacilli may be present which do not conform with recognized classification systems. Other types of lactic acid bacteria, including *Lactococcus*, *Leuconostoc* and *Pediococcus*, can be isolated from vacuum-packed bacon, but are only rarely present in significant numbers.

Lactic acid bacteria are of greatest numerical dominance in bacon of low pH value and low NaCl content, stored at temperatures below 10°C. In practice, this means the great majority of bacon and consequently this group forms the spoilage microflora. Patterns of spoilage vary from a general sourness to rancid and cheesy flavours. This may be accompanied by stickiness or slime formation. Lactic acid bacteria are little affected by smoke constituents and there is usually little difference in the ultimate spoilage pattern between green and smoked bacon.

Although species of *Lactobacillus* and *Carnobacterium* are the most common lactic acid bacteria spoiling pre-packed bacon, *Leuconostoc* is isolated in large numbers on rare occasions, usually where temperature abuse is suspected. The spoilage pattern is similar to that associated with other lactic acid bacteria, but significant quantities of gas may be produced.

Micrococci are able to grow in bacon by utilizing nitrate, or nitrite, as terminal electron acceptor. Growth is slow and under most circumstances the genus is unable to compete effectively with lactic acid bacteria. Micrococci have been involved in spoilage of high NaCl/high pH pre-packed bacon, including some continental types, where storage temperatures exceed 15°C. Spoilage involves slime formation, accompanied in some cases by proteolysis and lipolysis. Circumstantial evidence suggests that the proteolytic and lipolytic activity of *Micrococcus* is significantly reduced at low oxygen tensions. Coagulase-negative species of *Staphylococcus* can also be involved in spoilage of high NaCl pre-packed bacon, but only when storage temperatures exceed 25°C.

Gram-negative bacteria cause spoilage of vacuum-packed bacon under some circumstances. Smoked bacon is only very rarely involved, Gram-negative bacteria being sensitive to smoke constituents. The most common Gram-negative species is *Vibrio*, usually derived from the brine. It has been stated that *Vibrio* is favoured by low NaCl, high pH and high storage temperature. Despite this, *Vibrio* species isolated from bacon are able to grow rapidly at temperatures below 5°C in NaCl concentrations in excess of 5%, but are relatively sensitive to nitrite. A more likely interpretation, therefore, is that *Vibrio* tends to spoil bacon of low $NO_2$ content, possibly following depletion at high temperatures. Depending on strain, spoilage by *Vibrio* may involve proteolysis and $H_2S$ production is also common. Stickiness and slime formation, accompanied by a distinctive odour, may also occur.

Members of the Enterobacteriaceae, predominantly *Enterobacter* and *Hafnia* species, *Morganella morganii*, *Proteus mirabilis* and *Pt. vulgaris*, are amongst the spoilage flora of bacon stored at excessively high temperatures (>20°C), but are not significant under adequately controlled refrigerated storage. Spoilage is characterized by production of large quantities of hydrogen sulphide, which can be detected visually by an in-pack detector. A further member, *Providencia* (*Proteus rettgeri*), is involved in spoilage of vacuum-packed bacon stored at temperatures between 10 and 15°C. *Providencia* is able to metabolize methionine to methane thiol with production of 'cabbage odour'. The incidence of this type of spoilage appears to be increasing in summer months, although occurrence is sporadic. This has been attributed to a higher incidence of high pH meat and the failure of the industry to appreciate that modern bacon, of relatively low NaCl content, requires storage at temperatures below 5°C.

Relatively little is known of the spoilage microflora of bacon in modified atmosphere packs. Available evidence suggests a similar pattern to vacuum-packed bacon, although spoilage due to *Providencia* appears to have been largely eliminated.

Lactic acid bacteria, especially *Lactobacillus* and *Carnobacterium* are dominant in spoilage of pre-packed sweetcure bacon and the involvement of other bacteria is rare. Initial numbers on sweetcure bacon are low, as few as $10^2$/g being present on some occasions. The relatively low NaCl concentration and the lack of a pre-existing microflora mean that growth is usually relatively rapid, even when temperature is adequately controlled. On some occasions, however, a delay of up to 6 days has been observed during which little growth occurs. This delay is then followed by very rapid growth but the reasons for this phenomenon are unknown. Spoilage is by souring, with stickiness and slime formation.

*Vibrio* species are rare, but not unknown, in spoilage of sweetcure bacon. A particularly spectacular type of spoilage, which primarily affects bacon made with added sucrose, is due to dextran formation by some strains of *Vibrio*. This results in massive slime formation, rashers of bacon often being completely coated.

Members of the Enterobacteriaceae may also be involved in spoilage of sweetcure bacon. *Proteus* and *Providencia* are most commonly implicated, although cabbage odour does not appear to be a significant problem. The most common spoilage pattern is faecal odours and stickiness.

### (c) Dry-cured hams

Dry-cured hams, when properly made, are a stable product with a long storage life. External mould or yeast growth is common and within industry has previously been thought to be of little significance, or to be indicative of correct ripening and good quality. It is now accepted that mould growth is undesirable and there is concern over the possibility of mycotoxin production. A link has been made between consumption of mould-spoiled hams and some types of cancer in the Adriatic region of the former Yugoslavia. A large number of mould genera have been isolated from dry-cured hams but *Aspergillus*, *Cladosporium* and *Penicillium* are dominant. Potassium sorbate sprays are permitted in the US and some other countries to control mould growth.

Yeast growth can also be a fault on dry-cured hams, alternate patches of yeast and moulds being present on heavily spoiled products. Species present are generally the same as those found on Wiltshire bacon and include *Candida*, *Debaryomyces*, *Rhodotorula* and *Torulopsis*.

Internal spoilage of dry-cured hams is rare in commercial practice and is usually associated with exposure to high temperatures before a sufficiently high concentration of NaCl and curing agents is attained in deep muscle. Internal spoilage is often attributed to species of *Clostridium*, although in modern practice this type of spoilage appears to be rare. A wide range of clostridia have been isolated from spoiled hams, including *Cl. bifermentans*, *Cl. putrefaciens* and *Cl. sporogenes*. In extreme cases gas production was sufficient to disrupt the structure of the ham, although the significance of these findings to current practice must be questioned.

Some investigations have suggested that a miscellaneous group of bacteria is responsible for internal spoilage of hams. The normal internal microflora, for example, is *ca.* $10^6$ cfu/g and is dominated by lactic acid bacteria, especially *Lactobacillus* and *Pediococcus*, *Micrococcus* and *Staphylococcus xylosus*. Numbers of bacteria in spoiled hams are in excess of $10^8$ cfu/g, with species described as 'coryneform' (*Aerococcus*, *Arthrobacter*, *Bacillus* and *Sarcina*) being present in significant numbers in addition to members of the normal microflora. On other occasions, the spoilage microflora is dominated by *Providencia* and, to a lesser extent, species of *Proteus*. In any case considerable variation may be expected even with hams produced at the same plant.

Internal mould growth is very rare in traditionally cured hams but can be a significant problem with more modern cures. Mould growth, usually involving *Aspergillus* or *Penicillium*, is a common problem in deboned hams, in the cavity remaining after bone removal. Internal mould growth is also a serious problem in

restructured dry-cured hams and has limited development of this
type of product (see page 189).

### 4.4.4 Microbiological analysis

*(a) Routine analysis of Wiltshire and related immersion brines*

Direct microscopic enumeration, using a counting chamber,
permits gross changes in brine microflora to be readily detected
and remedial action taken. The accumulation of debris can make
direct counting difficult and phase contrast microscopy is preferred. Direct counting is a relatively crude method and should be
supplemented by colony counts. Colony counts of high NaCl environments are beset by difficulties and there is a large discrepancy
between microscopic and colony counts. Further, no single
medium will recover all types of micro-organism in the brine. A
medium containing 4–5% NaCl is most widely used and recovers
halotolerant bacteria as well as those halophilic bacteria involved in
bacon spoilage. Recovery of the dominant, more strongly halophilic bacteria requires a medium of higher NaCl concentration
(*ca.* 20%) and special techniques which are not appropriate to
routine monitoring.

Much of the current knowledge of the microbiology of Wiltshire
brines and their technical management stems from the work of Dr
Alan Gardner at the Ulster Curers' Association in Belfast. This work
demonstrated the level of control possible by means of relatively
simple techniques and suggested means by which more sophisticated control could be applied. A selective medium for *Vibrio* spp.
was devised, crystal violet–kanamycin agar, which, although
designed to indicate backwash contamination of brines, may well
have a wider role as a predictor of brine instability and a short
storage life of bacon. The value of other microbiological examinations of curing brines, such as counts for *Escherichia coli* I or
*Pseudomonas*, appear less well defined and it is possible that

---

* Halophilic bacteria can be highly sensitive to osmotic shock and it is necessary to
'think salt' whenever the organisms are handled. This means that not only must
cultivation media contain NaCl at an appropriate concentration, but also a similar
concentration must be present in dilution blanks, etc. Preparation of heat fixed
smears for microscopic examination is also complicated by the need to make cell
suspensions in a solution of NaCl rather than water. The smear must then be
desalted, a procedure which makes the Gram stain difficult and more complex
operations, such as flagella staining, virtually impossible.

control should be exerted through good manufacturing practice rather than through retrospective microbiological examinations.

## (b) Bacon and other uncooked cured meats

Microbiological analysis is most commonly undertaken at pre-packing and usually involves colony counts on non-selective media containing 4–5% NaCl. Enumeration of lactic acid bacteria may also be made and can be useful, especially with sweetcure bacon. Low pH value selective media containing high levels of acetate are not suitable and a medium such as the MRS medium recommended for other meats should be used. The crystal violet–kanamycin agar used for enumeration of *Vibrio* in curing brines can also be used with bacon. Enumeration of *Vibrio* is of particular value in summer months, especially as a predictor of spoilage in high pH value Wiltshire joints. The analysis is not, however, considered worth while for sweetcure bacon. Examination for other microbiological parameters, including *E. coli* and yeasts, is not considered of value under normal circumstances.

Routine microbiological examination of ripened dry-cured hams is not considered necessary.

EXERCISE 4.1

The relationship between the age of the brine and the flavour of bacon has never been systematically investigated. Indeed, there are few opportunities since mature brines are only discarded under extreme conditions. Following the establishment of a small Wiltshire curing operation on a greenfield site, however, it was found that for the first 10–12 curing cycles (length 3–4 days), the bacon was of pale colour and, while recognizably bacon, had a rather insipid 'porky' taste. The brine at this stage contained no more than $10^3$ bacteria/ml, which appeared to have been derived directly from the meat. The brine had a pH value of 6.6, was clear and of a yellow-green colour. The 'porky' taste disappeared over the next few curing cycles and the colour and flavour improved slowly until, after *ca.* 100 cycles, the bacon produced was considered to be of typical high quality Wiltshire flavour. At this stage, bacterial numbers in the brine had increased to $10^6$/ml, most of which were halophilic Gram-negative rods. The pH value of the brine had fallen to 6.2, a moderate amount of suspended material was present and the colour had darkened to a light brown. Levels of NaCl, $NaNO_2$ and $NaNO_3$ had been controlled to similar levels throughout.

Discuss the relationship between changes in the brine and the properties of the bacon. On the evidence presented, do you consider that the development of a brine microflora is important in the curing of Wiltshire bacon?

Assume you are given the opportunity to study a greenfield curing operation more fully than that described, and draw up an outline plan to determine the effect of chemical and microbiological changes in the brine and the quality of the bacon.

## EXERCISE 4.2

The relatively low levels of NaCl and $NO_2$ in sweetcure meats mean that vegetative pathogens, such as *Salmonella*, can survive for extended periods and may even be capable of growth. Although undesirable, this is of limited public health significance provided that cooking is adequate. Sliced sweetcure bacon, however, has been identified as being a potential risk product for *Listeria monocytogenes*. This stems from the resistance of *L. monocytogenes* to NaCl and $NO_2$, its ability to grow at low temperatures and the possibility of its surviving the light cooking often applied to thin sliced bacon. As far as is known, however, there are no epidemiological grounds for implicating sweetcure bacon in listeriosis.

Assess the basis for the identification of bacon as a high risk product for *L. monocytogenes*. To what extent do you consider the identification justified?

Develop a method by which the risk presented by sweetcure bacon can be analysed (quantitatively if possible). Apply this method to compare the risk presented by sweetcure bacon to that presented by pasteurized milk, cooked chicken (as a commercial product), Wiltshire bacon and cornflakes.

# 5

# COOKED MEAT AND COOKED MEAT PRODUCTS

---

## OBJECTIVES

After reading this chapter you should understand
- The various types of cooked meat product
- The basic technology of industrial-scale cooking
- The post-cooking handling of meat products
- The manufacture of specific types of cooked meat product
- The development of 'recipe' meals
- Quality assurance and control
- The physical structure of cooked meat products
- Chemical changes resulting from cooking
- Public health aspects of cooked meat products
- Microbiological spoilage of cooked meat products

---

## 5.1 INTRODUCTION

A very wide range of cooked meat products is available, from traditional products, such as pork pies and cooked sausages, to relatively recent developments, such as pre-prepared meat meals. From a socio-economic viewpoint, consumer convenience is the common thread linking these products and this may often be related to work patterns. Some of the earliest commercially cooked meat products, pies and pasties, were derived from the pastry-wrapped 'dinners' prepared for field workers, while modern 'ready-meals' are particularly popular in families where both adults work. It is notable, however, that while traditional products allowed manufacturers to make use of lower quality meat, and were accordingly cheap, ready-meals and other more recent developments tend to be premium cost products.

In industrialized countries, the continuing demand for convenience means that manufacture of cooked meat products is an increasingly important area of the meat industry. Performance in this sector is uneven. Sales of traditional products, including pies and cooked sausages, are static or slowly declining; such products are often perceived as old fashioned and even as 'food of the poor'. In contrast, sales of more recently introduced products are tending to increase rapidly. These include not only ready-meals but also pre-cooked meat joints of various types.

---

### BOX 5.1 **The baker's oven**

Precooked meat joints are seen today as an area of considerable opportunity for the meat industry. The concept is not new. In earlier times, the lack of adequate oven space in working class homes and the high cost of fuel meant that, on special occasions when joints were available, home cooking was not practical. In many areas it was customary to arrange for the joint, as well as home mixed cakes and puddings, to be roasted at the local bakery. At busy times, such as Christmas, it was necessary to devise elaborate precautions to ensure that meat was reunited with its correct owner and accusations of swopping, followed by non-festive bouts of fisticuffs, marred many a happy Christmas.

---

### 5.2 TECHNOLOGY

Cooking is a process in which, due to thermal treatment, chemical, physical and microbial changes occur in food finally leading to a palatable product. The wide variety of cooked meats produced means that a correspondingly wide range of technologies must be employed. The underlying functions are, however, the same in each case: a heat treatment which must be sufficient to kill vegetative pathogens, in addition to meeting technological objectives; cooling to prevent outgrowth of surviving bacterial endospores; and protection of the cooked product from contamination. In some cases, there may be an apparent conflict between the need to process for safety and the production of desired organoleptic properties. Under such circumstances safety is of paramount importance and must take precedence over all other considerations.

A knowledge of physical, chemical and biological changes as a function of cooking time and product temperature is required to enable optimal thermal processing conditions to be applied. This permits the use of cooking processes which maximize the quality attributes of the end-product, while minimizing undesirable effects, such as shrinkage. A cook-value (C-value) analogous to the F-value used in sterilization has been proposed:

$$C = \int t_0 10^{[T(t) - 100/Z_c]} \, dt$$

where C = cook-value in equivalent minutes at 100°C, T = product temperature, t = cooking time, $Z_c$ = necessary temperature rise (°C) needed for a 10-fold increase in reaction rate of a product property value.

The cook value concept is typically applied to a constant temperature environment. The $Z_c$ concept can be modified for application to change in product property rather than change in reaction rate constant. Thus $Z_p$ is the necessary temperature rise in °C needed for a 10-fold change in a property value.

## 5.2.1 Methods of cooking

### (a) Hot air

Hot air is a traditional means of cooking, and the most widely used on a domestic and catering scale. Hot air cooking is also widely used on an industrial scale. Ovens may operate on a batch basis, but conveyer ovens in which the product travels through the oven on a moving belt are universal in large-scale production. Air at up to 200°C is used and is usually heated by electricity or by gas burners. Heat transfer is only moderate, but can be increased by humidifying the air.

---

* A widely used guide to the heat treatment required is the reduction of numbers of *Salmonella* through 7 log cycles. This involves internal time/temperature combinations such as 57.2°C for 37 minutes and 60.0°C for 5 minutes. Such treatments provide an adequate safety margin with respect to *Salmonella*, but there is concern that *Listeria monocytogenes* may survive. In the US, there is a tendency to accept this possibility and to propose addition of preservatives to control its subsequent growth. Some microbiologists consider the use of preservatives to be contrary to good practice in cooked meat production. There is no doubt, however, that while preservatives should not be used as a substitute for good practice, their use is likely to improve significantly the safety of 'minimally processed' cooked meats, which are highly dependent on refrigeration for stability.

Smokehouse cooking is effectively a specialist form of hot air cooking. In basic form, kilns of the type used for bacon smoking are employed, although the operating temperature is rather higher. Heat is supplied by hot air, smoke being produced in a separate generator or sprayed into the chamber as a liquid extract. Alternatively, a hot air oven fitted with a means of introducing smoke may be used. In either case, careful control of humidity is required to prevent excessive drying.

### (b) Steam

Steam involves heating with saturated air at 100°C. Latent heat is given up at the meat surface as the steam condenses and heat transfer is therefore very good. The continuing presence of water at the surface, however, means that the temperature does not rise above 100°C and thus browning and other reactions associated with roast meat do not occur. Steam cooking can be used as the first part of a two-stage process for products of this type, the second stage involving radiant or hot air heating to impart the characteristic roast appearance and flavour.

Meat stews, soups and similar products of high water content can be heated by direct steam injection. This is an efficient process, but it is necessary to take account of the water produced by condensation. Heating by direct steam injection is also widely used in production of ultraheat-treated (UHT) soups and stews. In this case water of condensation is removed by flash cooling in a vacuum chamber.

Steam at pressures greater than atmospheric is generally associated with canning, but is also used in production of soups, stews, etc. Temperatures in excess of 100°C are obtained (typically up to 125°C), allowing for faster processing and shortening of the production cycle.

### (d) Hot water

Hot water at temperatures up to 100°C has relatively good heat transfer properties and is an efficient means of cooking (stewing, braising). There can, however, be a leaching of constituents, including flavour compounds, and the method is generally used only where the water is ultimately to be consumed with the meat and other ingredients. This is the case with soups, stews and pie fillings, which are often cooked in large, jacketed vats, fitted with a stirrer to assist heat distribution. In such cases, water is the heating medium for meat and other solid constituents, but the product as a whole is heated by conduction through the metal walls of the jacket from steam, or circulating hot water. A similar situation exists where heat exchangers, usually of the tubular type, are used to heat particulate meat products.

Some use is also made of hot water for cooking sausages in plastic shrouds and joints of meat in plastic bags. Heat transfer is reduced by the relatively poor conduction of the plastic. A similar process is used for cooking pâtés in metal moulds. Baths are used where the entire process involves hot water, but hot water sprays can be used in conjunction with hot air, or other forms of heating.

### (e) Hot fat or oil

Hot fat or oil are used for cooking (frying) at temperatures of 150–190°C and has very good heat transfer properties. The temperatures are sufficiently high for browning reactions to occur and there is absorption of the fat or oil used for heating. This imparts a characteristic flavour and texture to fried meats, which depends to some extent on the fat or oil used. Heated fats and oils rapidly become oxidized and off-flavours can readily be picked up by the meat being fried. It is usual practice to replace a part of the fat or oil at the end of each production cycle. Small-scale fryers operate on a batch basis, but large-scale production requires continuous frying.

### (f) Radiant heating

Radiant heating (grilling) primarily involves the infra-red portion of the spectrum. Heat transfer is very good and high surface temperatures are attained, with rapid onset of browning reactions and charring. For this reason, use is restricted to thin pieces of meat of even cross-section, or to heating the surface of larger products. Radiant heating, for example, may be used to 'brown' the surface of joints cooked in steam ovens. Electrical elements or flames can be used as the heat source; some processes involve a short passage through flames to char the surface deliberately.

## (g) Dielectric heating

Dielectric heating is a generic term which includes both microwave and radio frequency heating. For practical purposes, dielectric heating can be considered to be a volumetric process, which does not depend on heat transfer through a surface. This permits a significant improvement in heating rate by largely eliminating the temperature gradient.

Both microwave and radio frequency radiation are electromagnetic energy forms, with wavelengths between the audio range and the infra-red. Microwave is normally considered to be between 500 and 5000 MHz and radio frequency between 1 and 200 MHz. The major heating mechanism in microwave heaters involves the dipole effect, in which water molecules act as dipoles and oscillate in sympathy with the electric field reversals. This results in frictional heating, in proportion to the amount of water present in the food mass. A second heating mechanism, ionic conductivity, is of less importance in microwave heating but is of the greatest importance in radio frequency heating, especially at low frequencies. In either case, the ease with which heat can be generated is known as the loss factor and varies with a number of parameters including composition of the food, temperature and frequency of the applied field.

With respect to foods, the best known application of microwaves is the microwave oven, widely used domestically and in catering, primarily as a means of re-heating pre-cooked foods. Microwave heating is also used in industrial-scale food processing applications and has been used successfully, on a small scale, for cooking a wide range of meats, including sausages, pâté and bacon. There are doubts concerning uneven heating in heterogeneous products and in products of differing geometry. Despite this, a particularly valuable role is envisaged in the in-pack, post-cooking pasteurization of products such as ready-meals, where a high level of

---

* It is a common misconception that microwave ovens cook food from the inside outwards. The outer layers receive energy at the same rate as the inner mass and, provided that moisture content and other factors affecting heating are the same, will rise in temperature at the same rate. Immediate and significant heat losses can, however, occur at the surface through conduction and convection. This process continues after cooking is completed, resulting in a thermal gradient from the inner mass to the outer.

handling of cooked material is inevitable during manufacture (see page 253).

Radio frequency cooking has been most widely used for completing the baking of cereal products. More recently combination ovens have been developed which combine conventional heating with radio frequency heating in the same enclosure. Such ovens have applications with a number of meat products, including pies (see page 240).

### (h) Extrusion cooking

Extrusion cooking cannot be applied to conventional meat products, but is seen primarily as a means of upgrading meat of very low quality. Basic extrusion cookers consist of a pitched screw, or two parallel intermeshing screws, rotating within a barrel. The barrel may be heated or cooled, according to requirements. The operating characteristics of single and twin screw extruders are different, the latter now being the more widely used in most applications, including meat processing. Screws may be counter- or co-rotating, the latter design being the most popular. Meat and other ingredients, including water, enter at one end of the barrel and are carried through by the screws. There is a constriction towards the end of the barrel to provide compression, where a combination of mechanical work and heat converts the discrete food particles into a viscous dough. The meat then passes through the constriction, with accompanying pressure release, cooling and moisture loss, leaving the barrel through a die. Where high temperature cooking is used, the product is puffed by steam production on pressure release. Heating is primarily through the mechanical energy input used to drive the extruder screw, this accounting for 50–100% of the total energy input. Steam injection into the barrel may be used to provide additional heat input and heat transfer may take place through the heated or cooled walls of the barrel.

Extrusion cooked meat is primarily used as an ingredient in other products but there is interest in using it as a basis for a new type of restructured meat.

### 5.2.2 Cooling

It is unfortunate that there can be a tendency to regard cooling as incidental to the main technological functions in cooked meat production. This attitude can undoubtedly lead both to risk to public health, due to growth of surviving pathogens, and to poor quality due to overcooking. A target temperature of 10–12°C within 6 hours is generally considered safe, provided that the temperature is below 30°C in 2 hours and 20°C within 3 hours. Some strains of *Clostridium perfringens* and *Bacillus cereus* are able to grow, albeit slowly, at 6°C and a final temperature of 4°C should be attained within 24 hours.

Cooling usually commences immediately after cooking. Either chilled water or air may be used and it is common practice, where feasible, to incorporate cooling facilities into the cooking apparatus. Water, where used, should be chlorinated and of potable quality to avoid risk of contaminating the cooked product.

### 5.2.3 Prevention of recontamination

Prevention of recontamination, either from raw material or from other sources, including food handlers, is a major aspect of the safe production of cooked meats. Strategies must be built into the process design and extreme precautions are required in industrial-scale operations, which effectively means operating two physically separated factories. Routes by which meat may be contaminated after cooking are summarized in Figure 5.1 and means of control summarized in Table 5.1.

One of the most important precautions is the planning of operations to minimize hands-on operations. In some cases, however, handling is inevitable. These include the rebagging of cooked joints and the assembly of ready-meals. A further trend in some countries is production of a basic bulk packaged cooked meat product,

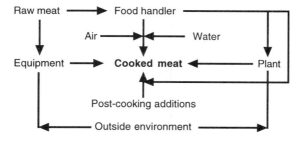

**Figure 5.1** Possible routes of contamination of cooked meat.

**Table 5.1** Precautions against the recontamination of cooked meat products

| Source | Precaution |
| --- | --- |
| **Raw meat** | Ensure strict physical separation of raw and cooked meat. |
| **Food handlers** | Persons working with cooked meat must have no contact with area of plant where raw meat handled. **All** persons must meet appropriate medical criteria (Table 5.2). Clean protective clothing and footwear must be worn. |
| **Post-cooking additions** | **Either** ensure brought-in ingredients conform to agreed microbiological specifications, **or** apply processing sufficient to ensure safety at plant. Ensure added ingredients handled and stored correctly. |
| **Equipment** | Separate equipment **must** be used for handling raw (including fermented) and cooked meat. Equipment should be cleaned and sanitized according to a predetermined schedule with an allowance for emergency cleaning if required. Food contact surfaces must be maintained in a condition permitting effective cleaning. |
| **Plant** | The fabric of the plant must provide adequate protection for the food and must be cleaned and maintained to minimize the risk of its becoming a reservoir of micro-organisms. Contamination of food by condensation or splashes from floors must be prevented. Drains must be correctly cleaned and maintained and flow must be from cooked areas to raw. (*Ideally* the drainage system for each part of the plant should be separate.) |
| **Water** | All water used for cooling, cleaning, etc. must be of potable quality and chlorinated. |
| **Air** | The general air flow in the plant should be from cooked to raw areas. Local air flows should protect product from contamination. Filtered air should be used for cooling. |
| **Outside environment** | Entry of wild animals, birds and insects must be prevented. The environs of the plant must be kept clean and protected from contamination by flood water. Outside clothing and footwear must not be worn in the cooked meat part of the plant. |

---

BOX 5.2  **The Judas window**

Restricting access to the cooked meat side of the plant is an important aspect of protecting the cooked product from contamination. Staff working in that area usually wear clothing of a different colour to those working in the raw meat side and, in some cases, access can only be gained by use of card operated locks and similar devices. Special arrangements must be made for engineers, etc. who may legitimately require access to both parts of the plant. Despite these precautions there is often a Judas, a person who has no authority to enter the plant but who, like Chesterton's postman, goes unnoticed through familiarity. Examples are known to all with experience of factory life, the man with the football pools, the girl with the cosmetics catalogue, the sports organizer and the inevitable collector for birthday and leaving presents. Although the actual risk presented by such visitors is probably small, since food handling is not involved, an important principle is being ignored. This, in turn, can lead to difficulty in enforcing other precautions.

---

which is opened and embellished at a retail outlet before repackaging. Handling of cooked meats in a retail outlet is considered to carry particular risk, since separation from raw meat prepacking operations may be difficult to impose. A further problem lies with the fact that training and experience of both management and staff will be relevant to a retail rather than a manufacturing environment. Any major hands-on operation involving cooked meat involves extreme precautions and a high level of supervision to ensure safety.

---

* The use of sterile, disposable gloves by persons handling cooked meats is a constant source of debate within the food industry. Gloves eliminate, in theory at least, any possibility of contamination from the micro-organisms of the skin. The major concern is *Staph. aureus*, which may be present on the healthy skin, although this site is considered less important in carriage than the nasal fossae. Gloves also inhibit undesirable habits such as nose-picking. Gloves may, however, lead to a false sense of security and the restriction of tactile response means that the wearer may not immediately appreciate the fact that soiling has occurred. Gloves require frequent changing and the resulting expense has led to situations where rationing of the number used was imposed. It should also be appreciated that the use of gloves does not obviate the need for frequent hand washing, or adequate protection of cuts, etc.

The value of medical examinations for food handlers has been much discussed, especially in the context of cooked meats and other cooked foods. Although regular examination is mandatory in some countries, the World Health Organisation considers that the general health of an employee is best assessed by a health questionnaire conducted by a suitably qualified nurse. Physical examinations and other tests, including stool examinations, could then be targeted to specific situations where doubt exists over the employee's suitability for work as a food handler. Although some employers continue to test stool samples routinely, this is not considered effective except in certain circumstances.

As part of initial training, and before food handling has commenced, the employee must receive basic hygiene instruction, including the special responsibility of food handlers and procedures for reporting illness, exclusion, etc. Exclusion is obviously required during illness and it is generally, although not universally, agreed that the risk involved in handling cooked meats is such that exclusion should continue, through convalescence, until clearance criteria are met (Table 5.2). During this period, excluded persons may be employed elsewhere, provided that, with the exception of *S. typhi* and *S. paratyphi*, solid stools are formed and personal hygiene is good.

Colonization of equipment is a major potential problem and can involve spoilage organisms as well as pathogens, such as *Salmonella* and *Listeria monocytogenes*. Sanitization of equipment for handling the cooked product, including conveyers, slicers, etc., is of prime importance and cleaning schedules should be planned to break the cycle of microbial development. This may involve a mid-shift cleaning, and in the US the Department of Agriculture insists that positive proof is provided that mid-shift cleaning is not required. The problem is complicated by the fact that wet cleaning procedures effective against *Salmonella* may be ineffective against *L. monocytogenes*. Attention must also be given to sanitization of cold stores, since these can be the major source of spoilage micro-organisms where cleaning of equipment is of a high standard. Cold stores and other low-temperature areas select for psychrotrophic micro-organisms, which are particularly serious with respect to product spoilage. In theory a build up of psychrotrophic pathogens, such as *L. monocytogenes* and *Yersinia enterocolitica*, may also occur in cold stores, although this does not appear to be a problem in practice.

**Table 5.2**  Criteria for clearance of food handlers after infectious disease

| | |
|---|---|
| *Salmonella typhi*<br>*S. paratyphi* | 12 consecutive negative stool samples over a 6-month period |
| Other salmonellas | None where personal hygiene standards are high, otherwise 3 consecutive negative stool samples |
| *Shigella* | 3 consecutive negative stool samples |
| *Vibrio cholerae*<br>(01 and non-01) | 3 consecutive negative stool samples |
| *Escherichia coli* | 3 consecutive negative stool samples |
| *Staphylococcus aureus*<br>   vomiting<br>   skin lesions | None<br>Until treated and healed |
| SRSVs[1]<br>Hepatitis A virus<br>Rotavirus<br>*Entamoeba histolytica*<br>Threadworms and<br>'pork' tapeworm | 2 days after diarrhoea ceases<br>7 days after appearance of jaundice<br>7 days after complete recovery<br>3 consecutive negative stool samples<br>Until treated |

[1]Small, round, structured viruses

In addition to sanitization of food contact surfaces it is necessary to control pathogens in the plant environment, including walls, floors, etc., which can otherwise act as continuing sources of contamination. In most cases soil removal is of prime importance, but use of a sanitizer as an adjunct is recommended in plants handling cooked meats.

## 5.2.4 Manufacture of the main types of cooked meat product

### (a) Whole and restructured uncured joints and poultry

Centralized cooking of meat joints initially involved little more than a scaled-up domestic process using hot air ovens. Cooking losses were high and the meat often of variable quality due to poor temperature control. High temperature differentials between the surface and the centre of the meat meant difficulties in obtaining an adequately cooked centre without overcooking the exterior.

Some roast joints are still cooked in this way in large catering or small industrial operations, but the vast majority of processes now use the 'cook-in' system in which the meat is cooked in plastic bags or film. Whole joints are usually deboned, pumped with a polyphosphate brine and tumbled or massaged for a short period. It is necessary not only to distribute the brine evenly but also to extract salt-soluble proteins into a layer of exudate on the outer surface of the meat so that the plastic packaging adheres closely to the meat. The exudate also 'sets' during heating and forms a skin which retains moisture within the meat. Dry seasoning may be added at this stage, along with an antioxidant to minimize development of warmed-over flavour. Many products of this type, as well as those re-formed from meat pieces, are of low fat and minimal NaCl content and of a tender, succulent texture. As such, they are regarded as 'new generation' meat products and considered to meet the requirements of the health and quality conscious consumer. From a technical viewpoint, they are also often considered as minimally processed products and rely on refrigeration for stability over shelf lives as long as 90 days. There is consequently increasing interest in the use of preservatives to control *Clostridium botulinum*, in the event of temperature abuse and to control post-process contaminants in the case of products which are handled after cooking. Various preservatives have been proposed including sodium lactate and propionate, nisin, bacteriocin-producing strains of lactic acid bacteria and bacteriocin-based preparations. Of these, lactate is of greatest interest and is a permitted additive to a number of types of meat product in the US. In addition to antimicrobial properties, lactate enhances the meat flavour and increases tenderness through its humectant properties raising the water-holding capacity. Sodium lactate is also considered more acceptable than many other preservatives, since the compound occurs naturally in meat.

An alternative approach is the Wisconsin process, in which the cooked meat is inoculated with lactic acid bacteria which, if the food is severely temperature-abused, will grow and cause spoilage through acidification. In this way, the food 'fails safe'.

Large joints are vacuum packed into plastic bags, although a film laminate may be used for smaller pieces. The packaging has been developed to withstand high temperatures and the bags consist of multi-layer structures formed by tubular extrusion. Common materials include nylon/EVA, polythene/polypropylene, polyester and co-extruded Surlyn/EVA. A Surlyn/nylon laminate is widely used for film material. Two basic variants are used: cook-in-ship, in which the meat is cooked, distributed and retailed in the same packaging, and cook-in-strip, in which the film used for cooking is removed after processing and the product repackaged. In this case there is no requirement for a gas barrier layer.

Before cooking, it is necessary to ensure that frozen meat has been totally defrosted. Joints are also sorted into batches of the same size and weight, to ensure the same degree of cooking. Cooking of bagged meat may be either in hot air ovens or in tanks of hot water, but hot water is preferred due to the superior temperature control and low temperature differentials. Early processes tended to use fairly high cooking temperatures to shorten the process. It is now common practice to use temperatures below 60°C, provided that the overall process is compatible with safety. This improves texture by degrading collagen without accompanying shrinkage and water loss. The low temperature differential obtained also allows the interior to be adequately cooked, without overcooking the exterior. Cooking to an internal temperature of 54.4°C for 135 minutes is a widely used process. The temperature of the heating water must be controlled to $\pm 2$°C to provide an adequate safety margin and the tank must be designed and loaded to permit adequate circulation. Meat packed in film laminate can also be cooked in hot water, a higher temperature of 75–85°C being used. Alternatively, high humidity air may be used, the moisture improving heat transfer and preventing the film cracking.

Following cooking, the roasts are cooled by either air or water and are transferred to a cold store. Cook-in-ship products usually receive no further handling, but cook-in-strip are removed from the cooking bags for further processing as a cooked product. This usually takes place at the manufacturing plant, but may also take place at retail outlets, where small vacuum-packaging machines are available for repacking. The process inevitably involves a very high level of handling and associated risk of contamination. Further processing may involve slicing, to produce individual retail pre-packs, or bulk portion-controlled packs for catering use. Meat may also be seasoned or embellished at this stage, the greater flexibility permitting a wider range of products to be manufactured. A secondary pasteurization (post-pasteurization) after repacking has been proposed to overcome problems due to contamination. *Listeria monocytogenes*, however, has been shown to survive a secondary pasteurization as severe as 96°C for 5 minutes. Microwave heating has been proposed as the basis of a more effective secondary pasteurization. Low-dose irradiation has also been used as an alternative secondary treatment, but on an experimental scale only.

Cook-in-ship meats may occasionally require rebagging due to accumulation of fluid (purge) in the bags. Special arrangements must be made for this process, to avoid risk of contamination.

Restructured meats may be formed in several ways (see Chapter 8, pages 372–373). The most common is to bind pieces of meat by proteins extracted in the presence of NaCl and phosphates, and there is also interest in use of the algin/calcium gelation system. Restructured meats were initially moulded into a roll, but three-dimensional moulding into a shape approximating that of a joint is possible. In some processes, cooking is by hot air in a reusable mould, but this process is inefficient and cook-in processing is now more common. A common system is to use a mould liner made from a heat-resistant film and to cook in either hot water or moist hot air. The end-product can be either retailed whole, or sliced and pre-packaged. In the latter case vacuum or gas packaging is now in general use.

Vacuum packaging is unsuitable for whole or parts of bone-in poultry. The birds (usually chicken) are portioned, if necessary, and then may be cooked by radiant heat in a rotisserie in small-scale manufacture, or by hot-air or steam ovens. Frying is also used and is particularly popular in the US. The birds may be injected with a brine of NaCl and phosphates before cooking and seasoning and colouring may be added to produce varieties, such as tandoori. Seasoning and colouring may also be added after cooking. Chickens are highly prone to drip and this, together with the high volume

---

* On a number of occasions it has been possible to relate outbreaks of food poisoning involving cooked meats to social and economic trends. Some outbreaks of food poisoning due to *Salmonella*, for example, can be attributed to consumer perception that undercooked (rare) beef is superior to fully cooked. In the case of cooked chickens, a number of incidents of *Salmonella* food poisoning occurred in north-west England during the 1980s. This stemmed from economic pressures placed by supermarkets on local butchers, who responded by installing rotisseries and supplying cooked chicken to market stalls and other outlets. In a number of cases, the butchers had neither the knowledge nor the facilities to ensure effective separation of raw and cooked chickens, problems being exacerbated by poor handling conditions on the market stalls.

throughput, means that precautions against cross-contamination must be particularly stringent.

A particular type of deboned cooked chicken is produced in continental Europe by cook-in processing. The chicken is vacuum packed in bags that have high barrier properties and it is then subjected to an intense heat treatment in hot water. Such products are given an extended life (>30 days) at a nominal storage temperature of 0–1°C. When originally imported into the UK, these products were claimed to be stable at ambient temperatures. The processing applied (the exact nature of which is proprietary) is not sufficient to ensure destruction of bacteria endospores, including those of *Cl. botulinum*, and such products are considered inherently unsafe. A superficially similar product is now available which has received a heat treatment equivalent to a 'botulinum cook'. Sales, however are limited by poor appearance and eating quality.

## (b) Pies and puddings

Pies and puddings are a diverse group of traditional meat products and some types, such as pasties, probably represent the earliest commercially manufactured convenience foods. Products of this type are often considered to have a 'downmarket' image and are also seen as unhealthy foods. Scope for product innovation is limited and largely restricted to variations in filling components, or use of different types of pastry. Direct non-meat equivalents of some types are available.

---

### BOX 5.3   Simple Simon

The baking of giant meat pies appears to have a fascination in parts of England and forms a periodic feature of festivities in various villages. It is quite impossible to ensure that such pies are either adequately cooked or cooled, resulting (in some cases at least) in a long history of foodborne illness. Despite this, the action of Environmental Health Officers in preventing the baking of a giant pie was condemned by the press as 'bureaucratic interference'. It is pertinent to ask those writers who object to even the most sensible measures to ensure food safety, if they would equally be prepared to travel in a car with defective brakes or in an unsafe aircraft.

---

Pork pies, a traditional English product, are suitable only for cold eating. Pork is the only meat used in pies of this type, although a variant contains whole hard boiled eggs. The meat filling of pork pies is solid and is prepared using a similar technology to that used for finely chopped sausage (see page 246), the minimum meat content (in the general case) being 25%. The degree of comminution varies, a coarse cut often being equated with high quality. Nitrite is added to some formulations at levels as high as 100 mg/l, although the prime purpose is to provide a pink colouration.

The meat filling of pork pies is baked in a shell of short pastry. The pastry is prepared by incorporation of finely divided fat into a paste of wheat flour and water. A small quantity of NaCl is added for flavouring. Bran may also be added in a few cases in attempts to exploit the market for high fibre products. The pastry is usually glazed, a solution of guar gum now being widely used for this purpose. Most pie pastry is prepared using cold water, but hot water, which gelatinizes some of the starch during mixing and results in a stiffer paste, is preferred for large hand-moulded pies. Pie manufacture is wasteful of pastry and large quantities of scrap are produced. This may be reworked into new pastry, provided that excessive quantities are not used and the scrap pastry has not been stored for excessive periods. Pastry supports rapid growth of micro-organisms, of which lactic acid bacteria and members of the Enterobacteriaceae are of greatest spoilage potential. Use of spoiled scrap pastry leads to a sour pastry of poor textural qualities. Off-odours and taints are also present and tend to partition into the meat filling.

During manufacture of pork pies, sheets of pastry, or pastry balls from a dough divider, are cut and formed to produce lids and bottoms. The bottoms are filled with meat mix and the lids are applied. Processing may be in stages, but large-scale manufacture of smaller pies is carried out in a single machine. It is usual practice to 'rest' the pie before baking, in order to reduce elasticity in the pastry and thus avoid shrinkage and toughening during baking. Alternatively, a relaxing agent such as cysteine, sodium bisulphite or soya derivatives can be added to the pastry, but there is often an adverse effect on texture after baking. Pies are baked to an internal temperature of 85°C. Hot-air ovens, which may be operated on either a batch or a continuous basis, are the most common, but ovens combining hot air with radio frequency heating are now being used. This reduces baking time to *ca*. 33% of that in a conventional oven and ensures both adequate heating at the centre of the meat filling and a correct degree of browning of the pastry.

The meat filling of pork pies shrinks during baking and it is customary to fill the space between the meat and the pastry with a jelly. Filling involves injecting a liquefied jelly through the lid of the partly cooled pie. Gelatin is most commonly used as jelly and carries a risk of contamination with *Salmonella*. It is therefore necessary to heat the gelatin solution to 90°C before use. The most suitable method is to use an on-line heat exchanger, heating the gelatin solution directly before injection. This is not feasible in small-scale manufacture, in which case the gelatin should be maintained at a temperature of 80–85°C before use. Considerable care is also required in sanitizing the equipment and utensils used in handling the gelatin.

Injection of jelly takes place after cooling of the pastry. To avoid risk of absorption by the pastry, the pie should not be moved until the jelly has set. This is inconvenient and agar, which sets at a higher temperature than gelatin, may be preferred. The risk of contamination with pathogenic micro-organisms is also lower, although full precautions during handling are still necessary.

A wide variety of hot-eating pies are made, including steak-and-kidney pie, chicken-and-ham pie and meat-and-potato pie. Non-meat equivalents, such as cheese-and-onion pie, are retailed in parallel with meat pies. Pies made with meat analogues are also available and include textured soya protein and Quorn™, a myco-protein derived from *Fusarium graminearum* grown on a glucose substrate in submerged culture.

Most hot-eating pies differ from pork pies in that the filling is cooked in bulk. According to the type of pie, the meat may be minced, diced or cut into strips, use also being made of re-formed meat and structured pastes. Structured pastes are made from comminuted meat, intensively worked with NaCl and water to produce a stiff, flowable paste. This is set by extrusion into hot water or a chamber heated by microwave radiation. More recently extrusion

* Gelatin is also a potential source of spoilage micro-organisms and growth of the thermophilic *Bacillus stearothermophilus* during holding at high temperature has been alleged. This suggests, however, that the gelatin was held for excessively long periods.

cooking has been used. Meat fibres or pieces are often added to the paste to provide texture. Following extrusion, the set paste is cut into various sized pieces.

Meat for pie fillings is usually cooked to a minimum temperature of 85°C, at atmospheric pressure in jacketed vessels. Pressure cooking is occasionally used but is uncommon. Meat requiring the longest cooking is added first, together with the bulk of the water and seasoning. Other meat, kidney and vegetables, etc., which require shorter cooking are added later. Thickeners are usually added to provide body to the filling. Various types of thickener are suitable, but starch-based are the most common. Thickeners are added late in cooking to avoid interference with heat circulation. It is essential, however, that the thickener is adequately cooked and the starch granules fully gelatinized. Additional ingredients can include lactose, dried milk powder or milk protein to improve body and texture, and caramel as a colouring agent. Carboxymethylcellulose is also used to prevent separation.

After cooking, the meat mixture is filled into bottoms made of short pastry. The meat should be cooled to 4°C if storage is required before filling. Filling may be a single stage operation, but with pies containing a large quantity of gravy, that ingredient may be added at a separate filling head. A lid is then placed on to the pie. The lid may be made of short pastry, although puff pastry is also common. This differs from short pastry in being made by laminating fat with short pastry, or a pastry of minimal fat content. This normally involves repeatedly folding and rolling out a paste of moderately plastic consistency. During baking, steam accumulates at the fat layers and then escapes, 'puffing' the pastry apart in thin layers. Antioxidants, such as octyl gallate, may be added to the pastry to minimize oxidative deterioration of the fat.

Pies for chilled distribution are baked at the producing factory. Baking is required only to cook the pastry, although heating of the filling inevitably occurs and is sufficient to kill vegetative bacteria. An initial temperature of *ca.* 150°C rising to *ca.* 175°C is used for

---

* Octyl gallate, like all alkyl gallates used in foods as antioxidants, has been associated with gastric irritation. Gallates are also considered to be potentially dangerous to persons who suffer from asthma, or who are allergic to aspirin. Octyl gallate and other gallates are not permitted in foods specifically intended for consumption by babies or small children.

short pastry, while baking of puff pastry involves a significantly higher initial temperature of *ca.* 235°C rising to *ca.* 245°C.

Although pork pies and the general type of hot-eating pies are the most widely made commercial products of this type, there are many variant and related products. Some are manufactured on a national basis, while others are strictly regional products.

Pasties and Forfar bridies are traditionally made by folding either short or puff pastry around the filling. This contains both meat and vegetables and is usually of stiffer consistency than most hot-eating pie fillings. The product may be baked as a whole, or the filling pre-cooked.

Scotch pies and traditional meat puddings, such as steak-and-kidney pudding, have a filling similar to that of hot-eating meat pies, but the pastry is a boiled paste. Modern steak-and-kidney puddings may be similar to steak-and-kidney pies, but of different shape.

Shepherd's pie has a minced meat filling, but is filled into a container (aluminium foil or heat-resistant plastic in the case of commercially made products) and topped with mashed potato. Reheating usually takes place in the container.

Sausage rolls have a filling of sausage meat in a pastry casing. The casing, which is open at each end, is usually made from puff pastry, but short pastry is also used. Vol-au-vents are technically similar, consisting of a cone of puff pastry and a filling of meat in a thickened sauce. Sausage rolls and vol-au-vents may be eaten either cold or after reheating.

A number of cold-eating pies are produced, in which the filling is cooked separately from the pastry. Many of these products are of large size and intended for serving by the slice at buffets, etc. The filling must remain firm after the pie is cut and usually consists of large pieces of meat in a matrix of gelatin or other gelling agent. Manufacturing practices vary somewhat, but a common procedure is to cook the meat in a minimum quantity of water and to add a preheated gelatin solution either just before, or immediately after, filling into the pastry base.

Rapid cooling is an essential stage in the manufacture of all types of meat pies. Vacuum cooling is highly efficient, especially for large pies, but equipment is expensive. Cooling by circulating air is most commonly used, but problems can arise due to contamination with mould spores derived from flour. Pies must, therefore, be protected during cooling, which should be carried out in a separate room with its own air supply. An effective arrangement is a cooling tunnel supplied with chilled, filtered air. Cooling conditions must be such as to avoid condensation under the lid of the pie.

Pies are usually packaged directly after cooling. Film overwrap is most common, although moisture-proof paper is used for traditionally made pork pies. Cardboard inserts are used in some cases to strengthen the pack and reduce crushing. Hot-eating pies, with a thin bottom, are often baked in foil trays, which remain with the pie until re-heating in the home. Care must be taken to ensure that pies are adequately cooled to prevent condensation on the inside of the film and enhanced risk of mould growth. Gas packing has been used for some types of pie, including sliced pork pie. In most cases, however, life is limited by staling and textural changes in the pastry and gas packing is of no benefit.

Meat pies are perishable products and refrigeration is required during distribution and retail display. Staling of the pastry proceeds more rapidly at lower temperatures and for this reason a storage temperature of *ca.* 7°C has often been recommended in the past. This represents a compromise between minimizing microbial spoilage and a rapid rate of staling. It is considered that microbiological considerations should take precedence and a storage temperature of 4°C should be used. The problem of staling is irrelevant with hot-eating pies, since the process is reversed during re-heating. Care must be taken to minimize temperature fluctuations during storage to prevent condensation under the lid of the pie, or between the pie and the wrapping film.

### (c) Cooked sausages

A wide variety of cooked sausages, many of which originated in continental Europe, are manufactured. From a technical viewpoint, these range from coarse comminutes, in which the structure is stabilized by meat binding (see Chapter 3, page 149) to liver sausage and similar sausages classified, in the present context, with spreadable meat products (see page 248). The most common type is the emulsion sausage. In many countries, emulsion sausages have a maximum permitted fat content of 50%

in the meat portion, while a minimum meat content of 70% is a common legal requirement.

Many types of emulsion sausage are available and there can be considerable variation in size and appearance. All are characterized by a very high degree of comminution, which results in most of the fat being present in 'free' form, the sausage structure being stabilized by a strong lean meat–NaCl mixture. This imparts characteristic properties on the cooked product, which may be demonstrated by the manner in which narrow diameter sausages, such as frankfurters and wieners, break with a distinct 'snap'. Emulsion sausages such as frankfurters, wieners, bierschinken and jagdwurst have a homogeneous meat filling. The German bratwurst and similar sausages consist of an emulsion matrix containing discrete pieces of meat, fat, spices, etc. In sausages such as plockwurst, the fat content is sufficiently high for separation to occur, forming a layer of free fat on the sausage surface (cf. pâtés, page 248). In most types, however, fat separation is a fault and must be avoided.

Low-fat cooked emulsion sausages are produced, the approach being similar to that with fresh sausages (see Chapter 3, pages 142–144). In the US, the '40% rule' stipulates that added water may be substituted for fat up to a combined total of 40%. The undesirable increase in firmness and cohesiveness associated with low fat products can be offset by increasing the proportion of added water. Higher water content, however, leads to problems with texture and purge during storage. To some extent this can be overcome by massaging the meat mix to increase protein–water and protein–protein interactions. Although this results in a less cohesive, softer and juicier sausage, there is likely to be increased purge.

Pork, including boar meat, is the most common meat used in emulsion sausages, although some types contain beef. In recent years, poultry meat, especially turkey, has become relatively widely used in frankfurter-type sausages. Mechanically recovered meat is used in some formulations, particularly those containing poultry meat. Phosphates may be added to improve binding, but are not permitted in Germany and some other countries. Salts, such as sodium citrate and lactate, may also be added to improve binding, without excessive saltiness. Dried milk powder and whey proteins have also been added for this purpose. Sodium nitrite is added as a preservative in many, but not all, emulsion sausages and, in some countries, antioxidants are permitted. The antimicrobial properties of sodium lactate may be exploited, by addition at a higher level than required for binding, in non-nitrite containing sausages. Sodium lactate is of particular use where the sausage is intended for slicing and prepacking.

Manufacture of the meat mix for emulsion sausages is similar to that of uncooked sausages. Traditional procedures remain in use in small-scale manufacture and involve pre-salting the meat by chopping coarsely with NaCl and holding for up to 2 days at 2–4°C. The meat is then chopped with the addition of water; fat and other ingredients are added and chopping is continued at high speed until a finely comminuted mass is formed. In modern large-scale practice, it is common to add all ingredients together and chop or mill at high speed to obtain the correct degree of comminution. Careful temperature control is required with either method and 22°C is considered the absolute upper limit at the end of the process. Many manufacturers control the temperature to a maximum of 18°C, or even as low as 15–16°C.

After comminution the meat mix is stuffed into either natural or synthetic casings. In frankfurter-type sausages, disposable cellulose casings are used and removed after cooking. With many other types of sausage, the casing remains until the point of consumption. Vacuum stuffing equipment is most effective. Hot water baths are used for cooking in a few cases, requiring water-impermeable casings or the use of plastic shrouds. Hot-air cooking is more common, however, and may be combined with smoking. Alternatively, the sausage may be smoked as a secondary process following the main cook. Liquid smokes are available but are not used in traditional practice. Sausages should be cooked to a centre temperature of 85°C. Lower temperatures, typically *ca.* 69–77°C, held for longer periods are acceptable provided that safety requirements are met (cf. roast joints, page 237). In the US a minimum temperature of 58.3°C is stipulated to ensure destruction of *Trichinella*.

In conventional sausage making, full automation is difficult due to the variable nature of casings, including synthetic types. A fully automated system has now been developed (Figure 5.2), based on the coextrusion of a sausage mix with a collagen dough to form the casing. This process can be used with any type of sausage, including fresh and fermented, but appears most suited to the

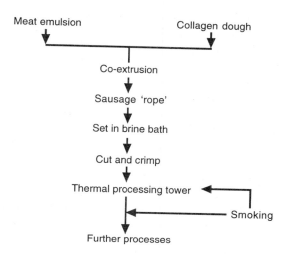

**Figure 5.2** Automated manufacture of emulsion sausages using co-extrusion. (Data from Stork Protecon-Langen, 1994, Non-stop sausage production, *Food Technology International Europe 1994*, 77–80.)

emulsion type. The basis of the process is the coextrusion equipment, which extrudes a 'rope' of sausage mix from the centre nozzle and a layer (casing) of spun collagen fibre from the peripheral jet. Preliminary setting of the casing is achieved by passing the 'rope' through a brine bath. This sets the collagen to provide sufficient mechanical strength to withstand mechanical stresses during subsequent operations. After immersion in brine, the sausages are crimped and cut by a series of blades to produce individual pieces of satisfactory appearance. Following cutting, the sausages undergo a thermal process to further strengthen the collagen skin by promoting cross-linking, while simultaneously removing excess moisture. This process is carried out continuously in a specially designed thermal processing tower. It should be appreciated that this stage is not a cooking stage, but designed to complete the encasement of the sausage. The tower can incorporate a smoking section but in most cases liquid smokes are preferred. These are sprayed on to sausages passing along a slatted conveyer belt, the sausages then being returned to the thermal processing tower for drying to induce colour fixation and stabilization. The sausages may then pass to continuous cooking, vacuum packing, etc., all of which is in conjunction with the initial coextrusion process and under the same computer-based control.

Following cooking, sausages should be cooled as rapidly as possible. Stripping the temporary casing from frankfurter-type sausages involves a significant amount of handling and special precautions should be taken against contamination at this stage. Decontamination by flash steam heating at 115–136°C for 30–40 seconds, followed by evaporative cooling, has been proposed as a means of minimizing risk. Such a process results in a 4 log cycle reduction in microbial numbers without an adverse effect on quality. Vacuum or gas packaging of small sausages is now common practice. In cases where the sausages are re-heated before consumption, boil-in-bag packaging is sometimes used. Large sausages may receive no packaging other than wrapping in greaseproof paper and placing in boxes for physical protection. The long storage lives required, especially for international commerce, means that vacuum packaging of single large sausages is now common practice. The sausages are usually packaged after cooking, but in recent years cook-in packaging has been introduced and this, in most cases, avoids risk of contamination at the processing plant. Packaging is normally removed before retail display.

Large cooked sausages may be portioned on demand in delicatessens, etc. or sliced and pre-packaged. Contamination with lactic acid bacteria during slicing is a common problem, which can lead to rapid spoilage (see page 290). Vacuum packaging, which often results in poor appearance due to the product being crushed, is still used, but gas packaging is now increasingly common. In the case of sausages imported into the UK, the combined bulk and pre-packaged storage life can be very long indeed and good temperature control at all stages is required to minimize spoilage.

### (d) Spreadable cooked meat products

The main spreadable cooked meat products are liver sausages, pâtés and some types of meat spreads. Legal requirements vary somewhat, but a common stipulation is a minimum meat content of 50% for liver sausage and 70% for other spreadable cooked meats. A feature of true spreadable cooked meat products is that the meat is at least partially cooked before comminution. This means that there is no strong lean meat–NaCl binding and the fat, which is almost entirely in the free form, is adsorbed on to meat particles. Above *ca*. 15% fat, separation occurs and a fat layer is allowed to form on the surface of the product (cf. sausages, page 245). It should be appreciated, however, that some types of pâté and liver sausage are intended for slicing rather than spreading, and in such cases the products resemble cooked sausages.

Pork is the most commonly used meat, although poultry meat is increasingly common, while mutton is used in major producing countries. Beef and, to a lesser extent, mutton spreads are traditional products in northern England, Scotland and Ireland. The eponymous liver sausage usually contains liver as *ca.* 50% of the meat content and extensive use may be made of jowls, rind, lung and other ingredients of high connective tissue content. These provide gelatin in the final product, which contributes towards texture. Pâtés usually contain *ca.* 25% liver, although higher levels are present in some cases.

---

### BOX 5.4 *Pâté de foie gras*

*Pâté de foie gras* is considered to be a pâté of exceptionally high quality. The pâté is a speciality product of Strasbourg and is produced from geese and ducks, which are quite literally force fed to enlarge the liver. This practice has attracted attention from animal rights campaigners and on January 1, 1994, Associated Press reported that a Washington-based group had been successful in persuading Air Canada to stop serving *pâté de foie gras* on its international services.

---

Pâté and some liver sausages, but not meat spreads produced in the UK, contain nitrite at an input level of up to 200 mg/l. The main function is pigment production, but there is also a powerful antimicrobial effect. Phosphates may also be added to increase water binding. Antioxidants are permitted in some countries.

The manufacture of all spreadable meat products shares a similar basic technology, although there may be considerable variation in detailed practice. The usual procedure is to cook rinds, etc. fully and then to cook lightly the other meat components, including liver. The cooked meat is then finely chopped with other ingredients in a bowl chopper. Liver sausages and similar products are filled into casings and cooked in moist air or hot water to an internal temperature of 75–85°C. Pâtés and meat spreads are filled into moulds or various sized bowls (terrines) and cooked to 75–85°C. Hot-air ovens have been traditionally used for baking pâtés, but the introduction of cook-in bags, which can also be used for liver sausage, means that cooking in hot water is increasingly common. Pâtés and meat spreads may also be packed in cans or glass jars.

Following cooking, spreadable meat products should be cooled as rapidly as possible. This can present a problem in the case of large pâtés filled into deep bowls. A common solution is the use of cold water sprays to lower the temperature immediately after cooking, followed by transfer to a blast chiller.

Although the use of cook-in vacuum packaging obviates risk of post-process contamination at the manufacturing plant, a number of types of pâté, particularly the large bowl type, are traditionally embellished after cooking. This may involve no more than a thin layer, or a glaze, of gelatin, but more elaborate embellishment is general with some types. Widely used additions include fruit, such as cherries, sprigs of herbs, spices and even pieces of tree bark. These materials are usually set in a layer of gelatin. Handling of gelatin should be subject to the same precautions that are employed in pie manufacture (see page 241). It is also necessary to ensure that embellishments are not themselves sources of either pathogenic or spoilage micro-organisms.

Where cook-in packaging is not used, spreadable meat products are usually vacuum packed before distribution. Large quantities of liver sausage and pâtés are sliced and pre-packed before retail sale but in most cases these are not products of the truly spreadable type.

### (e) Ready-meals (recipe products)

Ready-meals form a rather ill-defined category, which may be taken to include pre-cooked meat together with other ingredients such as rice, noodles and sauces. The product is formulated to comprise a complete meal, or its major component. The product is usually prepared as a single dish but, where components are incompatible with long-term storage, compartmented packaging may be used. The components are then mixed directly before re-heating and may include 'fresh' vegetables. Ready-meals may be seen as evolutionary developments of the more ambitious added-value raw meats. At retail level, ready-meals may be distributed at chill temperatures (*ca.* 2°C) or frozen. 'Ambient' ready-meals, which are fully retorted products, are also available.

---

### BOX 5.5   The Law of Diminishing Marginal Returns

Ready-meals are relatively expensive and the consumer must perceive good value. For this reason dishes such as Chicken Kiev, which are considered difficult to make, exciting and yet not too exotic, are highly popular. Sales of ready-meals are highest when an entire range is available and constant innovation is required to maintain interest. Despite this Chicken Kiev and other well established ready-meals, such as chilli con carne, continue to account for a very high proportion of total sales. New product development, although necessary, is expensive and may result in little overall sales increase. Further, as the market matures, the search for novelty in formulation becomes increasingly difficult and expensive mistakes become more likely. These include a ready-meal consisting of sausage, mashed potatoes and baked beans; hardly a product difficult to prepare, or to command a premium price.

---

Ready-meals (cook–chill) are widely used in catering, meals being prepared at a central location, chilled and re-heated directly before serving. Cook–chill has been established for many years in specialist situations, such as airline catering, but is now widely used in hospitals, schools, etc. The major advantage is cost reduction, although it has also been claimed that the higher level of control available in large-scale, centralized preparation both reduces the microbiological hazards associated with catering operations and improves quality. The use of cook–chill also allows a wider choice of menu to be provided.

*Sous-vide* may be considered to be a special type of cook–chill, in which products are par-cooked, vacuum packed and then subjected to a heavy in-pack pasteurization. In an alternative procedure the products are vacuum packed when raw and fully or partially cooked in the pack. The products are then refrigerated until use. At this stage the products are either warmed or fully re-heated to complete cooking. *Sous-vide* processing was first used by a French restaurant and its application is still greatest in restaurant catering. The process is claimed to provide higher quality than conventional cook–chill by preventing evaporative loss of flavour volatiles. There may also be a reduction in loss of nutrients through oxidation and leaching. Oxidative rancidity and aerobic spoilage are reduced, but the process is considered to be inherently

**Table 5.3**  Recommendations concerning the processing and storage of *sous-vide* products

---

*Sous-vide* Advisory Committee
    **Process**: Based on 6D reduction of *Cl. botulinum*[1] type E 80° C, 26
            min; 85° C; 11 min; 90° C, 4.5 min
    **Storage**:  Up to 8 days at 4° C or less

Advisory Committee for Microbiological Safety of Foods
    **Process**: Based on 6D reduction of psychrotrophic strains of *Cl.*
            *botulinum* type B[2]
            80° C, 129 min; 85° C, 36 min; 90° C, 10 min
    **Storage**:  10 days maximum

---

[1]*Cl. botulinum* type E considered to be more heat-resistant than type B. This is **not** general opinion.
[2]More heat-resistant psychrotrophic strains of *Cl. botulinum* type B are known. These would require up to 31 minutes at 90° C, which could cause 'unacceptable organoleptic damage'.

hazardous with respect to *Clostridium botulinum*. It is particularly alarming, especially in catering, that the risk presented by that organism is not fully appreciated and that organoleptic considerations should take precedence. Equally, it is alarming that a body involved in determining safety standards for *sous-vide* should be prepared to recommend re-heating as a means of inactivating any botulinum toxin present. Two further bodies have made recommendations concerning minimum heat processing and maximum life (Table 5.3), although there is doubt that these provide adequate safety margins, especially with respect to assumptions made concerning heat resistance of endospores of non-proteolytic strains of *Cl. botulinum*. It is notable that several studies on the safety of *sous-vide* make use of assumptions for which there appears little factual backing.

Chilled ready-meals for home consumption present no specific problems during processing. Any meat may be used, although chicken is most popular. This stems partly from the low cost of this meat, but there are also technological benefits. Chicken is easily formed into the desired shape and is of relatively bland taste. This means that added flavouring is readily taken up and that there is little conflict between the meat flavour and that of other ingredients. Pork is being increasingly used in whole meat products, but use of beef is generally restricted to mince. In this case, the type of beef used is dictated by the lowest cost compatible with desired quality. Cooking method used depends on the nature of the meat. Large pieces may be fried or roast, while small pieces or mince may be cooked with water in jacketed vessels. In the case of soups, or stews containing only small particles, cooking is possible by ultraheat treatment, which is usually followed by aseptic packaging. In the case of some ready-meals, the major product characteristics, in terms of consumer acceptability, are determined by non-meat ingredients. A very wide range of ready-meals is available and the number of ingredients is correspondingly large. This inevitably presents problems to the technologist during both development and production. Non-meat ingredients are usually cooked separately from the meat, the most appropriate method being chosen according to the nature of the ingredients.

Although the cooking of ready-meals presents no specific problems, provided that processes are properly validated and controlled, many types require a high level of handling after cooking. The significance of the potential for post-process contamination is increased by the very long shelf lives given to ready-meals. Preservatives are not present and growth of micro-organisms may be enhanced by the presence of non-meat ingredients. These problems are not insuperable, but require a very high level of supervision and control. Further, precautions to minimize risk during production of ready-meals should be devised as early as possible in product development. By this means, it is possible to reduce hands-on operations to a minimum. Consideration should also be given during formulation to devising means of reducing the growth rate of micro-organisms in the more susceptible non-meat ingredients, such as sauces. Reduction of the pH value by use of acidulants, or the $a_w$ level by use of humectants, for example, may be possible.

In established ready-meal operations, two areas have been of particular concern. The first involves the possibility of colonization of plant with environmental pathogens, such as *L. monocytogenes*. Surveys conducted at the end of the 1980s showed a high incidence of the organism in ready-meals, suggesting that precautions were not fully effective. There is evidence that not only does control of *L. monocytogenes* require more rigorous sanitization, but also that wet cleaning methods, suitable for *Salmonella*, are not effective against *L. monocytogenes*. The second area of concern is handling of ready-meals during retail display. A number

**Table 5.4** Conditions necessary for the sale of extended-life meat ready-meals

1. Avoidance of bacteriological contamination from machinery and proliferation due to temperature abuse during formulation and processing.
2. Meticulous design and validation of heat processing procedures.
3. Rigorous avoidance, or the subsequent elimination of post-process contamination; to be achieved by rapid post-process chilling and packaging under aseptic conditions or by post-packaging heat treatment by, for example, microwave processing.
4. Pre-shipment storage to be at no more than 1° C.
5. Distribution and display at food temperatures guaranteed never to exceed 3° C. **This is considered to be the most important factor of all.**
6. Sales to be made only of items which have never been exposed to ambient temperature or manipulation by potential customers.
7. Delivery to customers to be made only in well insulated bags, not accommodating other items.
8. Very clear ultimate date marking and the 'thermometer logo' in red and green with 3° C as the cut-off temperature.

*Note*: Modified from Mossel, D.A.A. (1989) *International Journal of Food Microbiology*, **9**, 271–94.

of recommendations have been made by experts, but in some cases it is doubtful that these can be applied in practice. It is recommended that ready-meals should be given special attention during retail handling and that all staff should be aware of potential difficulties with these products.

Special precautions for production, retailing and display of ready-meals are summarized in Table 5.4. The need for many of these precautions would be obviated by in-pack pasteurization of the end-product, possibly using microwave heating. Such a procedure is not currently general practice.

*(f) Other cooked meat products*

A number of cooked meat products exist that are either distinct from or intermediate between the main categories. Most are traditional products that utilize significant quantities of offal and low quality trimmings, etc. As such their importance, in earlier years, in providing meat in the diet should not be underestimated. Such products may be seen as 'food of the poor' and consumption has fallen as higher quality meat products become of lower relative cost. Many products of this type are now produced on only a small scale, or only in a restricted area.

*Black pudding (blood sausage)* is a product of northern England, although similar products are made elsewhere. Traditionally made black puddings contain no muscle meat, the main ingredients being blood (*ca.* 48%), pork fat (*ca.* 24%) and a cereal filler (*ca.* 18%). Fresh blood may be used directly from the slaughterhouse or obtained from outside sources. In the latter case, it is usually necessary to add an anticoagulant to prevent clotting before use. The blood is mixed with the fat and then with pre-cooked cereal filler. This may be barley flour, wheat flour or oatmeal. The azo dye, Black PN, may be added to overcome problems due to red colouration in the centre of the pudding. The cause of this problem is not known. The mix is filled into casings, which vary in size and shape, and cooked in a hot water bath to an internal temperature of 82–85°C. The end-product is bound by heat-set blood.

*White pudding* is a local speciality of Scotland, Ireland and a few parts of northern England. The pudding consists of a mixture of soaked oatmeal and beef suet, filled into casings and cooked. The oatmeal is responsible for binding the end-product.

*Brawn* and related products are made from ingredients that are high in connective tissue and low in fat. In traditional practice ingredients such as bones, rind and gristle, cheek meat, ears, feet and other trimmings of high connective tissue content are boiled in water to produce a stock of high gelatin content. Vinegar, or acetic acid, may be added to the water to enhance hydrolysis and gelatin formation. The stock may be clarified by egg white or blood plasma, the clarified stock usually being referred to as aspic. The stock is then used to form a jelly around pieces of precooked cured meat. In modern practice, it is usual to add an additional 1% gelatin to improve the strength of the jelly, or to prepare the entire product using powdered gelatin. In all cases precautions are required to ensure the microbiological safety of the stock or gelatin solution (see page 241).

*Saveloys and polonys* consist of a coarsely chopped filling containing *ca.* 70% meat (including fat, trimming, mechanically recovered meat and offals) and *ca.* 7% rusk, or other cereal, together with water and seasoning. The meat mix is filled into sausage casings and cooked to an internal temperature of 80–85°C by smoking, in moist air, or in hot water. The products are bound by a combination of weak meat binding and the cereals.

*Faggots* are made from a similar filling, but are formed into balls, covered in caul fat and roasted to a centre temperature of 75°C or higher. Alternatively, faggots are packed in trays with gravy.

*Haslet*, a dish of the midlands of England, is similar but of higher meat content (75% or higher) and made with pre-soaked rusks. Haslet is more strongly bound than other products of this type.

*Cooked rissoles and croquettes* are similar to the uncooked comminuted products of the same name (Chapter 3, page 139), but pre-cooked meat is used. There is consequently no meat binding and integrity of the product is entirely from gelatinized starch derived from rusk, potatoes, flour, etc. in the filling.

### 5.2.5 Quality assurance and control

Quality assurance is primarily concerned with ensuring that meat is correctly cooked, cooled and protected from post-mortem contamination. Application of HACCP is considered invaluable in cooked meat production. Cooking should be monitored by recording cooking times and temperatures of the heating medium and product. Suitably calibrated continuous thermographs or data loggers should be fitted to equipment. A number of methods have been proposed for verification of correct cooking. Tests used in the past were cumbersome and unsuitable for routine use in a manufacturing environment. More recently there has been much interest in enzyme inactivation as an end-point indicator of correct processing. Proposed enzymes include catalase and, more recently, N-acetyl-β-D-glucosaminidase. In each case the value may be limited by residual activity. The commercial Api-zym© system (which provides a simple means of semi-quantitative analysis of 19 enzymes) has also been used with some success. Determination of *Enterococcus* as an index for the more heat-resistant vegetative pathogens, such as *L. monocytogenes*, may also be of value.

Cooling should also be monitored, using temperature–time recording equipment where possible. Where batch processing is used and heated product must be physically transferred to coolers, this should be a specific responsibility of designated personnel. In air-cooled systems, the performance of air filters should be checked on a routine basis. Where water cooling is used, levels of chlorination should be monitored and effectivity verified by microbiological analysis.

Precautions against recontamination should be built into the manufacturing process and, indeed, the physical infrastructure of the building. A high level of observation is required by supervisory management to ensure that precautions are being taken and that no 'short circuits' exist between raw and cooked meats. 'Hands-on' operations require particularly close supervision. Ingredients added after processing require careful monitoring. A formal control system is required to monitor correct make-up and use of gelatin. Microbiological examination of the herbs, etc. that are used to decorate pâtés may be worth while but the value of such examinations should be assessed on the basis of the risk associated with each ingredient. A recently introduced and rather unusual pâté, for example, contains desiccated coconut as part of the decoration. This ingredient is a known source of *Salmonella* and clearance of each batch on a positive-release basis would be required.

In recent years kits based on ATP detection have become available which permit rapid detection of microbial contamination of cleaned surfaces. Such testing can be of value as part of the overall sanitization programme. It must be appreciated, however, that ATP monitoring cannot in any way substitute for a properly designed and implemented cleaning procedure. Further, possible variation from one part of equipment to another means that results from a small number of assays are not necessarily representative.

In addition to the key processing factors, it is necessary to control factors related to product quality. Selection of meat is usually on the basis of visual and microbiological examination, although water-holding capacity may also be determined, especially in the manufacture of emulsion sausage. There is relatively little inter-active production control, products being made to predetermined recipes and processing. As with fresh comminutes, however, on-line control of emulsion sausage manufacture is being developed and its importance is likely to increase in future. Parameters monitored are largely those applied to comminutes (Chapter 3, page 145), but physical parameters including viscosity and water-holding capacity of the emulsion mix are also of importance.

End-product testing is common practice with cooked meat products. Chemical analysis is primarily applied to ensure compliance with legal and compositional standards. Microbiological analysis is primarily concerned with overall status, although some micro-organisms may serve as an index of good practice. Product quality is still largely assessed on the basis of visual and organoleptic analysis, although there is increasing interest in objective determination of rheological parameters.

## 5.3 CHEMISTRY

### 5.3.1 Nutritional properties of cooked meat

The major effect of cooking is loss of thiamine. The extent is dependent on temperature and time of cooking and is greatest where cooking losses are high. Under these conditions as much as 80% may be lost, compared with *ca.* 45% at lower levels of cooking loss. Destruction of thiamine proceeds most rapidly at low pH values and losses are reduced where pH values are high. Some loss of riboflavin also occurs, but niacin is highly stable.

### 5.3.2 Flavour of cooked meat and meat products

The flavour of freshly cooked meat is believed to result from a fairly small number of compounds, although the total number of volatile compounds exceeds 1000. The importance of an individual compound depends both on its concentration and on its odour threshold, although compounds which cannot be identified individually may still play a role in determining the overall flavour.

The volatile flavour compounds of meat arise during heating through a large number of reactions between non-volatile compounds (Table 5.5). The role of lipids in particular has been contentious, although the negative effects of rancidity are well known.

**Table 5.5** Main types of reaction leading to formation of volatile flavour compounds during cooking of meat

1. Breakdown of individual substances.
2. Reactions between amino acids and sugars (Maillard reaction).
3. Lipid oxidation.
4. Interaction between lipid oxidation and the Maillard reaction.

It may be demonstrated, however, that thermal oxidation reactions during cooking, despite following similar pathways to autoxidation, contribute desirable flavours. The oxidation of phospholipids appears to be of considerable importance, while triacylglycerols play only a minor role. It is accepted that sufficient lipid is present within the muscle for development of flavour and that visible fat is not necessary.

In recent years it has been possible to identify a number of the key compounds responsible for the characteristic flavour of meat. The most important are a class of heterocyclic thiols and disulphides, which have been known for many years to have 'meaty' flavours and which have also been used as synthetic meat flavouring. The meaty character is determined by the degree of unsaturation and on the position of the thiol group. Furans and thiophenes with a thiol group in the 3-position tend to have a meaty aroma, but a thiol group in the 2-position results in burnt and sulphurous aromas. Meat-like aromas are most closely associated with a ring containing at least one double bond and a methyl group adjacent to the thiol group. In beef, a total of seven compounds have been identified as being of greatest importance in determining meat flavour (Figure 5.3). The first was 2-methyl-3-(methylthio)furan,

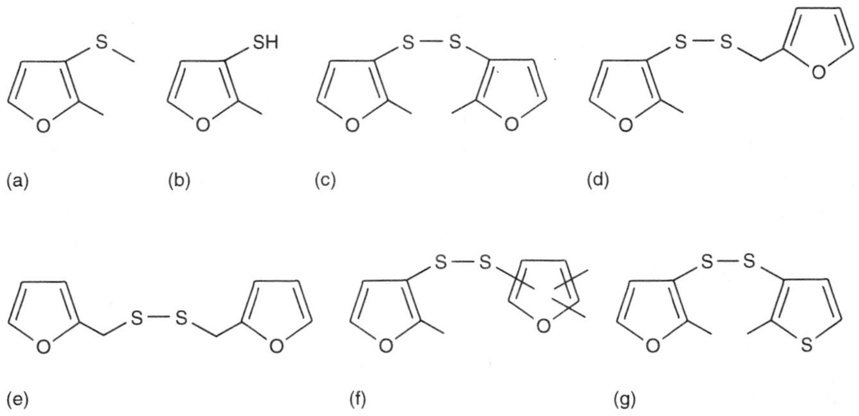

**Figure 5.3** Volatile compounds of key importance in determining the flavour of cooked meat. (a) 2-methyl-3-(methylthio)furun; (b) 2-methyl-furan-3-thiol; (c) bis(2-methyl-3-furyl) disulphide; (d)2-furfuryl 2-methyl-3-furyl disulphide; (e) bis(2-furfuryl) disulphide; (f) dimethylfuryl 2-methyl-3-furyl disulphide; (g) 2-methyl-3-furyl 2 methyl-3-thienyl disulphide.

followed by 2-methylfuran-3-thiol and bis(2-methyl-3-furyl) disulphide. Subsequently 2-furfuryl 2-methyl-3-furyl disulphide, bis(2-furfuryl) disulphide, dimethylfuryl 2-methyl-3-furyl disulphide and 2-methyl-3-furyl 2-methyl-3-thienyl disulphide have been identified. These four compounds are present at high concentration in the heart, which is reputedly very strongly meat flavoured after cooking.

Investigations showed that while some of the key compounds in beef are also of importance in chicken, there are significant differences. Bis(2-methyl-3-furyl) disulphide, for example, was present at only low levels, while much of the characteristic chicken odour was derived from unsaturated aldehydes such as deca-*trans*-2,*trans*-4-dienal, which are oxidation products of linoleic acid (cf. Chapter 1, page 31).

Flavour of stored cooked meat is modified by the rapid development of 'warmed-over' flavour (see below). To some extent a degree of warmed-over flavour may be accepted as part of the normal flavour profile of cooked meat. Many cooked meat products, of course, contain additional flavour sources. Some, such as NaCl and some spices, enhance the 'meaty' taste, while contributing flavour directly. In some meat products, including garlic sausage, the non-meat flavours may be dominant. Perception of meat flavour will also differ between hot-eating and cold-eating products.

### 5.3.3 Warmed-over flavour

The term 'warmed-over flavour' (WOF) was first used to describe the rapid development of oxidized off-flavours in refrigerated cooked meats. Stale or rancid flavours develop within 48 hours (and often earlier) during storage at 4°C. Warmed-over flavour is thus easily distinguished from off-flavours due to oxidative rancidity in frozen meat, fats and oils, which develops over a period of many weeks or months. It is now recognized, however, that WOF is not restricted to refrigerated cooked meats but may also develop in uncooked meats, especially following comminution, and in cooked meats maintained at high temperatures before serving.

The underlying cause of WOF is lipid oxidation. Animal lipids are considered to be fairly saturated and resistant to oxidation. Sufficient quantities of polyunsaturated fatty acids, however, are present in the phospholipid fraction of intramuscular lipids to

BOX 5.6  **Cold meat on Mondays**

Warmed-over flavour was scientifically recognized in 1958, but it seems likely that the phenomenon was recognized by consumers at a much earlier date. Increasing sales of pre-cooked meat products through retail outlets and increasing use of pre-prepared meals in mass catering means that the importance of WOF as a counter-indicator of acceptability is increasing. A survey in the US showed that while the food service industry felt there to be no major problem with WOF, consumers saw the matter rather differently. Critically, however, only 22.5% of those detecting WOF complained, other persons affected remaining silent but, presumably, disgruntled. (Cross, H.R. *et al.*, 1987, in *Warmed-Over Flavor of Meat*, (eds A.J. St. Angelo and M.E. Bailey), pp. 1–18, Academic Press, Orlando.)

permit a significant degree of oxidation. In beef, for example, 44% of the fatty acids contain two or more double bonds, compared with 3.5% of the triacylglycerol fatty acids. Similarly, in chicken meat, 31% of the phospholipid fraction contains three or more double bonds, compared with 1% of the triacylglycerol fraction. The overall level of unsaturation is highest in poultry which, for this reason, is most susceptible to WOF, followed by (in decreasing order) pork, beef and lamb/mutton. Lipid composition may also vary between sexes and between cuts of meat; and in non-ruminant animals it is also directly affected by dietary fats. 'Fishy' taints in turkey flesh, for example, have been attributed to oxidative reactions following the use of feed containing high levels of unsaturated fats. Rate of lipid oxidation, however, is modified by levels of the naturally occurring antioxidant, tocopherol. The rate of lipid oxidation in pork, for example, appears to be dependent on muscle tocopherol levels, while the greater susceptibility of turkey than chicken to oxidative rancidity has been attributed to lower natural levels of tocopherol.

Lipid oxidation involves the reaction of unsaturated lipids with oxygen to yield lipid hydroperoxides and the subsequent degradation of the hydroperoxides to a range of products, including alkanals, alkenals, hydroxyalkenals, ketones, alkanes, etc. These secondary compounds are the cause of the off-flavours associated with WOF.

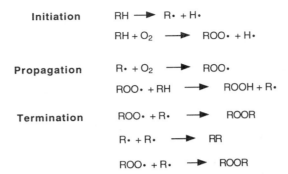

**Figure 5.4** Free radical mechanism of lipid oxidation. RH = unsaturated lipid; R· = lipid radical; ROO· = lipid peroxy radical; ROOH = lipid hydroperoxide.

Oxidation of unsaturated lipids, almost exclusively the initial substrates, is autocatalytic, in that the oxidation products catalyse the reaction, resulting in an increase in reaction rate as oxidation proceeds. This may readily be explained in terms of a free-radical reaction mechanism (Figure 5.4). The mechanism of initiation is less well understood, but appears to involve singlet oxygen. It has been postulated that singlet oxygen is formed by a photochemical reaction, involving light-mediated formation of an excited sensitizer, which reacts with oxygen to form excited singlet oxygen. The singlet oxygen reacts with unsaturated fatty acids to form hydroperoxides which decompose to free radicals. These, in turn, enter the lipid oxidation chain reaction (Figure 5.5). Singlet oxygen reactions probably continue throughout lipid oxidation, but it is likely that, once hydroperoxides are formed, the free radical reaction dominates. It has also been suggested that phagocytic cells are involved in initiation of oxidation through $H_2O_2$ production during phagocytosis and generation of the superoxide anion radical by leucocytes.

* Molecular oxygen exists in its ground or triplet state, while both lipids and hydroperoxides exist in the singlet state. Hydroperoxide formation from triplet oxygen and lipids would require both a total change in electron spin and an activation energy of 35–65 kcal/mole and is not considered possible. These barriers do not exist for singlet oxygen, which exists with unpaired electrons and is thus highly electrophilic. (Lillard, D.A., 1987, in *Warmed-Over Flavor of Meat*, (eds A.J. St Angelo and M.E. Bailey), pp. 41–67, Academic Press, Orlando.)

$$S + H\nu \longrightarrow 1_S^* + 3_S^*$$

$$3_S^* + 3_{O2}^* \longrightarrow 1_{O2}^* + 1_S$$

$$1_{O2}^* + RH \longrightarrow ROOH$$

$$ROOH \longrightarrow \text{free radicals}$$

**Figure 5.5** Formation of singlet oxygen and reaction with lipids. S = sensitizer; H$\nu$ = light; $1_S$ = singlet state sensitizer; $1_S^*$ = excited singlet state sensitizer; $3_S^*$ = excited triplet state sensitizer; $3_{O2}^*$ = ground state triplet oxygen; $1_{O2}^*$ = excited singlet state oxygen; RH = unsaturated fatty acid. (Rawls, H.R. and van Santen, P.J., 1970, *J. Am. Oil. Chem. Soc.*, **47**, 121–8.)

Hydroperoxides are very unstable and rapidly degrade to secondary reaction products by a free radical mechanism (Figure 5.6). Free radicals formed during lipid oxidation react with other meat constituents, including proteins, amino acids, enzymes, vitamins and pigments. Secondary reaction products can also take part in browning reactions with proteins.

Metals, especially iron, are important catalysts of lipid oxidation. Both initiation of oxidation and degradation of hydroperoxides are catalysed by metals, free radical generation during the second reaction being of greatest importance. Iron is relatively insoluble at physiological pH values and is present in association with either low or high molecular weight compounds. Low molecular weight (LMW) or 'free' iron is primarily associated with amino acids,

**Hydroperoxide cleavage**

$$R\text{--}C(\text{--}OOH)H \longrightarrow R\text{--}C(\text{--}O\cdot)H + OH\cdot$$

**Free radical reactions**

$$R\text{--}C(\text{--}O\cdot)H\text{--}R \longrightarrow R\text{--}CHO + R\cdot$$

$$R\text{--}C(\text{--}O\cdot)H\text{--}R + RH \longrightarrow R\text{--}C(\text{--}OH)H\text{--}R + R\cdot$$

$$R\text{--}C(\text{--}O\cdot)H\text{--}R + R\cdot \longrightarrow R\text{--}C(=O)\text{--}R + H$$

**Figure 5.6** Degradation of hydroperoxides. (Keeney, M., 1962, in *Lipids and Their Oxidation*, (eds H.W. Schultz, E.A. Day and R.O. Sinnhuber), pp. 79–108, AVI, Westport.)

nucleotides and both inorganic and organic phosphate. Iron in this form is now considered to be of greatest importance in catalysis of lipid oxidation. High molecular weight (HMW) iron is associated with membrane lipids and proteins, such as haemoglobin, myoglobin, ferritin and the ferritin degradation product, haemosiderin. Some sources of HMW iron, including haemoglobin, myoglobin and ferritin, are able to catalyse lipid oxidation, but others have no role unless the iron is released into the LMW iron pool. Factors contributing to release are likely to increase the catalysis of lipid oxidation in meats.

The concentration of LMW iron increases in processed and stored muscle, but the sources are not fully understood. In poultry and the major meat animals, between 42 and *ca.* 77% of iron is water soluble. This component consists largely of LMW iron and iron bound to myoglobin, haemoglobin and ferritin. Release of bound iron from these sources is at least partly responsible for the increase in LMW iron. A further 23–58% of iron is associated with water-insoluble components of muscle. These include haemosiderin, contractile protein, collagen and membrane lipids. There has been doubt concerning the availability of iron associated with insoluble compounds, but it is now known that release can occur, especially at higher pH values (increasing over the range 5.0–7.0) and in the presence of ascorbate.

Several mechanisms have been proposed to explain the role of haem iron in catalysis of lipid oxidation. Formation of a coordinate complex between the haem compound and lipid hydroperoxides, followed by a homolytic scission of the peroxide bond, is favoured by many workers. Alternatively, haem proteins may act as a sensitizer in formation of singlet oxygen.

Although WOF is known to occur in uncooked meats, heating has a pronounced accelerating effect. Release of iron from haem-containing compounds occurs during heating due to oxidative cleavage of the porphyrin ring. Less iron is released, however, than would be expected from theoretical considerations and other mechanisms may be involved. Released iron may also be ionically bound to coagulated proteins. Haem proteins themselves may be more active initiators of oxidation when denatured by heating. A further possibility is increased exposure of lipoproteins to oxygen and catalysts as a consequence of loss of integrity of muscle membranes and breakage of lipoprotein complexes.

It is generally agreed that meats cooked to an internal temperature of 70°C for 1 hour develop WOF most rapidly. At 80°C and higher temperatures, heating leads to formation of antioxidants, which significantly delay oxidation. Maillard browning products are among the compounds postulated as antioxidants. Meat products containing higher levels of carbohydrates, such as some types of meat loaf, are relatively resistant to oxidation, possibly due to the presence of a higher level of Maillard browning products. Addition of glucose to cooked meat products has been proposed as a means of increasing Maillard browning.

For many years, lipid oxidation reactions have been considered to be non-enzymatic. It is now well recognized that muscle microsomes contain lipid oxidizing systems, which may be involved in development of WOF. At the same time, free fatty acids produced by phospholipase A activity may exert a protective effect against oxidation. The significance of enzymatic reactions has not, however, been fully elucidated.

Although many of the factors governing development of WOF are intrinsic, the process can be affected by extrinsic factors of relevance to meat processing. Sodium chloride has a well known pro-oxidant effect. Several mechanisms may be involved, including an increase in ionic iron release. The effect of NaCl does not appear to be temperature dependent and increases over the range 0.5–2.0%.

In contrast to NaCl, nitrite at levels as low as 50 mg/l is a powerful inhibitor of WOF, which is not a significant problem with nitrite-containing products (see Chapter 6, pages 304–306). Ascorbic acid plays a dual role in lipid oxidation and development of WOF. At levels up to 250 mg/l, ascorbate catalyses oxidation, possibly through enhancing release of bound iron. At levels of *ca.* 500 mg/l, lipid oxidation is inhibited. At these concentrations, ascorbate may either change the balance between ferrous and ferric iron or act as an oxygen scavenger.

---

* The antioxidant effect of rosemary (*Rosmarinus officianalis*) has been recognized empirically for many years. An extract, oleoresin rosemary, is commercially available and contains the phenolic compounds carnosol, rosmanol, rosmaridiphenol and rosmariquinone. Phenolic compounds derived from rosemary also have activity against Gram-positive bacteria, including *Listeria monocytogenes*.

Antioxidants are widely used in control of WOF, the most effective being free radical chain terminators, such as butylated hydroxytoluene, butylated hydroxyanisole, α-tocopherol and phenolic derivatives of plants such as rosemary. Metal chelating agents, including phosphates, EDTA and citrates, also have antioxidant properties but are less effective and do not entirely prevent oxidation. A common strategy for preventing lipid oxidation is the use of antioxidants in conjunction with vacuum packaging. It has been suggested that vacuum packaging the meat before cooling would retain natural antioxidant systems in fully reduced form and be a highly effective means of preventing lipid oxidation.

### 5.3.4 Colour of cooked meat products

In non-nitrite containing meat products that have been adequately cooked, the red colour of raw meat changes to off-white, grey and brown hues, depending on the nature of the muscle. The behaviour of sarcoplasmic haemoproteins is the most important factor determining colour of cooked meat, but additional browning involves reactions between the amino groups of proteins and reducing sugars of other carbonyls.

Myoglobin, in pure solution, is denatured by heating to 85°C, although in meat the haemoproteins begin to coagulate at *ca.* 65°C, co-precipitating with other muscle proteins. The globin is denatured but the haematin nucleus remains intact, although separation from the globin is usual. Colour complexes thus consist of denatured haemoproteins in which the haematin nucleus is bound to a denatured protein. The denatured protein may be one of several present and not only globin.

The red globin haemochromogen, in which the iron is in the $Fe^{2+}$ form, may be formed by denaturation of either myoglobin or oxymyoglobin, while the brown globin haemichromogen, in which the iron is in the $Fe^{3+}$ form, may also be formed directly from metmyoglobin. Globin haemochromogen is highly susceptible to oxidation and thus globin haemichromogen is present in much greater quantities.

In general terms, the colour of cooked meat will be determined by the extent of globin haemichromogen formation and the quantity of undenatured myoglobin (including oxymyoglobin) present. Beef cooked at an internal temperature of 60°C will be red in colour and that cooked at 60–70°C pink. At higher temperatures the colour is predominantly grey-brown, although a residual pink colour, due to residual undenatured myoglobin, may be retained in beef and other meats at cooking temperatures as high as 76°C.

Pink colouration is also relatively common in meat products, especially pork and turkey, heated to an extent that effectively all of the myoglobin is denatured. This is perceived as a fault by consumers who complain that the meat is undercooked or, less commonly but sometimes correctly, that nitrite is present. A number of possible causes exist and it is necessary to distinguish between pink colouration as a continuing phenomenon, affecting all producers and caused by endogenous factors, and sporadic outbreaks, affecting individual producers and caused by exogenous factors.

Reduced cytochrome $c$ is believed to be responsible for a residual pinkness in turkey and pork muscle cooked at temperatures of 85°C and above. It has been postulated that cytochrome $c$ is released from the inner membranes of mitochondria following degradation during cooking. The depth of the pink colour is determined directly by the cytochrome content of the muscle. This is usually low, but may be increased by factors such as preslaughter stress. Reduced cytochrome $c$ is slowly oxidized during refrigerated storage, corresponding to a gradual fading of the pink colouration.

A separate phenomenon has been observed in roast pork joints cooked at 82°C and involves a development of pinkness during 12 days storage in a vacuum pack at 2°C. In this case, the pink colouration, which fades after exposure to air, was attributed to globin haemochromes, or related non-nitrosyl haemochromes.

'Pink spot' is a relatively common problem affecting cooked turkey roll, in which patches of pink discolouration appear on the freshly cut surface of the roll. The pink pigment is thought to be a reduced nicotinamide-denatured globin haemochrome complex, formation of which is promoted by reducing conditions and reversed by oxidizing conditions.

Sporadic problems of pink colouration have also been attributed to a number of causes. Contamination with nitrate, which is subsequently reduced to nitrite, or nitrite itself appears to be most common. Particular problems may occur where cured cooked meats are produced in the same premises as uncured. Nitrate and nitrite, however, may be derived from many sources. Water is important in some areas, while a particular problem faced by one plant involved nitrite leached from packaging adhesive. Sporadic pink discolouration of cooked meats has also been attributed to contamination with ammonia from a leaking refrigeration unit and carbon monoxide from engine exhaust fumes or gas-fired ovens. Oxides of nitrogen may also be involved in the latter cases.

### 5.3.5 Structure of cooked meat products

#### (a) Whole muscle products

At the start of cooking meat has a flacid feel. During the cooking process the most obvious changes are the shrinkage in muscle volume, with a consequent loss of fluid, and the development of a rigidity absent in raw meat. In simple terms a change has developed in the texture of the meat concomitant with the change from a raw to a cooked product. Texture is probably the most important factor determining eating quality and the extent of change varies considerably from muscle to muscle.

Shear force required to break tissue is a convenient objective measure of texture and, in general, correlates well with taste panel assessments of that parameter. Increase in shear force with increasing temperature occurs in two distinct phases (Figure 5.7). The first occurs at 45–50°C and the second at 65–70°C. Prolonged heating at temperatures of 70°C or higher eventually leads to a decrease in shear value.

Increase in shear force value at lower temperatures (40–45°C) has been attributed to denaturation of the myofibrillar proteins, actin and myosin, coagulation of which leads to formation of a semi-rigid gel. Shrinkage of the actinomyosin occurs within the endomysial sheath, which is unaffected at lower temperatures. In raw meat, this sheath serves to constrain the myofibrils from swelling and is under a degree of tension. Denaturation of actinomyosin leads to a release of tension and fluid is forced out of the space between the endomysium and the denatured myofibrils, accounting for the observed loss of fluid at these temperatures.

The higher temperature (65–70°C) increase in shear value is attributed to shrinkage of the perimysial collagen. The fibres are dena-

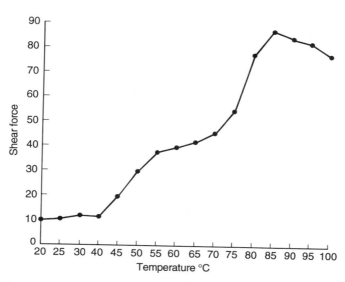

**Figure 5.7** Increase of shear force with increasing temperature.

tured and may be seen to change from an opaque, inelastic fibre with characteristic banding pattern to a swollen fibre with elastic properties. Denaturation is accompanied by shrinkage and a tension is generated against the muscle fibres, within the bundles surrounded by the perimysium. Fluid released on denaturation of the actinomyosin is expelled by shrinkage of the perimysium. The extent of shrinkage and loss of fluid depends on the nature and extent of the intermolecular cross-links which stabilize the perimysial collagen fibres (see Chapter 1, pages 20–21). For this reason the extent of shrinkage is greater in older animals.

Heating for prolonged periods at temperatures above 70°C eventually causes a reduction in shear value, probably due to cleavage of peptide bonds in the molecule. Extensively cross-linked collagen appears to be able to regenerate the native structure when cooled to *ca*. 20°C (Ewald reaction). As a consequence the residual strength of fibres binding the muscles is increased, resulting in measured shear force being greater at 20°C than at 70°C.

The residual strength of fibres binding the muscle together contribute to toughness of meat, in addition to tension generated by the thermal shrinkage of perimysial collagen. Tensile properties of cooked meat along the fibre axis are about ten times greater than the force required to separate the bundles transversely. This indicates the weakness of the perimysium responsible for lateral binding of the muscle fibres. Studies of the fracture mechanics of cooked meat have shown that the endomysial–perimysial bonding is the first to fracture under load and is thus the weakest point. On increasing the load further the individual muscle fibres break, followed finally by the perimysial fibres.

In an overall view, it may be seen that the mass of meat is a function of the denatured myofibrillar proteins, but that texture is determined by the collagen fibres of the perimysium. Two effects are involved, the compression of the muscle bundles during collagen shrinkage and the binding of the muscle bundles due to the residual strength of the denatured collagen fibres. In each case, the effects are determined by the nature and extent of collagen cross-linking.

### (b) Formation of heat-set protein gels

The texture, after cooking, of many types of comminuted and re-formed meat products depends on the formation of a heat-set protein gel. Myofibrillar proteins are partially solubilized during comminution in the presence of NaCl. A thick sol is formed, which on heating sets to form a gelatinous structural network of proteins, which impart structure to the entrapped water.

Formation of the gel follows a sequence of events initially involving thermally induced conformational changes in the proteins, followed by exposure of hydrophobic groups and, finally, gelation. The gel formed is irreversible and results from interaction of protein molecules, heated above their denaturation temperature. The underlying mechanism is heat-catalysed interprotein cross-link formation. The properties of the gel are also affected by factors influencing the ability of the gel to form cross-links (Table 5.6). Although gel formation involves thermal denaturation of proteins, other factors leading to denaturation (including low pH value, dehydration, or freeze denaturation) may interfere with cross-linking and result in a weak gel structure.

Many types of muscle protein can form heat-set gels, but myosin is recognized as being of greatest importance in meat products. Actin is also important through formation of F-actomyosin and also appears to play a direct role in increasing gel strength. Other myo-

**Table 5.6** Factors affecting formation of heat-set muscle gels

| | |
|---|---|
| pH value | Lowest $E_a$ at 5.5 |
| Ionic strength | Fine-structured gels at 0.25 M KCl, coarse-structured at 0.60 M KCl |
| Protein level | Critical concentration is *ca.* 2 mg/ml. Shear modulus increases with the square of protein concentration |
| Temperature | Higher shear modulus and greater elasticity after heating at 44–56° C than at 58–70° C |
| Muscle type | Slow (dark) muscles give firmer but weaker gels than fast (light) muscles. Strength of gels usually depends on myosin content |
| Rigor state | Strength of fast muscle gels greater post-rigor. No effect with slow muscle |
| Washing | Increases gel strength with chicken meat. Improves texture of restructured cardiac muscle, but not skeletal muscle |

*Note*: Data from Aguilera, J.M. and Stanley, D.W. (1993) *Food Reviews International*, **94**, 527–50.

fibrillar proteins may be involved under some circumstances, but are unlikely to play a major role, since proteins such as tropomyosin do not undergo conformational changes at the temperatures used for heat-setting.

Native myosin is a protein assembly consisting of several subfragments. The most important of these are paired globular heads and a fibrous tail, linked by a trypsin-sensitive hinge region (see Chapter 1, pages 14–15). Both the head and tail subfragments can form a heat-set gel, but the characteristics are different and the two subfragments play a different role. Formation of a myosin gel is thus the result of two separate reactions. The first, which takes place at 30–45°C, involves a partially irreversible helix-to-coil transition of the tail subfragment and subsequent network formation. The second reaction takes place at temperatures above 50°C and involves irreversible aggregation of the head subfraction, which contributes to the final three-dimensional matrix structure of the gel. Aggregation of the head subfraction has been attributed to disulphide interchange, but hydrophobic interactions are probably also involved. F-actomyosin increases gel strength by cross-linking the tail subfractions of free and actin-bound myosin. Optimal gel strength is obtained when the ratio of myosin to F-actomyosin is *ca.* 4:1.

*(c) Structure of meat emulsions*

The term 'emulsion' is used to describe meat products in which fat is finely dispersed within a matrix of meat protein. Proper stabilization of such systems is required to prevent separation of fat and water during cooking. There has been considerable debate concerning mechanisms of stabilization and whether or not a classical emulsion is involved.

Finely comminuted (emulsion) meat products may be described as a complex mixture of muscle tissue, fat particles, water, spices and solubilized proteins which are held together by a variety of attractive forces. The NaCl-soluble myofibrillar proteins, myosin and actomyosin, are of prime importance as the major structural components and also form an interfacial protein film (IPF) around fat globules. Two well established theories currently exist to explain the stability of meat emulsions: the emulsion theory and the physical entrapment theory. It should be appreciated that these theories are not mutually exclusive and that both mechanisms may be involved.

The emulsion theory proposes that in the uncooked product, myofibrillar proteins are attracted to, and concentrated at, the fat globule surface, forming a stabilizing membrane. Localized frictional forces during comminution are believed to result in the formation of a thin layer of melted fat at the surface of the fat globule, to which undenatured myosin is adsorbed. The heavy meromyosin head of myosin is of relatively high surface hydrophobicity and it is probable that the molecule is orientated at the interface so that the head faces the hydrophobic phase and the light meromyosin tail the aqueous phase. The overall result is the formation of a monomolecular layer of undenatured myoglobin around fat globules. In uncooked meat emulsions, protein:protein interactions lead to the binding of other proteins to the myosin monolayer. The emulsifying properties of myosin, however, appear to be of key importance in stabilization of meat emulsion products before cooking.

An interfacial protein film also surrounds the fat globules in cooked meat emulsion products and a role has been suggested for stabilization during cooking. The IPF undergoes considerable change during cooking and, in the cooked product, is penetrated by small pores. Fat exudes from the pores which, it has been suggested, act as a 'vent' or 'safety valve', allowing the thermal expansion of fat during cooking (see below) while largely maintaining the integrity of the IPF.

The physical entrapment theory proposes that stability is derived through physical entrapment within the three-dimensional matrix of a heat-set protein gel (see above). This theory suggests that uncooked meat emulsion products are partly stabilized by a large quantity of intact fat cells. Support for this theory is obtained from both photomicrographic evidence and physical analysis and some meat scientists consider that the emulsifying properties of myofibrillar proteins are incidental to other stabilization mechanisms.

Instability of emulsion meat products continues to be an infrequent but potentially serious problem in commercial meat processing. It has been thought for a number of years that stability is related both to the thickness of the IPF and to the integrity and density of the protein matrix. Fat channels have been noted in unstable emulsions and have been attributed to formation of a weak IPF together with aggregation of the protein matrix. Each of these factors predisposes to fat and water loss.

More recent studies have shown that both stable and unstable emulsions contain two types of fat globule, with rough and smooth protein coats respectively. The rough type, which are more prevalent in unstable emulsions, contain pores in the coat, which are often of large size. The rough portion of the protein envelope may be unevenly distributed around the globule and globules with both rough and smooth coats are present in both stable and unstable emulsions. In these cases, pores are concentrated in the rougher portions of the coat. The rough coat is continuous with the protein matrix and appears to consist of thick, dense protein.

The smooth type of fat globule is prevalent in stable emulsions. Smooth globules are themselves of two types: small, with a relatively thick coat and a few tiny pores, and larger, with a thin coat and several pores. It appears that fat globule morphology is a major determinant of emulsion stability. Morphology, in turn, is dependent on the amount and type of protein forming the IPF, as well as IPF thickness.

A number of meat research workers consider that, while physical entrapment can account for stability of uncooked emulsions, it is difficult to envisage this mechanism being effective in the cooked product, unless the fat is localized within a membrane. This is a consequence of the sequence of events during cooking, where fat liquefies at 35–50°C before a significant amount of gelation has occurred amongst the matrix proteins. The fat will also be in the liquid state at higher temperatures (50–70°C), when most of the matrix proteins have undergone gelation. In either case, large pools of free fat would be expected to spread throughout the matrix during cooking, a phenomenon that does not occur with either stable or unstable batters.

It appears obvious, therefore, that the IPF plays an important role in stabilizing fat globules during cooking and preventing coalescence. Physical restriction of fat globules does occur, but this is due to binding to the protein matrix rather than physical entrapment. This probably results from protein:protein interactions between the IPF and matrix proteins, the IPF proteins possibly being part of the total protein network. Evidence for this possibility has been obtained by use of transmission electron microscopy, which demonstrates continuity between the matrix and the protein coats of fat globules at various points on their circumference. Scanning electron microscopy has indicated that some fat globules, especially those of smaller size, are connected to the fat matrix by thread-like protein strands.

Fat exudation occurs during cooking from both rough and smooth coated globules. As noted above this has been thought of as a 'safety valve', and some support for this contention has been obtained in detailed studies of the phenomenon. Exudation does not occur from very small globules of either type. In larger globules, however, exudation is universal and, in some cases, involves multiple exudation, in which newly exuded globules are themselves releasing fat. In this case, however, the protein coat of the parent and exuded globule is continuous.

A mechanism has been proposed which suggests that, during cooking, expanding fat pushes out the IPF at weak points to form

---

* Fat exudation and coalescence are distinct phenomena. The two may be distinguished by transmission electron microscopy, which shows that exuded fat totally lacks a defined spherical shape. Incomplete IPF residues are present in the fat and there are numerous interconnections between fat pockets. (Gordon, A. and Barbut, S., 1990, *Food Structure*, 9, 77–90.)

stable, round appendages. Insufficient protein is present to 'repair' the gap in the parent globule, thus leading to pores remaining in the coat. This process is repeated until smaller, more stable globules are formed, or until the temperature is sufficiently high to gel the proteins. Stable emulsions contain globules which show several small, uniform pockets of exuded fat. In contrast, unstable emulsions contain globules which show large exudations at weak points. In such cases, formation of fat channels and coalescence is facilitated.

The protein layer surrounding thinly coated fat globules is probably simple in structure, consisting of a monomolecular layer of myosin. The protein layer surrounding thickly coated globules is of considerably greater complexity. Pores appear to consist of a complex, convoluted series of tunnels, extending into the globule. The protein coat also appears to be multi-layered. There is some variation in the structures proposed by different researchers, but the general concept involves a thin internal layer coating the fat. This is bound through a diffuse region to a third internal layer, similar to that coating the fat. These three layers have been described as a thermodynamically favoured lipid bilayer-type structure. The outer layer is a very thick, relatively diffuse coat.

Although small globules consist only of fat, some of the larger have a relatively complex internal structure. The architecture of this structure varies, but in some cases, appears to impart additional stability to the globule. The internal structure consists of protein, which interacts with the IPF. The origin of the internal structures is unclear.

## 5.3.6 Chemical analysis

Standard methods can be applied to analysis to ensure compliance with legal and compositional standards (see Chapter 3, pages 152–153). Texture and cohesiveness are important with cooked meats and simple penetrometer devices (see Chapter 3, page 153) are suitable for objective assessment of these parameters in products such as emulsion sausages. In the case of cooked meat joints, shear force measurement using the Warner–Bratzler method may be more suitable. Alternatively an Instron tester fitted with 'jaws' to determine force encountered during chewing often gives very good results, although application is limited by cost.

Meat species determinations are difficult with cooked meat due to denaturation of the proteins. Special immunological techniques have been developed, which have had some degree of success. Such methods tend, however, to be cumbersome and isoelectric focusing is often considered the method of choice.

## 5.4 MICROBIOLOGY

Cooked meats as a whole, when correctly processed and stored, are highly stable products. The critical processing stages (adequate cooking, rapid cooling, prevention of recontamination and storage at low temperatures) are well defined and should be well understood. Despite this, microbiological problems continue to occur and, while outbreaks of food poisoning naturally receive most attention, problems due to rapid spoilage continue. In many cases, food poisoning may be attributed to unacceptable practice in catering or small industrial operations, and stems from underlying ignorance and/or inadequate facilities. Conversely, the very large scale of some operations can itself lead to difficulties of maintaining control while, especially in the case of spoilage, problems can be exacerbated by the very long storage lives required by retail chains.

### 5.4.1 Cooked meats as an environment for micro-organisms

It has become customary to consider cooked meats which do not contain nitrite as an ingredient separately from those containing nitrite. This distinction should be regarded with caution, however, since in some cases nitrite is only added at a level sufficient to obtain the desirable pink pigmentation. In other cases, levels of $NaNO_2$ and NaCl are sufficiently high to inhibit the outgrowth of endospores, including those of *Clostridium botulinum*, but have little effect on the development of the vegetative spoilage micro-

* The effect of nitrite in inhibiting the outgrowth of endospores in cooked meat products may be illustrated by comparing pork pies containing nitrite in the filling with otherwise similar pies containing no nitrite. At storage temperatures above 15°C, germination and outgrowth of *Bacillus* endospores occurs relatively rapidly in pies containing no nitrite and, depending on the species present, visible spoilage of the meat filling usually occurs after 4–7 days. In contrast, in pies with an input level of 75–100 mg/l $NaNO_2$, germination does not occur over any reasonable holding period. The presence of nitrite, however, has no effect on germination and growth of *Bacillus* in the gelatin layer within the pie and only a minimal effect at the meat:gelatin, or meat:pastry interface.

flora or growth rate of non-halotolerant pathogens, including some *Salmonella* serovars. In contrast some cooked meat products, including some emulsion sausages and pâtés, contain nitrite and, in some cases, NaCl added at sufficient levels to inhibit markedly the outgrowth both of endospores and the growth of vegetative micro-organisms (cf. cooked cured meats, Chapter 6).

In a general sense, cooked meat products containing insignificant levels of added inhibitors support rapid growth of micro-organisms, refrigeration being of prime importance in control of micro-organisms. There are, however, differences which can be attributed both to inherent variations between products of different types and to the application of different processing methods to products of the same type. The relatively high moisture content of the filling of many hot-eating pies, such as steak-and-kidney for example, supports more rapid growth than the filling of pork pies. Similarly a significant degree of drying occurs during oven baking of meat joints and growth of micro-organisms is usually significantly slower than on the surface of joints produced by cook-in processes.

In addition to dehydration, other changes occurring during cooking may affect microbial growth. It has been suggested, for example, that Maillard products have an antimicrobial effect, as well as acting as antioxidants (cf. page 265), and that microbial growth rate is lower where Maillard products are present at high concentrations. Similarly, antimicrobial properties have been attributed to volatile breakdown products of sulphur-containing amino acids and lipids. It is necessary, however, to place the antimicrobial properties of compounds formed during heating in context. It seems likely that, at most, such compounds modify the growth patterns of micro-organisms on cooked meat and may exert some selective pressure. The overall effect, however, is probably very limited and those micro-organisms important in the microbiology of cooked meats, including pathogens (such as *Cl. botulinum*) and spoilage organisms, are able to grow very rapidly indeed if temperature control is inadequate. For this reason cooked meat products which are effectively totally dependent on refrigeration are considered significantly less safe than their nitrite-containing counterparts.

Concern over the safety of non-nitrite containing products of low NaCl content, such as roast beef and some types of sausage, has led to the use of sodium lactate as a preservative. Sodium lactate, at a level of 2–3% is effective in preventing toxigenesis by

*Cl. botulinum*. The salt is also effective against many vegetative micro-organisms, Gram-positive bacteria (including *Listeria monocytogenes*) being more susceptible than Gram-negative bacteria. In general, however, lactate is considered to be effective against bacteria capable of growing at a pH value of 6.5, in the presence of NaCl at an $a_w$ level of 0.95 or below. The effect of lactate is specific and not due to reduction of pH value or $a_w$ level. In contrast to bacteria, yeasts are resistant even at levels of 10% or above. Various other preservatives have been suggested, including other organic acids, but results appear variable.

In addition to ingredients added primarily as preservatives, an antimicrobial function has been ascribed to a number of ingredients used for other technological purposes. Notable amongst these are phosphates, but antimicrobial properties have also been claimed for various antioxidants as well as for herbs and spices, including rosemary, garlic and black pepper, more usually considered a problem as a source of bacterial endospores. Although antimicrobial properties can often be demonstrated in pure culture experiments, the effects in cooked meat products appear limited. The results of different experiments may also be contradictory.

As with uncooked meat products (see Chapter 3, page 141), there is disagreement as to whether use of the algin/calcium system as a means of binding meat pieces reduces or increases microbial growth rates. Experimental work has suggested that the presence of alginates and kappa carrageenan increases the survival of *L. monocytogenes* during cooking.

### 5.4.2 Cooked meats and food poisoning

#### (a) Whole and restructured joints and poultry

Food poisoning resulting from undercooking is a continuing problem with these products. During the 1970s a number of outbreaks of salmonellosis occurred in the US due to consumption of undercooked roast beef. The underlying cause appeared to be a consumer desire for beef of 'rare' appearance, the cooking processes used being inherently unsafe. This problem appears to have been overcome by the introduction of cooking at time/temperature combinations of proved safety (see page 226). Since the 1970s, a series of outbreaks of salmonellosis has occurred over a wide area of north-west England and north Wales, in which undercooking of

BOX 5.7 **A needle in the . . .**

Epidemiological investigation of geographically widespread outbreaks of food poisoning, where the vehicles of infection are widely eaten foods and which involve commonly isolated micro-organisms, poses particular difficulties. This is illustrated by an outbreak, involving 39 known cases, which occurred in a wide area of north west England and north Wales during the late spring of 1991. The causative organism was *S. typhimurium* DT 193, currently the most common *S. typhimurium* phage type in the UK, and the vehicle of infection was under-processed cooked ham, from a single small supplier. The existence of the outbreak was detected through phage typing and antibiotic resistance patterns, which identified a cluster of distinctive isolates of *S. typhimurium* DT 193, resistant to sulphonamides, trimethoprim and furazolidone. Identification was made possible by the practice, in England and Wales, of forwarding isolates of *Salmonella* to the national reference laboratory. The value of this practice is well illustrated by comparing the successful detection of this outbreak with experiences in the US, where isolates are not routinely sent to a reference laboratory and where an outbreak of *S. typhimurium* food poisoning was only linked to inadequately pasteurized milk after more than 16 000 culture-confirmed cases. It is a matter of concern that the practice, in England and Wales, of forwarding cultures to the national reference laboratory may become less common due to organizational changes in the National Health Service. (Thornton, L. *et al.*, 1993, *Epidemiology and Infection*, **111**, 465-71.)

joints has been one of the contributory factors. Roast pork and ham have been primarily involved and in some cases it has been possible to trace the source of the *Salmonella* to outbreaks of infection at individual pig farms. No less than 550 known cases resulted from consumption of roast pork and ham during the 1970s, while in the summer of 1989, 206 persons were infected in a single outbreak involving these meats. In many cases the meat was cooked in a small operation, operated in conjunction with a butcher's shop, where process control was poor and control of cross-contamination difficult, even when overall management was good.

*Salmonella typhimurium* DT 193 was the causative organism in a number of outbreaks, including that involving 206 known cases. The organism survived cooking, the relatively large number of organisms present being a possible contributory factor alongside defective equipment. Secondary contamination occurred, either directly from raw meat to cooked, or between contaminated and uncontaminated cooked meats. The ultimate source of the organism was an infection amongst pigs at the producing farm.

Poultry also has a high incidence of contamination with *Salmonella*. Undercooking is rarely a problem with chickens but problems do occur with turkeys, including an outbreak in which 61 persons are known to have suffered *S. kedougou* infection.

Concern remains over the possibility that *L. monocytogenes* may survive processes validated for *Salmonella*, even when these are correctly applied. Undercooking of commercially processed chickens has been implicated as being a cause of sporadic listeriosis in the US, although in this example doubt has been cast over the interpretation of epidemiological data. Cooked chickens have also been associated with listeriosis in the UK, although this probably resulted from post-process contamination.

Reported cases of food poisoning due to either inadequate cooling or post-process contamination continue even in industrial scale production of cooked joints and poultry. In some outbreaks it has appeared that virtually every aspect of good practice has been ignored. The situation is particularly serious in some catering operations where direct contamination of cooked chicken with drip from raw birds has been reported. In many cases the need to separate cooked poultry from raw has either been totally misunderstood, or ignored. Poor temperature control has often been identified as a contributory factor, including the sale of cooked chickens from unrefrigerated market stalls.

---

* The hands of inexperienced chefs have been shown to be an important vehicle in transmission of *Campylobacter* from raw chickens to cooked foods. Adequate drying of hands, as well as washing, is considered to reduce significantly the possibility of transferring *Campylobacter*, an organism which is highly sensitive to dehydration. It is notable, that while food hygiene recommendations invariably, and correctly, refer to hand washing, the importance of drying is completely overlooked.

Most reported cases of food poisoning due to post-process contamination of cooked joints and poultry have involved *Salmonella* but *L. monocytogenes* has been isolated from a wide range of cooked meats, including roast beef and ham. Some evidence suggests that *L. monocytogenes* is derived from environmental sources (cf. pâté, page 284) rather than raw meat, and plant colonization may be involved. Cross-contamination of cooked poultry from raw is also considered to play a large and possibly a major role in sporadic *Campylobacter* infections in the UK. There is also an obvious risk from other pathogens associated with raw meat including *Yersinia enterocolitica* and Vero cytotoxin-producing *Escherichia coli*.

*Staphylococcus aureus* can readily be isolated from cooked joints and poultry, although numbers are usually no greater than $10^2$ cfu/g. Significantly higher numbers, in the order of $10^4$–$10^5$ cfu/g are unacceptable and indicate a high level of contamination and/or poor temperature control. *Staphylococcus aureus* is usually derived from human sources, although poultry was considered to be the source in an outbreak of food poisoning involving an enterotoxin A-producing strain, which contaminated cooked chicken. Sliced products are especially prone to contamination with *Staph. aureus* from human sources. On a number of occasions, outbreaks of enterointoxication involving large numbers of cases have been associated with special events such as weddings, dances and christening parties. This usually results from the advance preparation of meals based on sliced cooked meats held at ambient temperatures for extended periods. In this context, it has been noted that too often socializing takes precedence over refrigeration.

Although *Staph. aureus* can readily be isolated from vacuum- and gas-packed sliced cooked meat, in small numbers, enterointoxication is rare. This has been attributed to suppression of enterotoxin at low oxygen tensions. A further possibility is inhibition of growth of *Staph. aureus* by lactic acid bacteria, which almost invariably are present in large numbers in commercial pre-packs. In the case of poultry, cooked turkey meat has been found to be a poor medium for enterotoxin production.

Slow cooling of cooked large joints of meat and whole large chickens or, more usually, turkeys is a common cause of *Clostridium perfringens* food poisoning. Problems rarely occur in large-scale production and catering operations are most commonly involved. The usual underlying cause is inadequate or insufficient refrigeration and in some small catering operations the cooling and storage of cooked meat at ambient temperature is not uncommon.

### (b) Pies and puddings

Commercially produced pies and puddings generally have a good safety record. The major exception is the association of pork pies with salmonellosis due to the post-cooking addition of gelatin. Gelatin is both a potential source of *Salmonella* and a good medium for growth of the organism. In the past, outbreaks due to use of contaminated gelatin were a common cause of salmonellosis. Staphylococcal enterointoxication has also been reported as a consequence of gelatin contamination.

In recent years, the precautions necessary for safe handling of gelatin have been more rigorously applied and problems significantly reduced. A relatively large outbreak of salmonellosis that occurred in northern England during the early 1980s, however, was attributed to the use of contaminated gelatin in pork pie manufacture. This outbreak was notable for the total mishandling of gelatin, including the use of domestic teapots as dispensing utensils. Resulting adverse publicity led to closure of the producing factory and the bankruptcy of the owners.

Undercooking of commercially made pies is rare due to the relatively severe nature of the cooking and the resulting large safety margin. An outbreak of salmonellosis has, however, been reported which resulted from undercooking of large pork pies. The pies were cooked in a batch oven together with smaller items, including sausage rolls. As a result, the process usually applied for the pies was shortened considerably to avoid overcooking the sausage rolls.

Commercially made hot-eating pies have only rarely been implicated in food poisoning. Hot-eating pies made in catering establishments have, however, been implicated as a cause of *Cl. perfringens* enteritis. In such cases the pies were held at a temperature suitable for rapid growth of the organism for an extended period. Growth of *Cl. perfringens* has also been noted in commercial pie filling held without cooling after the bulk cooking stage. In this case, the vegetative cells were killed either during the baking of the pastry of the filled pie, or during cooking in the home – a fortuitous tyndallization process. Germination and outgrowth of other endospore-forming bacteria may also occur during holding at high temperatures. This normally results in spoilage, but four outbreaks of botulism in the US resulted from gross mishandling of beef or chicken pot-pies. In each case the pies had been produced as frozen products, heated in the home and then stored for an extended period at or above ambient temperature. Strains of diarrhoeal syndrome *Bacillus cereus*, *B. licheniformis* and *B. subtilis* have also been responsible for food poisoning following consumption of hot-eating pies that had been held for long periods at high temperatures. Emetic syndrome *B. cereus* food poisoning may have resulted from consumption of a shepherd's pie, but growth probably occurred in the potato-based topping rather than in the meat filling.

Occasional cases of *Staph. aureus* food poisoning have been reported in which hot-eating pies have been implicated. Growth of the organism and toxin elaboration appear to have occurred during prolonged holding of the meat filling after cooking. The toxin then survives any subsequent heating.

### (c) Cooked sausages

In recent years, the safety record of cooked sausages has been good, despite the fact that some cooking processes are marginal for safety. In many cases, the presence of $NaNO_2$ or, less commonly, sodium lactate does provide an additional safety margin, especially as any surviving vegetative cells are likely to be sub-lethally damaged. In the US, however, undercooked turkey frankfurters have been implicated as a cause of *Listeria monocytogenes* infection.

In a small number of cases, cooked sausages are known to have been responsible for *Staph. aureus* food poisoning. The products had been sliced and packed in an air-permeable film, it being assumed that contamination occurred during post-process handling and that temperature abuse occurred during storage. *Clostridium perfringens* food poisoning has been reported following slow cooling of sausages of a nitrite-containing type and the outbreak was unusual in that *Cl. perfringens* is sensitized to $NaNO_2$ by sub-lethal heat treatment. It appears, however, that the sausages were of incorrect formulation and that input nitrite levels were very low.

## (d) Spreadable cooked meat products

The safety record of spreadable cooked meat products is generally good, although sporadic outbreaks of food poisoning have been reported, involving a range of pathogens. Undercooking of pâtés, especially those of the large bowl type, can occur but the problem usually involves survival of *Enterococcus* and heat-resistant strains of *Lactobacillus* rather than pathogenic micro-organisms. The heat resistant *Salmonella senftenberg* has been isolated from chicken pâté, although food poisoning was not reported, possibly because this serovar is of low infectivity to humans.

Surveys in the UK have indicated that *L. monocytogenes* was present in 10% of pâtés sampled in 1989 and 4% of those sampled in 1990. The difference was largely due to a particularly high incidence of the organism in pâtés from one manufacturer during the 1989 survey. Although the possibility of *L. monocytogenes* surviving cooking cannot be discounted, evidence suggested that contamination occurred after cooking. Particular risk was associated with service delicatessen counters, where pâté is sliced manually on demand (cf. *Staph. aureus*). There has been no recognized outbreak of listeriosis involving pâtés, but the products have been implicated on epidemiological grounds in Australia and the UK.

Gelatin, used in the post-process embellishment of some pâtés, is an obvious potential source of *Salmonella*. No outbreaks have been reported, however, despite obvious mishandling of gelatin on some occasions. Embellishments such as herbs and fruit slices have long been recognized as potential sources of spoilage micro-organisms. More recently, herbs have been identified as a potential source of *L. monocytogenes* and possibly other environmental pathogens, including *Aeromonas*.

Pâtés have been implicated in *Staph. aureus* food poisoning resulting from post-process contamination and temperature abuse. On one known occasion the pâté had been sliced and vacuum packed at a central plant. The pack was a leaker and growth of *Staph. aureus* occurred during storage at *ca.* 12°C. This case was unusual in that the pâté was visibly spoiled, despite the fact that *Staph. aureus* was present in almost pure culture. The major problem, however, lies with service delicatessen counters (cf. *L. monocytogenes*), where large pâtés can be subjected to repeated handling during a long period of storage under poor temperature control.

Slow cooling can be a problem in the manufacture of large pâtés. In some cases, cooling is deliberately slow to produce unspecified improvement in texture. An obvious hazard exists with respect to *Cl. perfringens*, although the presence of $NaNO_2$ does reduce risk. For this reason, pâtés involved in outbreaks of *Cl. perfringens* food poisoning have usually been made without nitrite. Diarrhoeal syndrome *Bacillus cereus* and *B. subtilis* have also been implicated in food poisoning following consumption of pâté. An unidentified species of *Bacillus*, which produced characteristic 'rocket' colonies on nutrient agar plates, may also have caused food poisoning in liver sausage.

### (e) Ready-meals

Although ready-meals, produced for retail sale, have been identified as high risk products, no outbreaks of food poisoning have been reported which can be attributed to these products. Concern over safety may be a positive factor in that managements are acutely aware of potential hazards. It is also true that, in many cases, ready-meals are produced in purpose-built premises, constructed to a higher specification than many plants producing conventional cooked meat products. Despite this, surveys have suggested that a relatively high incidence of ready-meals are contaminated by *L. monocytogenes*, although numbers present have generally been very small. With the exception of low temperature, there are no constraints on the growth of *L. monocytogenes* or other psychrotrophic pathogens and for this reason it is considered that both storage temperature and length of shelf life should be strictly controlled.

Cook–chill meals are of particular concern in that persons in situations such as hospitals and homes for the elderly may be particularly vulnerable. *Listeria monocytogenes* has been isolated from hospital cook–chill meals in the UK, the risk appearing to be greater where meals are purchased from external suppliers rather than prepared on the premises.

In-flight meals served by airlines are effectively cook–chill products and have been associated with a number of incidents of food poisoning. With meat dishes, *Staph. aureus* has been most commonly involved, but *Salmonella* and *Shigella* have also been implicated. The cause in each case appears to have been post-process contamination at point of preparation and inadequate temperature control during storage.

### 5.4.3 Spoilage of cooked meat products

*(a) Whole and restuctured joints and poultry*

To a considerable extent, the spoilage of joints and poultry depends on the processing technology. Cook-in-ship joints, for example, in which post-process contamination is eliminated, are extremely stable when stored at less than 5°C. Psychrophilic and psychrotrophic strains of both *Bacillus* and *Clostridium* are known, but appear to be of little, if any, importance in this context. Spoilage due to these organisms does, however, occur at storage temperatures above 7°C and is often rapid at temperatures of 15°C and above. The spoilage pattern varies according to the dominant organism present, but typically involves souring, off-odours, slime and gas production. Germination and outgrowth of endospores is delayed by preservatives such as sodium lactate.

Marginal underprocessing may result in survival of *Enterococcus* and heat-resistant strains of *Lactobacillus*. These bacteria are usually, but not invariably, unable to grow at temperatures below 5°C, although growth is often rapid at temperatures above 7°C. In many cases, spoilage potential appears to be low and typified by weak acid production and slime formation. On rare occasions gas is produced by heterofermentative species of *Lactobacillus*.

Cook-in-strip joints are inevitably contaminated after removal from the cook-in bag. The degree of contamination tends to reflect the extent of handling and is greatest where slicing is involved. In most cases the product is repackaged in vacuum or gas packs for distribution and retail sale. The spoilage microflora is dominated by species of *Lactobacillus* and *Carnobacterium*, although these organisms may be present in only very small numbers immediately after repacking. Spoilage typically involves acid production and a sweet/sour odour. Growth of lactic acid bacteria in these and other cooked meat products is indicated by the milky appearance of free moisture in the pack. This is due to precipitation of proteins by lactic acid and is usually apparent before overt spoilage. In non-nitrite containing products, a fairly clear relationship exists between bacterial numbers and spoilage, visible signs appearing when colony counts are in the order of $10^8$ cfu/g. The relationship is less clear when nitrite is present. Visible signs of spoilage may be detected when numbers are $10^6$-$10^7$ cfu/g, but equally products with microbial numbers in excess of $10^8$ cfu/g may remain unspoiled. *Brochothrix thermosphacta* is an occasional cause of spoilage of vacuum-packed cooked meats. Spoilage involves a characteristic cheesy odour.

A variety of micro-organisms cause spoilage of cooked meats which are unpackaged, or packed in air-permeable film. *Pseudomonas* and other oxidase-positive, Gram-negative, rod-shaped bacteria are common in uncured products stored at less than 5°C, producing slime and off-odours. Environmental strains of the Enterobacteriaceae are also common and may become dominant at higher storage temperatures. Common genera include *Citrobacter, Enterobacter* and *Klebsiella*. *Serratia* may also be present and red-pigmented strains can cause rapid and spectacular spoilage of cooked meat stored at temperatures in excess of 15–18°C. Formation of purple slime on roast beef has been attributed to '*Chromobacterium violaceum*' (probably *Janthinobacterium lividum*) but this condition is very uncommon.

The surface of roast poultry and meat joints cooked unbagged in hot air ovens can be subject to considerable drying. Under these conditions moulds may develop during extended storage at low temperatures. A number of genera may be involved but the most common are *Penicillium, Mucor, Aspergillus, Cladosporium* and *Thamnidium*. Yeasts, especially *Candida, Monilia* and *Torula*, may also be present in significant numbers on both meat and fat surfaces.

### (b) Pies and puddings

The meat filling of most pies and puddings is protected from post-process contamination by the pastry casing and, when correctly made, the only micro-organisms present are those that survive cooking. These products are thus highly stable at temperatures below 5°C. In the case of cold-eating pies, such as pork pies, germination and outgrowth of endospores in the meat filling is rare, especially if nitrite is present. Large *Bacillus* colonies, however, sometimes develop at the interfaces between the meat and gelatin, the gelatin and pastry, or the meat and pastry. On some occasions a single giant colony develops and may cause revulsion and even fear in the consumer.

Gelatin can be a source of endospores of *Bacillus* and, if mishandled, vegetative spoilage organisms. Discrete colonies sometimes develop in the gelatin layer and proteolytic micro-organisms occasionally cause liquefaction. Mould spores occasionally enter the pie during addition of the gelatin and a hyphal mass may develop between the gelatin layer and the pastry lid.

The filling of hot-eating pies supports rather more rapid microbial growth, although these products are also highly stable at temperatures below 5°C. Spoilage at higher temperatures is usually by *Bacillus* spp. and involves taints, acid and, sometimes, gas production. *Clostridium* is less often involved in spoilage, despite the presence of endospores of this genus. On rare occasions, the filling may be spoilt but no micro-organisms are present. This has been attributed to extended storage at high temperature leading to extensive growth in the bulk filling, the micro-organisms then being killed during the second heating. The thermophilic *B. stearothermophilus* has allegedly been involved in spoilage of this type, although firm evidence is not available.

Meat pies and puddings of all types can be spoilt by mould growth on the outer surface of the pie. In many cases, the pastry is baked to a low moisture content and spoilage of this type is relatively uncommon. In film-wrapped pies, however, condensation of water between the film and the pastry can lead to rapid mould growth. This results either from wrapping the pie before cooling is completed or from temperature fluctuation during storage. Mould spoilage of puddings filled into a boiled paste, or of shepherd's pie topped with mashed potato, is a much more common problem and yeast growth may also occur. Spoilage of the topping of shepherd's pie by species of *Bacillus* (cf. food poisoning by *B. cereus*, page 283) has also been known.

### (c) Cooked sausages

In whole sausages, the filling is protected by the casing and only bacteria which have survived cooking are involved in internal spoilage. Survival of vegetative organisms is rare, although *Enterococcus* and heat-resistant strains of *Lactobacillus* do occasionally survive, especially in marginally processed large diameter sausages.

As in other cooked meats, the spoilage potential of these bacteria is low. The most common spoilage pattern involves acid production and weak taints, although gas production by heterofermentative strains of *Lactobacillus* may occur. The most spectacular spoilage, however, occurs when hydrogen-peroxide forming strains of *Lactobacillus* are present, although this is more commonly associated with post-process contamination. Hydrogen peroxide, produced during aerobic growth, in the absence of catalase (which is inactivated during cooking), oxidizes the pigment, denatured nitrosyl myoglobin resulting in a grey-green or, less commonly, a yellow or white discolouration. Hydrogen peroxide is not produced in the absence of oxygen, and in some cases the classic 'green ring' is observed, marking the inward diffusion of oxygen. Potentially, any member of the lactic acid bacteria may produce $H_2O_2$, but the property is most commonly associated with *Enterococcus*, *Lactobacillus viridescens* and *Pediococcus*.

Germination and outgrowth of surviving endospores of *Bacillus* and *Clostridium* can occur, especially in sausages containing low levels of nitrite. In practice, however, this is usually only a problem where the product has been grossly temperature abused. Spoilage often involves massive gas production, with disruption of the filling and swelling of the casing or vacuum pack. Gas production may be accompanied by acid production and/or, especially in the case of *Clostridium*, foul smells. On some occasions, growth of *Clostridium* results in blackening of the meat filling. This phenomenon appears to be related to hydrogen sulphide production but the mechanism is not fully understood.

Mould and yeast growth on the outer surface of sausage is a potential problem during extended storage at 0–1°C, or if condensation is present. Moulds involved are those commonly associated with meat products, including *Mucor*, *Penicillium* and *Cladosporium*. Yeast growth leads to visible slime formation, which can be very extensive. A wide range of yeasts may be involved, including *Candida*, *Cryptococcus*, *Debaryomyces*, *Galactomyces*, *Pichia*, *Rhodotorula* and *Torulaspora*.

Slicing and pre-packing cooked sausage in either vacuum or gas packs is common practice in many countries. The type of packaging has little effect on the microflora present, although gas packing can have advantages with respect to product appearance. In all cases, the dominant spoilage flora comprises lactic acid bacteria, predominantly *Carnobacterium* and *Lactobacillus*. Numbers of up to $10^8$ cfu/g are not uncommon at point of consumption, although there is usually little effect on organoleptic quality beyond a limited souring. A further increase in numbers, however, results in slime production and off-odours. Greening due to $H_2O_2$ production is a persistent problem with pre-packs from some plants. In the case of vacuum packs, greening develops when the sausage is exposed to air and may be in discrete patches, reflecting uneven contamination. The original source of $H_2O_2$-producing lactic acid bacteria is often thought to be fermented sausages, handled in the same packing lines. It seems likely that colonization of the plant occurs.

### (d) Spreadable cooked meat products

The overall spoilage pattern of spreadable cooked meat products resembles that of cooked sausages. Large bowl pâtés are particularly prone to undercooking and the consequent survival of *Enterococcus* and heat-resistant strains of *Lactobacillus*. Outgrowth of endospores during the slow cooling of some pâtés and liver sausages can also be a significant problem. A problem involving a fiery red colouration in the centre of liver sausage has been attributed to competition for oxygen between *Bacillus* and haem pigments resulting in an excess of reduced nitrosylmyoglobin. It is not known why liver sausage can be particularly prone to this problem and other factors may be involved.

Embellishments added to some types of pâté after cooking can be a significant source of spoilage organisms. Herb sprigs, for example, can be an important source of moulds and fruit slices can be a source of yeast. The extent of contamination is best appreciated with pâtés in which the embellishments are set in a gelatin topping. In these circumstances the gelatin acts as a culture medium in which discrete colonies of contaminating micro-organisms develop. Gelatin itself can be an important source of spoilage micro-organisms as well as pathogens. Spoilage micro-organisms may develop as discrete colonies or in a film at the meat:gelatin interface.

Liver sausage can be particularly prone to spoilage by surface mould growth. Yeasts may develop, either as slime or as discrete colonies. Yeast growth can also be a problem on the outer fat layer of some types of pâté.

As with cooked sausages, souring and occasionally gas production by lactic acid bacteria is the predominant form of spoilage of sliced pre-packs. Greening appears to be relatively rare with both pâtés and liver sausage. Pâtés, however, are prone to discolouration at low pH values. An unusual case of gas production in pre-packed pâtés resulted from synergistic gas production by a species of *Lactobacillus* and an unidentified yeast.

### (e) Ready-meals

In most cases, spoilage of ready-meals results from post-process contamination. Patterns of spoilage are often related primarily to the non-meat constituents and can be extremely variable. Under most conditions, Gram-negative rod-shaped bacteria, especially *Pseudomonas*, are of greatest importance. Lactic acid bacteria, primarily *Lactobacillus*, can also be involved in spoilage, especially where the product contains sauces of reduced pH value. Yeasts have also caused problems in this situation.

Ready-meals may contain significant numbers of *Bacillus* spp., derived from non-meat ingredients such as flour, starches and spices, as well as from the meat. These are normally present as endospores and are not significant in spoilage unless gross temperature abuse occurs.

## 5.4.4 Microbiological analysis

Microbiological analysis of cooked meat products presents no particular problems but selection of analyses does require care. There is a tendency to attempt to determine too many types of microorganism, rather than too few.

According to circumstances and type of product, analysis may be made for *Staph. aureus*, *Salmonella* and *L. monocytogenes*. In the case of *Staph. aureus*, it has been argued that microbiological examination serves little purpose and that effort is better concentrated on process control. To a large extent this is true, especially since *Staph. aureus* is not a major problem with commercially produced meats and since contamination is sporadic and unlikely to be detected by examination. At the same time, cases are known where laboratory examination has led to early detection of contamination with *Staph. aureus*, possibly resulting from colonization of plant. For this reason, examination for *Staph.*

*aureus* is a routine in some laboratories as part of a 'belt and braces' approach. This is considered acceptable, provided that laboratory examination does not distract from the need for a high level of process control. Baird–Parker medium, or a derivative, is recommended for detection of *Staph. aureus*. Enrichment is not required.

Examination for *Salmonella* has been recommended for pre-cooked roast beef, following outbreaks of salmonellosis in the US. In this context examination for *Salmonella* has been described as verification of the cooking process. The suitability of *Salmonella* as an index for correct processing of roast beef, or indeed any cooked meat product, must be questioned. Despite the high incidence of *Salmonella* in poultry, and sometimes pork, the organism is by no means universally present in raw meat and its absence cannot therefore be taken to ensure the adequacy of a process. Use of *Enterococcus* as an index organism would probably be more reliable, although in the long term chemical indices, such as enzyme inactivation (see page 256) would be preferred. In addition, the statistically based sampling scheme required to ensure 'absence' of *Salmonella* is extensive and requires a very significant laboratory input. It is considered that this effort would be better used in ensuring adequate processing in combination with simpler and more reliable methods of verification. Where examination for *Salmonella* is considered necessary, a three-stage method is required involving pre-enrichment (resuscitation), enrichment and selective plating. Standard media, such as buffered peptone water (pre-enrichment), RV broth (enrichment) and brilliant green agar (selective plating), are suitable. Rapid and automated methods may also be used but require validation in each individual circumstance.

In recent years, examination for *L. monocytogenes* has become routine in some situations. Concern has been greatest with ready-meals and where products are prepared for some multiple retailers. The value of such examinations is controversial, especially where methods are used which do not differentiate between *L. mono-cytogenes* and other *Listeria* species. Some microbiologists also consider that examination for *L. monocytogenes* itself is of little value in the absence of a simple method for differentiating virulent from non-virulent strains. In any case it is considered unfortunate that a continuing preoccupation with *Listeria* has diverted atten-tion from more immediate problems, especially *Salmonella*. In many, if not all, cases there appears to be little justification for the 'blockbuster' approach adopted for analysis for *L. monocytogenes* and it is considered that resources would be better employed in ensuring adequate control of processing and product handling. Individual circumstances vary and it is the responsibility of each microbiologist to assess the risk from *L. monocytogenes* and determine the level of analysis required. Any testing should be in conjunction with a pathogen control programme and it is also necessary to be quite clear what action is to be taken if *L. monocytogenes* is isolated during routine testing. Where analysis is undertaken, a two-stage enrichment, selective plating procedure is required. A selective enrichment broth, incubated at 30°C for 48 hours, is most suitable for regular testing, although additional serovars may be recovered if low temperature enrichment at 4°C for 1 week is used in parallel with selective enrichment. Oxford agar is a suitable selective plating medium but, like other types, it cannot differentiate *L. monocytogenes* from other species. Biochemical testing of presumptive isolates is required for this purpose. Automated and rapid methods are available but many detect all species of *Listeria* rather than *L. monocytogenes*.

Examinations for *Cl. perfringens* and other potential pathogens are not considered necessary on a routine basis. An indication that products have been cooled slowly, under conditions likely to favour growth of *Cl. perfringens* and *B. cereus*, can be obtained by plating on blood agar incubated at 37°C. Colonies of *Bacillus* species are easily recognized by experienced microbiologists. This test is not considered necessary on a routine basis, but can be useful if problems arise.

With the possible exception of ready-meals (components of which may receive considerable post-cooking handling), examination for 'coliforms' and *E. coli* are not considered to be of value. The use with ready-meals should be assessed according to circumstances.

In addition to examinations for pathogens, some manufacturers routinely determine a wide range of potential spoilage organisms, some of which are extremely unlikely. In general terms, and again with the possible exception of ready-meals, 'total' viable counts are considered to have little value as a predictor of spoilage. In the case of pre-packed product, analysis is often made for lactic acid bacteria, but the value of this is also subject to doubt, despite the fact that lactic acid bacteria are usually dominant in spoilage. This stems from there being sometimes little relationship between numbers of lactic acid bacteria initially present and spoilage. This is usually a problem when initial numbers are low. High initial numbers of lactic acid bacteria usually indicate an unacceptable level of post-cooking contamination and a high potential for spoilage.

## EXERCISE 5.1

It has been claimed that treatment of cooked pork at very high (unspecified) pressure results in a product which has a similar appearance and flavour to ham and which could be the basis for 'a family of new generation meat products'. Discuss, from both a technical and marketing viewpoint, the stages in developing a range of pressure-treated cooked meats from concept to production. What particular problems are likely in terms of microbiological and chemical stability? What are the respective merits of marketing such products as additive-free cured meat analogues, or as an entirely new product?

EXERCISE 5.2

You are technical manager of a medium-sized company producing meat-based ready meals, primarily for supermarket own-
brands. Your company is faced with financial problems resulting from a combination of pressure on margins from the
major customers, high interest payments and exorbitant direc-
tors' fees. The factory is situated in an area of high unemploy-
ment and, in the past, it has been possible to maintain
adequate staffing levels despite low wages. This situation has
been changed by the opening of a Japanese-owned computer-
assembly operation and your management is faced with
increasing difficulties in obtaining staff and a very high staff
turnover. You are concerned over the difficulties in maintaining adequate quality assurance standards in the absence of a
stable and motivated workforce and alarmed at a number of
resulting incidents which have compromised product safety.
Senior management refuse to contemplate increasing wage
rates and have proposed to overcome problems of staffing
shortages by ending the strict demarcation between handlers
of raw and cooked products and the institution of 'floaters',
who can be transferred at short notice between the two areas.
These persons will receive as little as possible extra training,
but will be subject to a more rigorous initial medical examination.

Discuss the implications of the use of 'floaters' with respect to
public health risks. Do you consider that a system of this type
can ever be truly safe in a large-scale manufacturing environment? To what extent, if any, is a more rigorous initial
medical examination likely to reduce risk?

### EXERCISE 5.3

'Many foods with pH values and $a_w$s able to support growth of *C. botulinum* are not heated sufficiently to guarantee an adequate margin of safety. It seems inconceivable that all these inadequately treated foods are stored under perfect temperature control, and none is abused, before being eaten. Why is there not more botulism?' (Dr T.A. Roberts, Institute of Food Research, Reading Laboratory, speaking at the 1994 Summer Conference of the Society for Applied Bacteriology, Heriot-Watt University, Edinburgh). Consider carefully Dr Roberts' comments. Do you agree that more botulism might be expected? Assess the various factors which determine the probability of botulinum toxin being present in cooked meat products. Are we just lucky to avoid botulism on a large scale or do you think other factors are involved? As Dr Roberts went on to ask: 'Have we got it wrong?'

# 6

# COOKED CURED MEATS

---

## OBJECTIVES

After reading this chapter you should understand
- The different types of cooked cured meats
- Processes used in their manufacture
- Quality assurance and control
- The nature of the pigments in cured meat
- The role of nitrite in inhibiting development of warmed-over flavour
- Public health aspects of cooked cured meats
- Spoilage of cooked cured meats

---

## 6.1 INTRODUCTION

Cooked cured meats are a long-established product, some types of which have undergone considerable change in recent years. In the past, cooked cured meats have been characterized by a dry texture and distinctive salty taste, in some cases with heavy smoke overtones. Products of this type are still made by traditional processes, especially in central Europe, and can be very strongly flavoured indeed. In the US and elsewhere, however, the trend for a number of years has been to moist products of relatively low NaCl content and very little flavour. The descriptor 'plastic' is not inappropriate to some examples of this type of product, which from a technical and microbiological viewpoint are very similar to uncured cooked meats such as roast beef.

Cooked cured meats, such as ham and tongue, were previously staples of the traditional English high tea and are still widely consumed in buffets and as sandwiches, etc. Scope for new product development is limited, but there has been a significant growth in products made from poultry rather than the more common pork (see pages 301–302). Poultry-based cooked cured meats initially gained popularity, especially in the US, as being 'healthier' than pork-based, but in some cases they are now preferred in their own right.

## 6.2 TECHNOLOGY

### 6.2.1 Definition of cooked cured meats

The definition of cured meats is the same as that used in Chapter 4, pages 168–169.

### 6.2.2 Cooked ham, shoulder, etc.

Methods of cooking vary but are essentially the same as those used for uncured cooked meats (see Chapter 5, pages 226–230). Hot air is widely used and in the case of traditional European products is often combined with a heavy smoking. Ham cured by the Wiltshire process is also usually cooked in hot air. Hot water is now widely used for joints and may involve either cook-in-ship or cook-in-strip processing. Joints of this type are made from pieces of meat, tumbled or massaged with a polyphosphate containing brine, before bagging.

Traditional ham boiling is still practised on a small scale. This involves suspending Wiltshire-type hams, usually wrapped in muslin stockinette, in vats of hot water at a temperature approaching 100°C. The length of the process is based on an internal temperature of 70°C being attained. In most cases, it is a single-stage process, but use is still made of an older, two-stage procedure that involves a preliminary cook, followed by manual derinding of the ham while warm and then a finishing cook. This process carries a high risk of contamination with *Staphylococcus aureus* at the derinding stage and growth and enterotoxin formation can occur before inhibitory temperatures are reached during the second cook.

Sweet flavours have traditionally been used to offset the saltiness of ham and some traditional types are coated with a glaze of honey, or molasses in water. The glaze contributes flavour and promotes an attractive appearance. For honey-roast ham produced by modern technology, a major part of the glaze is likely to be a solution containing D-glucose and D-fructose (the sugars of honey). Some types of traditional ham contain cloves or juniper berries inserted in the outer layer of meat, while added-value variants of modern hams containing fragments of black peppercorns have also been produced.

---

### BOX 6.1  **The longest day**

Spam® is a canned, commercially sterile product, the name being derived from 'spiced ham'. At times of food shortage during the second world war, Spam and similar products, imported from the US formed a significant part of the UK meat intake. Sales of Spam have continued to the present day and the product has undergone something of a marketing revival as a consequence of nostalgia stimulated by the 50th anniversary of the invasion of German-occupied France by Allied forces.

---

Ham and similar products present particular problems due to photo-oxidation of pigments during the first 24 hours of display. This occurs in the presence of even small quantities of oxygen and means that conventional vacuum packaging is not adequate. In some cases a film containing metal foil is used, an illustration of the product being printed on the top surface of the pack. Although effective, the method is expensive due to the cost of the film. It is also disliked, and even considered dishonest, by some consumers. Two other methods involve use of film of very low oxygen permeability, pulling an initial vacuum of >95% (versus 90% normally used) and holding the product in the dark for 4 days to permit depletion of oxygen by reducing systems in the meat. Alternatively sophisticated packaging equipment may be used to flush the meat with $CO_2$ and then seal the pack with a $CO_2$ over-pressure. Neither of these methods can be considered satisfactory. More recently the possible use of 'active packaging' has been investigated. Complete elimination of photo-oxidative discolouration has been claimed by including a sachet of Ageless® GM-50, which scavenges oxygen remaining in the pack while concomitantly releasing $CO_2$.

### 6.2.3 Canned hams

Substantial quantities of canned ('pasteurized') ham are produced and used primarily for slicing, either for pre-packing or for immediate use in catering. The usual pack is a large rectangular can, or sometimes a large D-shaped can. Although the ham is packed in double-seam cans, the process applied is by no means a commercial sterilization. Internal temperatures of 65–75°C are most commonly used, a typical process being a minimum centre temperature of *ca.* 68°C attained by cooking in water at *ca.* 83°C for several hours. The temperature difference between the heating medium and the meat must be as low as possible to minimize cooking losses and jelly formation. Cans of this type may be classed as 'semi-preserved' and have a shelf life of at least 6 months at temperatures below 5°C.

---

### BOX 6.2  **Christmas cracker**

'Canned' hams, especially those packed in D-shaped cans, may be stored incorrectly at ambient temperatures. This stems partly from the association of canned products with ambient temperature storage, but also from confusion with canned hams which have undergone a full retorting process. These are usually of small size, but large D-shaped cans are also used. The ham after processing tends to lose its characteristic texture and there are usually large quantities of jelly present. Sales seem to be concentrated at Christmas and such hams are common in gift hampers and as raffle prizes.

---

## 6.2.4 Cooked cured poultry products

In recent years a wide range of cooked cured poultry products have been produced and while some have not been succesful, others have achieved sales greater than their pork-based equivalents in some markets. Cooked cured poultry products are normally produced using technology similar to that applied to some types of cooked ham. The process involves multi-needle injection, tumbling, restructuring and cooking. A brine of low NaCl content is used, especially with chicken products, to avoid a harsh metallic taste and sugars are also present for this purpose. Polyphosphates and, in some cases, binding agents, such as lactose, are present to enhance brine take-up and to minimize cooking losses. The finished product tends to be bland in taste and so brines may contain spices and flavour enhancers. Alternatively the product may be smoked, a liquid smoke usually being applied. Cooked cured poultry products are normally retailed in sliced pre-packed form, or as re-formed 'joints'. Small quantities of whole birds, halves and bone-in joints are also produced, some of which receive a very heavy smoke.

### 6.2.5 Tongue

Significant numbers of tongues are cured. Cattle (ox) tongue is most common in the UK but large quantities of lamb and sheep tongue are cured in Australia. Brines of similar composition to Wiltshire bacon brines are used, although nitrate is often omitted. Large tongues are usually artery pumped but multi-needle injection is widely used for smaller tongues. Yield is lower when multi-needle injection is used, although this may be offset to a small degree by addition of polyphosphates to the brine. Tongues are then placed in immersion brine for 2–3 days at *ca.* 1°C. There is normally no maturation period.

Tongue is traditionally retailed as a cooked meat, although small quantities are prepared for home cooking (see Chapter 4, pages 190–191). Cooking usually takes place immediately after curing and involves simmering at just below 100°C for *ca.* 2.5 hours for ox tongue or *ca.* 3.5 hours for sheep tongue. The skin and hyoid bone is then removed and the tongue packed into cans, usually with addition of gelatin. Process is to an $F_o$ value of 0.1–0.7. The vast majority of tongue is processed in bulk packs for slicing at point of sale or, more usually, pre-packing.

### 6.2.6 Bacon

Bacon is retailed as a raw commodity, although some types receive heating during processing. There is also small-scale production of cooked bacon for special purposes.

#### (a) 'Pasteurized' bacon

Small quantities of 'pasteurized' bacon are manufactured in which heating, usually during smoking, is sufficient to destroy vegetative micro-organisms. Pasteurized bacon is intended to have a long storage life in conditions where temperature control is inadequate and is primarily intended for special purposes, such as military or expedition use. The product may be vacuum packed after heating or packed in cans before processing. The eating quality is usually considered to be low.

### (b) Pre-fried bacon

Pre-fried bacon is a catering product designed for use in sandwiches. The product is fully cooked (cf. microwave bacon, Chapter 4, page 189) by frying and is not reheated before use. The product is vacuum packed for distribution in bulk packs. Quantities produced are currently very small and there appear to be unresolved problems due to poor texture.

## 6.2.7 Quality assurance and control

Quality assurance and control procedures for the initial manufacture of cooked cured meats are the same as those for the raw products (see Chapter 4, pages 198–199), while procedures for cooking and post-cook handling are the same as those for other cooked meats (see Chapter 5, pages 256–258). The exception is canned (pasteurized) hams, where canning technology is involved. It is necessary to ensure that air is evacuated to the correct level to ensure adequate heating and, of course, that processing is correctly applied. Post-process contamination appears to have been responsible for cases of staphylococcal food poisoning (see page 309) and seaming faults appear to have been a major contributory factor. Seam parameters should be checked at regular intervals through production and a graphical means used to detect at an early stage any significant deviation from the norm. It is also necessary to ensure that cans are protected from contamination at the critical stage immediately after removal from the retort when 'every can is a potential leaker'. This requires monitoring chlorination levels in cooling water, ensuring contact surfaces are correctly sanitized and that handling is minimized.

---

* In the canning industry as a whole, incubation of processed cans and examination for gas production are often used as a means of verifying correct processing. Specifications for canned hams may include this procedure, despite the fact that canned hams are not fully retorted products and that bacterial growth (and gas production) at elevated temperatures is therefore not neccesarily indicative of incorrect processing. Even in fully retorted products the validity of gas production as an index of bacterial growth and therefore non-sterility is dubious and incubation tests failed to detect contamination of cans of corned beef due to *Escherichia coli* and other members of the Enterobacteriaceae growing by nitrate respiration rather than fermentation. As a result the cans were released, followed by the 1964 Aberdeen typhoid outbreak.

## 6.3 CHEMISTRY

### 6.3.1 Nature of the pigments of cooked cured meats

The nature of the cooked cured meat pigment, nitrosylhaemochromogen, is of particular interest but remains enigmatic. Nitrosylmyoglobin, the pigment of uncooked cured meat, is formed as described in Chapter 4 (pages 202–203). During cooking of cured meat the protein of nitrosylmyoglobin is denatured and detached from haem. At the same time, a second mole of nitrite is incorporated. Assuming that nitrosylhaemochromogen is, as strongly suggested, a mononitrosyl complex, the second mole of nitrite must be bound to the denatured protein. The nature of the protein:nitrite complex is not known, but it is possible that nitrosylmyoglobin contains a protein radical which, on denaturation, reacts with an oxide of nitrogen or with nitrite itself. Further nitrosation reactions probably occur during prolonged heating, but may not be relevant to commercial practice.

As with uncooked cured meats, nitrihaemin may be formed as a consequence of excess levels of nitrite (nitrite burn). Green discolouration can also be a problem in cooked cured meats due to oxidation of the porphyrin ring by strong oxidizing agents, such as $H_2O_2$. This results in production of choleglobin, while further oxidation ruptures the porphyrin ring with production of a verdohaem complex, which is also green. Ultimately the haematin pigments may degrade to yellow or colourless pyrrole fragments (bile pigments). The most usual cause is production of $H_2O_2$ during growth of lactic acid bacteria (see page 311).

### 6.3.2 Role of nitrite in inhibiting the development of warmed-over flavour

It is well known that development of warmed-over flavour is very slow in cured cooked meats compared with uncured. In the past it was not known if nitrite itself or a reaction product of nitrite was the actual antioxidant, but it now seems probable that a number of nitrite-derived compounds are involved, acting through several different mechanisms and affecting both initiation and propagation of lipid oxidation.

Cooked cured meats contain different pools of nitrite-modified compounds, the most important of which are nitrosated cytochromes

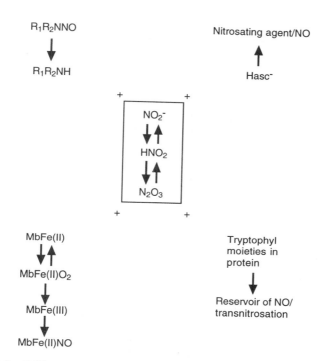

**Figure 6.1** Different pools of nitrite-modified compounds in cured meat and their dynamic interchange. (After Skibsted, L.H., 1992, in *Chemistry of Muscle-based Foods*, (eds D.E. Johnston, M.K. Knight and D.A. Ledward), Royal Society of Chemistry, London.)

and nitrosated tryptophyl moieties on the proteins, nitrosyl pigments and the unidentified nitrosating derivatives of ascorbic acid (see Chapter 4, pages 201–202). These appear to exist in a dynamic interchange system with each other, with residual nitrite and with $N_2O_3$ (Figure 6.1). The practical significance is that depletion of one group due to oxidation or involvement in antioxidative reactions will, under favourable circumstances, be followed by reformation of the antioxidant through transnitrosation reactions.

* It has sometimes been suggested that transnitrosation reactions occur only very slowly at the low temperatures used in commercial storage of cured meats. Isotopic labelling experiments, however, have shown that radical exchange reactions, such as the nitric oxide exchange in nitrosomyoglobin, have a low activation barrier. This indicates a low-temperature dependence for at least some types of transnitrosation reaction. If this finding is applicable to transnitrosation reactions in general, it suggests that reformation of active antioxidants in cooked cured meat does occur during storage at low temperature.

The early stages of lipid oxidation are generally considered to be of greatest importance in the ultimate development of warmed-over flavour. Nitrite-modified compounds may, therefore, be considered to be of greatest significance in affecting the initiation of oxidation by metal-ion catalysis and subsequent chain propagation.

### (a) Role of nitrite-modified compounds in protection against metal-ion catalysis

Iron(II) is known to be particularly active in generation of free radicals, although other metal ions have also been implicated:

$$Fe^{2+} + H_2O_2 \rightarrow Fe^{3+} + \cdot OH + OH^-$$

The antioxidative properties of $NO_2$ have been attributed to its function as a metal-ion chelator, although little is known of the nature of the nitrite iron complexes formed. There is evidence that chelation alone cannot account for the ability of $NO_2$ to inhibit the initiation of lipid oxidation and it is possibly an indirect effect resulting from interaction of $NO_2$ with haem pigments. This is attributed to its role, in conjunction with ascorbic acid, of preventing formation and accumulation of metmyoglobin. The lipids are thus protected against metmyoglobin-catalysed peroxidation.

### (b) Role of nitrite-modified compounds in chain breaking

Reactions yielding non-radical products require interaction between the radicals involved and other free radicals. There is strong evidence that nitrosylmyoglobin acts as a chain terminator. The nature of the products is not definitely known. It seems likely, however, that myoglobin and $NO_2$ are formed and regenerate the antioxidant nitrosylmyoglobin:

$$ROO\cdot + 3MbFe^{II}NO + 2H_2O \rightarrow ROH + 3MbFe^{II} + 3NO_2^- + 3H^+$$

---

* Nitric oxide is believed to modulate the reactivity of both haem and non-haem iron and to act as a scavenger during peroxidation of lipids. Although NO has a short life in the presence of oxidants, it is stabilized by binding in nitrosylmyoglobin and forms a readily mobilized free-radical buffer. (Kanner, J. *et al.*, 1991, *Arch. Biochem. Biophys.*, **289**, 130–8.)

## 6.4 MICROBIOLOGY

### 6.4.1 Cooked cured meats as an environment for micro-organisms

Cooked cured meats vary considerably as a medium for microbial growth. In the case of many modern products which are moist and of low levels of NaCl and $NO_2$, there is little difference from uncooked cured meats and only the more sensitive Gram-negative bacteria are inhibited. In such cases extrinsic factors (refrigeration and vacuum packaging) are responsible for stability.

Wiltshire-type and other more traditional hams have a higher level of stability as a consequence of higher levels of curing ingredients and a lower water content. Products manufactured today, however, are less stable than their historic counterparts. There is a selective effect favouring NaCl-tolerant bacteria, including *Staphylococcus aureus*, which is also favoured in many circumstances by the absence of a competitive microflora. Storage at low temperatures and in vacuum packs modifies the developing microflora and favours lactic acid bacteria. A few central European products that contain a high level of curing ingredients (and are heavily smoked and undergo a considerable drying during cooking) are highly shelf-stable and support the growth only of halotolerant strains of *Micrococcus*, yeasts and moulds.

For a number of years the stability of cooked cured meats with respect to outgrowth of endospores, including those of *Clostridium botulinum* was attributed to the 'Perigo factor'. This is a compound formed during autoclaving in nitrite-containing laboratory media, which has been shown to have powerful inhibitory effects against outgrowth of clostridial spores. These findings were extrapolated to cooked cured meats, the Perigo factor being (it was believed) a major factor in determining safety. A considerable amount of work was carried out in attempts to determine the

---

* Although deposition of phenolic compounds from smokes was an important factor in the past in determining the stability of both raw and cooked cooked meats, the effect of most modern smoking procedures is limited. Antimicrobial activity has been claimed for a number of liquid smoke preparations, but these claims must be treated with extreme care. It is certainly possible to demonstrate antimicrobial activity by *in vitro* assay of liquid smoke preparations, but it is rarely possible to detect any effect on the growth of micro-organisms on the meat.

nature of the Perigo factor and its inhibitory mechanism. Subsequently, it has been realized that the Perigo factor is of considerably greater importance in artificial media than in meats. Inhibition of *Cl. botulinum* stems primarily from the combined effects of NaCl, $NO_2$ and refrigeration, although polyphosphates can also be involved. The situation is complex and the polyphosphate curaphos 700, which at a concentration of 0.3% reduced toxin formation in a high pH value (6.3–6.8) cooked cured meat model system, slightly increased toxin formation in the same system at pH 5.5–6.3. Ascorbate and erythorbate (isoascorbate) may also be involved in inhibition of *Cl. botulinum*, although some of the evidence is conflicting.

### 6.4.2 Cooked cured meats and food poisoning

#### (a) Staphylococcus aureus

Cooked cured meats such as ham and tongue are classic vehicles of staphylococcal food poisoning. In the past a number of cases resulted from growth of *Staph. aureus* during two-stage ham boiling (see page 299). This process is considered inherently unsafe, but is now carried out on only a very small scale and for a number of years the most common cause has been contamination at slicing followed by storage at high temperatures. *Staphylococcus aureus*, favoured by the relatively high NaCl content, is able to grow very rapidly in air at high temperatures and critical levels of enterotoxin can be synthesized in as little as 6–8 hours. General use of refrigeration at retail level and the use of vacuum packaging for pre-packed products (see below) has reduced the incidence of staphylococcal intoxication due to ham, etc. although a substantial number of cases still occur. These are largely associated with catering, the classic scenario being a buffet meal on a summer day where the food is prepared and laid out several hours in advance of consumption. It is particularly unfortunate that these conditions often arise on special occasions, such as wedding

* *Staphylococcus aureus* food poisoning has affected airline passengers on several occasions, several types of food being involved. In a number of cases symptoms have been reported to be unusually severe, with a high percentage of those affected requiring hospital treatment after landing. Various reasons have been suggested, including the stress of flying, mass hysteria and the conditions of the atmosphere in the aircraft cabin. It should, however, be appreciated that although symptoms of *Staph. aureus* intoxication are generally considered to be 'mild', as many as 10% of those affected (in non-airborne outbreaks) seek hospital attention.

receptions. Staphylococcal food poisoning has also been associated with the ham filling of commercially prepared sandwiches, which had been stored at room temperature on a hot day. Airline catering has been involved and cooked ham used in preparation of an omelette was responsible for an outbreak of staphylococcal food poisoning which affected 197 passengers on a charter flight. Outbreaks of this type can be avoided by refrigeration. In outbreaks involving meals on commercial catering premises, it is likely that a major cause is preparation of buffets, etc. too far in advance, even when adequate refrigeration is available for bulk storage. This is usually to reduce the number of staff involved and in such situations it is not unfair to say that profit takes precedence over refrigeration.

*Staphylococcus aureus* is only rarely involved in food poisoning due to consumption of centrally sliced and vacuum packed cooked cured meats. This may be attributed, in part, to the lower level of handling associated with automated packing and to improved refrigeration. There is also an effect of vacuum packing that may be due to the inability of *Staph. aureus* to synthesize enterotoxins in conditions of reduced oxygen tension. Vacuum packing also favours the growth of competitive lactic acid bacteria, especially in relatively low NaCl products, and this is likely to be another significant factor.

Canned (pasteurized) hams were involved in a number of outbreaks of *Staph. aureus* food poisoning during 1989. The majority were processed in Romania and it is likely that incorrect seam sealing led to recontamination in a processing environment of generally poor hygiene standards.

### (b) Salmonella

*Salmonella* has been responsible for a number of outbreaks of food poisoning involving cooked cured meats. An outbreak in the UK due to *S. falkensee* during 1989, for example, caused 83 known cases. In this outbreak and in others, there was no specific connection with cooking of cured meats. On some occasions, uncured cooked meats have also been involved, problems arising from undercooking and/or cross-contamination. Cross-contamination can be a particular problem in small operations where separation of cooked and uncooked meats is physically difficult. *Salmonella* is unable to grow on Wiltshire-type and other more traditional products, but multiplication is possible on at least some types of sweetcure ham if temperature abuse occurs.

### (c) Listeria monocytogenes

In 1990, a major product recall was necessary due to a high level of contamination by *L. monocytogenes* on vacuum-packaged sliced ham. No cases of illness were known to have resulted, but political controversy followed allegations of unneccesary delay by the UK Department of Health. *L. monocytogenes* is relatively resistant to NaCl and $NO_2$ and can grow on some types of cooked ham held at low temperature. There are no known cases of listeriosis due to consumption of cooked cured meats and the risks are not specific to cured products.

### (d) Yersinia enterocolitica

A family outbreak of yersiniosis due to *Y. enterocolitica* serovar O3 occurred in Hungary following consumption of a cooked cured meat product, disznosajt. The product was locally made on a small scale and post-cooking contamination from raw to cooked product was the likely cause. Symptoms were predominantly extra-intestinal, including sore throats and fever.

### (e) Other bacteria

A number of other bacteria have been implicated in food poison-ing following consumption of cooked cured meats. Most frequent were members of the *Proteus–Providencia* group and *Enter-ococcus* (Group D *Streptococcus*). Symptoms tended to be mild and poorly defined and the validity of many reports has been doubted. The number of cases reported has fallen very con-siderably in recent years, which may be due to widespread use of refrigeration.

---

* The pathogenicity of *Enterococcus* has been a matter of debate for many years. The bacterium was originally considered to be a cause of mild gastroenteritis, but this became widely questioned and it was subsequently thought that *Enterococcus* was of little importance as a pathogen in normal circumstances. More recently *Enterococcus* has been recognized as an opportunistic pathogen in serious extra-intestinal disease, including endocarditis. Its importance as a nosocomial pathogen is also increasing with wider usage of broad-spectrum β-lactam antibiotics.

## 6.4.3 Spoilage of cooked cured meats

The vast majority of cooked cured meats are retailed as vacuum or modified atmosphere pre-packs. With occasional exceptions, the spoilage microflora is derived from post-process contamination. In many cases spoilage patterns are similar to those of uncured cooked meats. The microflora is dominated by species of *Carnobacterium* and *Lactobacillus*, spoilage being due to acid production and a 'sweet/sour' odour. In the later stages of spoilage this is accompanied by stickiness and slime formation. There may also be a fading of colour. The relationship between spoilage and numbers of bacteria present is not clear. Spoilage of ham has been noted when bacterial numbers (as determined by colony counts) were *ca.* $10^6$ cfu/g but, conversely, an identical product containing in excess of $10^8$ cfu/g can be organoleptically sound.

*Brochothrix thermosphacta* has been associated with a 'cheesy' odour in cooked cured meats, although spoilage by this organism is less common than in uncured meats, due to its relative sensitivity to NaCl and $NO_2$. Species of *Vibrio* may also be present in ham pre-packed close to the curing environment, but not where packing is at a remote site. *Vibrio* species have been associated with production of sulphide odours, but this type of spoilage appears rare and occurs only after significant temperature abuse. Temperature abuse also appears to be involved in production of 'faecal' odours by *Proteus* and *Providencia* although it seems likely that additional, unknown factors are operative.

'Greening' of the meat pigment is a problem restricted to cooked, nitrite-containing products. Discolouration occurs after exposure to oxygen on opening the pack and results from production of $H_2O_2$ by lactic acid bacteria under aerobic conditions. The cooked cured meat pigment, denatured nitrosylhaemochromogen, is oxidized by $H_2O_2$ (see page 304) resulting in fading to a grey-green or, less commonly, yellow or white colour. There appears to be considerable strain variation with respect to the quantity of $H_2O_2$ produced and thus the greening potential. It is possible that fermented meats, such as salamis, are a major source of strains with high greening potential in plants that pre-pack a wide range of cooked meats.

Canned, pasteurized (semi-preserved) hams are stable for *ca.* 6 months at storage temperatures below 5°C. Heat treatment does not kill endospores of *Bacillus*, or *Clostridium*, but outgrowth does not usually occur except under conditions of temperature abuse. There have, however, been unsubstantiated accounts of spoilage of low NaCl ham by psychrotrophic strains of *Clostridium*, apparently similar to *Cl. laramie*.

A number of processes applied in pasteurization of hams are marginal for the more heat-resistant strains of *Lactobacillus* and *Enterococcus*. Spoilage by *Lactobacillus* involves a similar pattern to pre-packed ham and souring is predominant. Greening due to $H_2O_2$ production by heat-resistant lactobacilli may occur but it is rare and spoilage of this type normally results from contamination of the ham after the can is opened. Heterofermentative species of *Lactobacillus* occasionally produce visible quantities of gas, but this is also rare. *Enterococcus faecalis* ssp. *liquefaciens* has weak proteolytic activity and can liquefy the gelatin on the surface of canned ham. In some cases, the organism survives only in the centre of the can and produces sufficient proteolytic enzymes for partial degradation of the muscle structure. This results in 'soft core' defect, which may be present in the absence of other detectable deterioration. At the same time *Enterococcus* in particular can attain high numbers ($>10^8$ cfu/g) in canned hams and apparently have no organoleptic effect.

### 6.4.4 Microbiological analysis

Methods used for uncooked cured meat are adequate, although analysis for halophilic species of *Vibrio* is not considered worthwhile. It is common to include analysis for *Staph. aureus*. Direct plating using Baird–Parker medium is recommended; enrichment is not considered worth while unless food poisoning is suspected. Canned hams are usually examined for *Enterococcus*. Kanamycin–aesculin–azide agar is probably the most satisfactory medium. Enrichment is not considered neccessary.

## EXERCISE 6.1

It has been stated that it is pointless to examine cooked cured meats (or other foods) for *Staph. aureus* since it is the enterotoxin and not the organism which is of importance. To what extent do you feel this statement is justified, bearing in mind the current available methodology for detecting staphylococcal enterotoxins. Discuss the significance of detecting the following numbers (cfu/g) of *Staph. aureus* in sliced, vacuum-packed ham:

$2 \times 10^2$ coagulase-positive;
$5 \times 10^3$ coagulase-positive;
$7 \times 10^5$ coagulase-positive;
$6 \times 10^6$ coagulase-negative.

# 7

# FERMENTED SAUSAGES

---

## OBJECTIVES

After reading this chapter you should understand
- The nature of the various types of fermented sausage
- The technology of their manufacture
- The role of lactic acid bacteria during fermentation
- The role of species of *Micrococcus* and *Staphylococcus* during fermentation
- Quality assurance and control
- Chemical changes occurring during fermentation, drying, etc.
- The nature and origin of the flavour and aroma
- Microbiological hazards associated with fermented sausages
- Spoilage of fermented sausages

---

## 7.1 INTRODUCTION

Fermented sausages are generally defined as consisting of an emulsion of meat and fat particles, NaCl, curing agents, spices, etc., which have been stuffed into a casing, fermented (ripened) and dried. In many cases, the finished sausage should be microbiologically stable at ambient temperatures.

Fermentation appeared to develop as a means of enhancing the production of dried meats, drying being the earliest form of preservation. The first use of fermentation is lost in the mists of antiquity, but the process is thought to have originated in China some 2000 years ago. The use of salt and nitrate came many years later, probably around the 13th century, the term salami (*salame*) being derived from the Latin *sale* – salt. In Europe, manufacture of fermented sausages first developed around the Mediterranean, subsequently spreading north and west. Little is known of fermented sausage production before *ca.* 1700. The technology was introduced into Hungary in 1851 and spread to the US with immigrants from central Europe.

---

### BOX 7.1  **I could eat a horse**

A distinct and high quality fermented sausage (salami) was developed in Hungary and consisted of pure pork meat stuffed into a horse or donkey intestine. The horse-loving British disapproved of this use of their erstwhile friends and the myth persists today, especially amongst children, that all salamis are made from donkey and horse meat. Horse meat is, however, used in some traditional eastern European products.

---

In Europe, manufacture of fermented sausages was largely a small-scale enterprise using traditional labour-intensive methods, but in the US, the development of the large-scale Chicago meat industry meant that a high level of automation was applied by the early years of the 20th century. In contrast to cheese, however, understanding of the principles of meat fermentation developed only slowly and it was not until the 1940s that the first attempts were made to put the fermentation process on a scientific basis, following the development of starter cultures. Even today there is a considerable craft element and in some countries the fermentation remains a spontaneous and uncontrolled process.

Fermented sausages are produced from a wide range of meat species, over a wide geographical range. A very large variety of types has developed, over 350 being produced in Germany alone, and while many are only minor variants, classification is difficult. A common basis for classification is the length of processing, final water content and final $a_w$ level. Using these criteria, three categories may be described: spreadable; sliceable, short processed; and sliceable, long processed (Table 7.1). If required, these categories may be subdivided on the basis of further criteria including fineness of chop, diameter of sausage, application of smoking and use of moulds in ripening.

Alternatively, fermented sausages may be classified simply as dry or semi-dry. Semi-dry sausages include both spreadable and sliceable, short processed. The division is not arbitrary; it has public health

**Table 7.1** Classification of fermented sausages

---

**Spreadable**
Process length: 3-5 days
Final $H_2O$ content: 34-42%
Final $a_w$ level: 0.95-0.96
Examples: Teewurst, frische Mettwurst (Germany)

**Sliceable, short processed**
Process length: 1-4 weeks
Final $H_2O$ content: 30-40%
Final $a_w$ level: 0.92-0.94
Examples: Summer sausage (US), Thuringer (Germany)

**Sliceable, long processed**
Process length: 12-14 weeks
Final $H_2O$ content: 20-30%
Final $a_w$ level: 0.82-0.86
Examples: Salamis (Germany, Denmark, Hungary), Genoa (Italy),
Saucisson (France), Chorizo (Spain)

---

*Note*: Based on Rocca, M. and Incze, K. (1990) *Food Reviews International*, **6**, 91-112.

---

### BOX 7.2 **Spiced dainties**

Fermented sausages are traditional products which offer little scope for development of variants or value-added products. Attempts have been made, on a small scale, to introduce new flavourings, but these have been unsuccessful. This results partly from incompatibility of added flavours with the intrinsic flavour of fermented sausages and partly from the conservative tastes of consumers. An exception is the successful marketing of 'salami-snack' products in the UK. These are small diameter, dry sausages, stable at ambient temperatures when vacuum packed in a foil envelope to prevent mould growth and reduce fat oxidation. The snacks are available strongly flavoured with a number of spice combinations. Salami-snacks are popular as an alternative to other savoury snacks, such as potato crisps (chips), but may also be preferred by young people as an alternative to sweets.

implications in that *Trichinella* is able to survive in semi-dry but not dry sausage. In the US, it is mandatory that semi-dry sausages which contain pork should be heated to a temperature of 58.3°C to inactivate the parasite. Similar legislation exists in some other countries.

It should be appreciated that, while fermentation and drying are inextricably linked in the products of the Mediterranean countries, central and northern Europe and the US, some Asian products are only minimally dried and depend entirely on low pH value for stability. These products include Mam, a traditional Thai fermented beef sausage.

## 7.2 TECHNOLOGY

The processing of all fermented sausages shares a basic technology (Figure 7.1). Many of the products are very similar, although there can be significant differences in the processing applied by different manufacturers to the same product. Some care is required when dealing with the less well known fermented sausages, since the same name may be applied to two distinct products.

### 7.2.1 Ingredients

*(a) Meat*

The initial mix used for manufacture of fermented sausages contains 50–70% lean meat. Any meat species may be used, choice being dictated by tradition and, in some cases, religious considerations. Pork is most widely used in northern, central and eastern Europe, the US and China, but elsewhere mutton and beef are more common. Chicken meat is also used in the manufacture of fermented sausages. Meat of any type should be of good quality and free from visible blemishes, such as blood splashes. The main factors affecting suitability are water-holding capacity, pH value and colour. Where pork is used, the initial pH value should be in the range 5.6–6.0. This assists the initiation of fermentation and ensures adequate fall in pH value. Dark, firm, dry meat is not suitable, but pale, soft, exudative meat may be used in manufacture of dry fermented sausages at 20% and, possibly, higher levels.

In many traditional processes, deboned meat is allowed to drain in cooling chambers before processing. Draining is thought to

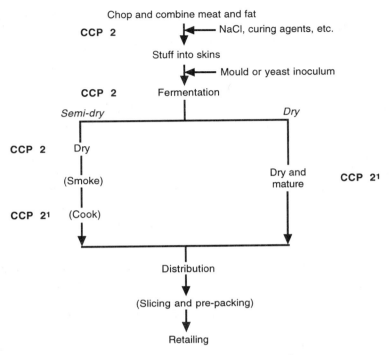

**Figure 7.1** Basic processing of fermented sausage. [1] Both drying during manufacture of dry fermented sausages and cooking during manufacture of semi-dry are intended to ensure inactivation of *Trichinella* but not bacterial pathogens.

improve the quality of the fermented sausage, although this is often disputed and the process involves significant extra refrigeration costs. Meat from old animals is often preferred and is considered to produce the highest quality products.

Meat should also be of good microbial quality to reduce competition at the onset of fermentation. This is of particular importance in the case of spreadable salamis, where high microbial numbers in the meat tends to result in an inconsistent end-product. This, together with poor functional properties, has limited the use of mechanically recovered meat in fermented sausages, though it has been used in production of spreadable sausage without adversely affecting the microbiological status of the end-product. Special precautions are necessary to prevent fat separation and to improve spreadability, these involving pre-emulsification with soy protein isolate or sodium caseinate. Although many fermented sausages are of high fat content, it is usual to add fat as a separate ingredient and the use of high fat meat is not recommended.

### (b) Fat

Fat is an important ingredient of fermented sausage and can comprise as much as 50% after drying. Fat oxidation can lead to rancidity and reduce the shelf life of the finished sausage. It is, therefore, important to use fat which is of high melting point and which has a low content of unsaturated fatty acids. Pork back fat is very widely used, since this has a low content of the poly-unsaturated linoleic and linolenic acids (8.5% and 1.0% of total fatty acids, respectively), which are highly prone to autoxidation. Soft fatty tissue from pigs fed on a diet high in unsaturates, however, causes colour and flavour defects and rapidly autoxidizes.

Fat of any type should not have been subjected to storage conditions which initiate early deteriorative changes. In some countries, the addition of antioxidants with the fat is permitted. Tertiary butyl hydroquinone and butylated hydroxyanisole are most widely used, but are under suspicion of causing adverse toxicological reactions. Increasing use is being made, where permitted, of ascorbyl palmitate (6-O-palmitoyl-L-ascorbic acid), nature-identical antioxidants, such as synthetic $\alpha$-tocopherol, and naturally occurring substances, such as 'extracts of natural origin rich in tocopherols'. The herb, rosemary, and spices such as garlic and mace also have antioxidant properties. Ascorbic acid may be added as an inhibitor of fat oxidation, but in many cases the prime role lies as a curing adjunct and in maintaining good colour.

### (c) NaCl and curing agents

NaCl is usually added to the mix at a concentration of 2.5–3.0%, which lowers the initial $a_w$ level to *ca* 0.96. Higher quantities are added to some Italian salamis, which may contain more than 8% in the dried product. In combination with sodium nitrite at a concentration of up to 150 mg/kg and at reducing pH values, this forms a powerful inhibitory system. NaCl also plays a role in solubilizing proteins, which form a sticky film around individual meat particles, while nitrite is of importance in determining the colour of fermented meats (see page 343) and retarding lipid oxidation. Although NaCl and $NO_2$ contribute to the stability of the finished sausage, the prime role in most cases is to create a selective advantage for lactic acid bacteria in the early stages of fermentation and to suppress the growth of undesirable micro-organisms. The role of nitrite in maintaining the stability of the end-product is greatest in semi-dry sausages of relatively high water content and $a_w$ level. Potassium nitrate may be added to fermented sausages in addition to, or in place of, sodium nitrite. Where used alone, levels of up to 600 mg/kg are added, but such quantities are often considered excessive. It is desirable to include a nitrate-reducing micro-organism in the starter culture to ensure sufficient nitrite is present.

Neither nitrate nor nitrite are added to the mixes of some traditionally made products, such as the Spanish *chorizos*. Small quantities of nitrate are derived from other ingredients, including garlic and paprika.

### (d) Starter culture

The use of starter cultures is almost universal in production of semi-dry fermented sausages and is widespread practice in production of dry fermented sausages in countries such as Germany. In other areas starter cultures are less commonly used, although their value is being increasingly recognized. Starter cultures mainly consist of lactic acid bacteria and are added to improve consistency and control of fermentation (see below). Where potassium nitrate is added without sodium nitrite, it is common practice to include a nitrate-reducing strain of *Micrococcus* or *Staphylococcus* in the starter culture. *Staphylococcus* may also be used in the absence of lactic acid bacteria.

### (e) Carbohydrate

Carbohydrate is often, but not invariably, added to the initial sausage mix. The purpose is to ensure that sufficient fermentable substrate is present to enhance the growth of lactic acid bacteria and the production of organic acids. The quantity and type of

---

* Many of the *Lactobacillus* present in meat fermentations possess both a nitrate reductase and a nitrite reductase. Two types of nitrite reductase may be present, haem-dependent and haem-independent. The end-product of the haem-dependent nitrite reductase is $NH_3$, whereas the haem-independent nitrite reductase produces $N_2O$ and $NO$, which contribute to colour formation. In either case, activity is limited and insufficient for manufacture of satisfactory sausages.

carbohydrate added represents a balance between the need to establish an effective lactic fermentation and the need to avoid an over-rapid fall in pH value. Glucose, which is rapidly metabolized, is usually added in combination with a more slowly metabolized oligosaccharide. Carbohydrate is usually added at a level of 0.4–0.8%. In sausages containing nitrate, but not nitrite, lower levels are used to reduce the rate of formation of organic acids and thus avoid early inhibition of the nitrate-reducing microflora. Conversely, up to 2% glucose is added to some semi-dry sausages manufactured in the US, to ensure rapid pH fall.

Results of a systematic study of the course of commercial meat fermentations has cast some doubt on the effect of adding carbohydrates. The acidification rate in a German sausage, made without an effective starter culture, was the same in mix containing either 0.8% or 1.6% sugar. Added sugars were either not utilized or only partially utilized. In contrast, sugars added at a level of 0.2% were fully utilized during fermentation of an Italian sausage. Excessively high levels of carbohydrates can lower the initial $a_w$ level sufficiently (in addition to the lowering by NaCl) to reduce the rate of fermentation.

In some countries, milk powder may be added as a source of lactose and potato flour as a source of starch. The prime role, however, is as a meat extender (see below) and problems may be encountered during drying as a result of the water-binding properties of these additives.

### (f) Acidulants

Acidulants are added to ensure a rapid reduction in pH value in the early stages of fermentation. The use of acidulants is considered by some experts to be essential for safety in fermented sausages made without the addition of starter cultures. Acidulants are often used in conjunction with starter cultures in spreadable sausages, where a particularly rapid fall in pH value is required. In other products, however, the use of both acidulants and starter cultures is often thought to lead to poor quality. Glucono-δ-lactone, which hydrolyses to gluconic acid, is preferred as acidulant by many manufacturers and may be added at levels of up to 0.5%. Glucono-δ-lactone is considered to interfere with nitrate reduction and aroma formation in dry sausages and is not generally used in these products. Direct acidification by addition of lactic or other organic acids is used in manufacture of some types of fermented sausage. The consistency of the mix is changed, due to coagulation of meat proteins, and this causes handling difficulties when filling into casings. This problem can be overcome by encapsulating the organic acids in a coating, such as partially hydrogenated vegetable oil. The coating is formulated to melt and release when the temperature is raised after stuffing. Release at room temperature is common, but in some cases the coating does not melt until high temperature drying at as high as 60–65°C. Glucono-δ-lactone may also be encapsulated and this avoids problems due to premature hydrolysis in the event of machine breakdown, or other delays.

### (g) Other ingredients

Spices of various kinds are added to the mix of most types of fermented sausage. These include black pepper, garlic, paprika, mace, pimento and cardamon. Chinese fermented sausages contain a particularly wide range of spices and some types are flavoured with wine. In addition to antioxidants, L-glutamic acid is permitted as a flavour enhancer in some but by no means all countries.

Meat extenders are permitted in some countries. Vegetable proteins, predominantly soya, have been most commonly used. Isolated soy protein has been used at levels of up to 5%, but adverse effects on quality result at levels greater than *ca.* 2%. Isolated soy protein has humectant properties and permits some reduction of the NaCl content. As an alternative to vegetable protein, blood products have been used. The final products are of acceptable quality, although changes in the pattern of fermentation may result in organoleptic differences with conventional formulations.

### 7.2.2 Preparation of the unfermented sausage mix

Although the texture of fermented sausages can vary widely, the vast majority of mixes are of the emulsion type. Two special factors must be taken into account: the need for the sausage to lose water

---

* Some spices, including red pepper and mace, have been found to stimulate organic acid production by lactic acid bacteria. This has been attributed to the provision of manganese, for which lactic acid bacteria have a high requirement. Manganese is required for enzyme activity, including that of the key fructose 1,6-diphosphate aldolase.

readily during drying and the high fat content of the mix. The lean meat is cut relatively coarsely at a temperature between $-4°C$ and slightly above $0°C$, to avoid strong binding of water. The fat is also cut while still frozen at a temperature of *ca.* $-8°C$. This minimizes smearing and the coating of meat particles with a layer of fat, which limits water loss during drying.

NaCl, curing agents and other ingredients are usually added after preparation of the basic meat/fat mix. Care must be taken to ensure even distribution of NaCl, etc. Pre-cured meat may be used and in some cases a mixture of uncured and cured meat is allowed to equilibrate before the fermentation proper.

Before fermentation commences, the mix is usually stuffed into a casing. Either natural or artificial (fibrous collagen) casings can be used, but the type can markedly affect quality of mould-ripened sausages. Comparisons of sausages made with different types of casing have shown, for natural casings, significantly greater colonization and penetration by *Penicillium* and enhanced growth of yeasts. Mould-ripened sausages made with natural casings are of better flavour and aroma and ripen more evenly. By implication, however, spoilage strains of moulds and yeasts may develop more readily where natural casings are used for non-mould ripened sausages. Irrespective of type, the casing must allow evaporation of water, permit penetration of smoke (where used) and follow shrinkage during drying. Care must be taken during stuffing to ensure that the casing is adequately filled to avoid quality faults. Mechanical stuffing gives most satisfactory results, but auger-type equipment can lead to fat smearing and so vacuum stuffers are preferred. These have the additional advantage of reducing the amount of air present and ensuring correct pigment formation. The mix must be maintained at $0-1°C$ during stuffing to minimize fat smearing. A few types of fermented sausage are not stuffed into casings but are pressed into moulds.

### 7.2.3 Inoculation with mould or yeast

Growth of mould or yeast on the outer casing of many dry fermented sausages plays an important role in development of organoleptic characteristics. Moulds or yeast also play a role in suppression of undesirable bacteria, protection of the sausage from light and oxygen and in producing catalase. Traditionally the micro-organisms have been adventitiously derived from the factory environment (in many cases a factory-specific monoculture develops). Reliance on adventitious inoculation, however, can lead to inconsistent quality, while the development of a mycotoxigenic mould is always a possibility. For these reasons, inoculation of the sausage with a pure culture of a non-mycotoxigenic strain is now common practice. In most cases the sausage is inoculated directly after stuffing, although inoculation may be delayed until after the commencement of drying.

Commercially available mould cultures are species of *Penicillium*. *Penicillium nalgiovense* is most widely used, but cultures of *P. expansum* and *P. chrysogenum* are also available. *Debaryomyces* is the most commonly used yeast culture, and a strain of *Candida* is also available. Mould cultures are supplied as freeze-dried spore suspensions, and yeast cultures as freeze-dried cells. Suspensions are prepared in vats of water into which the sausages are dipped. This is a simple and effective means of inoculation, but vats and ancillary equipment must be rigorously sanitized to prevent contamination with moulds from the environment.

### 7.2.4 Fermentation

In a narrow sense, fermentation can be considered to be that stage at which active growth and metabolism of lactic acid bacteria is occurring, with an accompanying rapid fall in pH value. It must be appreciated, however, that other important changes occur during this stage. Further, growth of lactic acid bacteria usually continues during drying and smoking of semi-dry sausages. In the case of dry sausages, fermentation occurs concurrently with the initial stages of drying and the activity of microbial enzymes persists even when conditions do not permit growth of the micro-organism.

### (a) The role of micro-organisms during fermentation

As in production of other fermented foods, such as cheese, fermented milks, dill pickles and sauerkraut, lactic acid bacteria play a key role. *Pediococcus*, homofermentative species of *Lactobacillus*

---

* Studies with Italian salamis have shown that the most commonly isolated moulds from spontaneously inoculated sausage belonged to highly mycotoxigenic species. *Penicillium verrucosum* var. *cyclopium* was most common on sausages produced on a small scale, while *Aspergillus candidus* was most common in large-scale production. (Grazia, L. *et al.*, 1986, *Food Microbiology*, **3**, 19–25.)

and, to a much lesser extent, *Lactococcus* are of prime importance. Heterofermentative lactic acid bacteria are undesirable due to production of gas and atypical flavour compounds and, in the case of some species of *Leuconostoc*, slime formation. In practice, heterofermentative lactic acid bacteria, although common in some types of fermented sausage, are usually present only in relatively small numbers. Some strains of the homofermentative *Lb. sake* also produce slime. While this appears to have no significant effect on the quality of the sausage at levels of $10^7$ cfu/g, it is considered undesirable.

The major role of lactic acid bacteria is production of organic acids, primarily lactic acid, from carbohydrates, via the Embden–Meyerhof pathway. This lowers the pH value and contributes to the inhibition of undesirable micro-organisms. The lowering of the pH value is also an important factor in reducing the water-holding capacity of proteins and thus in ensuring that drying proceeds correctly.

In recent years, there has been considerable interest in the production of inhibitors (in addition to organic acids) by lactic acid bacteria from sausage fermentations. In some cases, inhibition involves relatively non-specific mechanisms, such as hydrogen peroxide production, reduction of the redox potential and competition for nutrients, especially amino acids and the vitamins niacin and biotin. Many lactic acid bacteria also produce bacteriocins which are often highly specific in activity. Most interest obviously lies in bacteriocins inhibitory to pathogenic micro-organisms, such as *Listeria monocytogenes*. Bacteriocins produced by *Pediococcus* are effective against *L. monocytogenes* before significant fall in pH value occurs, although activity is enhanced at lower pH values.

Species of *Micrococcus* and coagulase-negative *Staphylococcus* are important in some types of fermented sausage in reducing nitrate

---

* The genus *Leuconostoc* is recognized as being heterogeneous and comprises three distinct genetic lines, *Leuconostoc sensu stricto*, the *Leuconostoc paramesenteroides* group (which includes a number of 'atypical' lactobacilli) and the single species, *Leuc. oenos*. It has been proposed that *Leuconostoc paramesenteroides* and related bacteria should be reclassified in a new genus, *Weissella*. This new genus includes a number of 'atypical' lactobacillus, including *Weisella confusus* (*Lact. confusus*) and *W. viridescens* (*Lact. viridescens*). *Leuconostoc*-like isolates from Greek fermented sausage have been found to form a distinct genetic line and have been placed in a new species, *Weissella hellenica*. (Collins, M.D. *et al.*, 1993, *Journal of Applied Bacteriology*, 75, 595–603.)

to nitrite. Many bacteria present in fresh meat, such as *Pseudomonas*, are also capable of nitrate reduction but are unable to grow during fermentation. Nitrite is a major constituent of the inhibitory system of fermented sausages and is also important in producing the characteristic pigmentation. Catalase production by the organisms is important in preventing discolouration due to peroxide-forming strains of lactic acid bacteria and in reducing fat oxidation. Some species of *Lactobacillus* have a significant level of catalase activity. *Micrococcus* and *Staphylococcus* are thought to be important as a source of lipolytic and proteolytic enzymes during ripening. The relative role of enzymes from micro-organisms and those derived from meat in flavour production is a matter of debate (see pages 340–343) but *Staphylococcus carnosus* does appear to make a positive contribution to the flavour of fermented sausages. Growth of some members of the indigenous meat microflora before significant lactic acid production during fermentation also appears to improve quality of dry sausages.

## (b) Natural fermentations

In some cases, natural fermentations rely entirely on the presence of lactic acid bacteria in the raw meat. Lactic acid bacteria are invariably present but initial numbers are very low, unless the meat has been stored for some time in vacuum packs. Initial conditions select against the dominant Gram-negative microflora of meat and favour Gram-positive bacteria, including *Micrococcus* and both coagulase-positive and coagulase-negative species of *Staphylococcus*, as well as lactic acid bacteria. There is evidence that initiation of the lactic fermentation involves a succession commencing with members of the Enterobacteriaceae followed by *Enterococcus* and finally by *Lactobacillus* and *Pediococcus*. A

---

* Bacteriocins are toxins produced by a range of bacteria. The majority are peptides, which attach to specific sites and inactivate sensitive cells, usually of Gram-positive bacteria. Bacteriocin production in many species of bacteria is plasmid-encoded, but this has not currently been demonstrated in *Lactobacillus*, except *Lb. acidophilus* and *Lb. sake*. Bacteriocin-producing lactic acid bacteria appear to be common in sausage fermentations and it is assumed that their presence enhances safety. There is also considerable interest in the use of bacteriocins to control pathogens, such as *L. monocytogenes*, in non-fermented meat products and other foods. Further work, however, is required to clone the gene encoding bacteriocin production and to identify the responsible plasmid. (Garriga, M. *et al.*, 1993, *Journal of Applied Bacteriology*, **75**, 142–8.)

wider range of bacteria is able to develop where nitrate is added in the absence of nitrite and this apparently improves the quality of dry fermented sausages. Where the fermentation proceeds correctly, growth of lactic acid bacteria is rapid and levels of $10^6$–$10^8$ cfu/g are present after 2–5 days fermentation. The corresponding fall in pH value results in the death of *Pseudomonas* and other acid-sensitive Gram-negative rods within 2–3 days, although more acid-tolerant genera, including *Salmonella*, may persist for extended periods. Numbers of lactic acid bacteria tend to decline after reaching the maximum, although in mould-ripened sausages a second period of growth often occurs after *ca.* 15 days, coincident with a rise in pH value due to metabolism of lactate. Delays in initiating the lactic fermentation and slow fall in pH value favour the growth of *Staph. aureus* and the possibility of enterotoxin production. Poor flavour may also result from the growth of other undesirable micro-organisms.

Attempts to improve the reliability of the fermentation originally involved the process of 'backslopping', in which a new mix was inoculated with material from the previous production cycle. Backslopping is still used and does increase the reliability of fermentation, but the process cannot be considered to be fully satisfactory. In the first place, lactic acid bacteria in the backslopped material may be in a physiologically attenuated state and unable to initiate a new fermentation rapidly. Secondly, the uncontrolled nature of the backslopping process can mean that lactic acid bacteria with undesirable characteristics, such as peroxide formation, become dominant, with consequent adverse effects on quality of the sausages.

Species of *Lactobacillus* are the dominant lactic acid bacteria in natural sausage fermentations. The most commonly occurring species are *Lb. bavaricus*, *Lb. curvatus*, *Lb. farciminis*, *Lb. plantarum* and *Lb. sake*. *Lactobacillus sake* is probably of greatest overall importance, followed by *Lb. curvatus*. Undesirable species, such as the $H_2O_2$-producing *Lb. viridescens*, may also be present in significant numbers. *Pediococcus* is usually secondary to *Lactobacillus*, but is numerically dominant during some fermentations. *Pediococcus damnosus*, *Pd. acidilactici* and *Pd. pentasaceous* are all of importance. *Lactococcus* and *Leuconostoc* are usually present in small numbers only, although large numbers of *Leuconostoc* have been associated with poor quality sausages.

## (c) Fermentations with starter cultures

The course of fermentations initiated by starter cultures of lactic acid bacteria is essentially the same as that of successful natural fermentations, although lactic acid bacteria may achieve domination rather earlier. Although strains of lactic acid bacteria selected for use as starters in fermented sausages must fulfil a number of criteria (Table 7.2), the ability to compete successfully with the indigenous bacteria of the meat is a prerequisite. Competitive ability, however, is influenced by a number of factors in addition to the intrinsic properties of the starter strain (Table 7.3).

**Table 7.2** Criteria for suitability of lactic acid bacteria as starter cultures in sausage fermentation

1. Must be able to compete effectively with indigenous lactic acid bacteria.
2. Must produce adequate quantities of lactic acid.
3. Must be tolerant of NaCl and able to grow in a concentration of at least 6%.
4. Must be tolerant of $NaNO_2$ and able to grow in a concentration of at least 100 mg/kg.
5. Must be able to grow in the temperature range 15–40° C, with an optimum in the range 30–37° C.
6. Must be homofermentative.
7. Must not be proteolytic.
8. Must not produce large quantities of $H_2O_2$.
9. Should be catalase-positive.
10. Should reduce nitrate.
11. Should enhance flavour of the finished sausage.
12. Should not produce biogenic amines.
13. Should not produce slime.
14. Should be antagonistic to pathogenic and other undesirable microorganisms.
15. Should be tolerant of, or synergistic with, other starter components.

* Studies of the course of a German sausage fermentation, showed that the starter culture, a strain of *Lb. plantarum*, showed no growth, presumably due to inability to compete with intrinsic lactobacilli. Lack of competitive ability may also be the reason why starter strains developed in northern Europe tend to be ineffective when used in fermentations in southern Europe. Competitive ability is, however, difficult to predict and strains of *Lb. curvatus* and *Lb. sake* have been isolated from sauerkraut which were effective competitors against the meatborne microflora during sausage fermentations. (Marchesini, B. *et al.*, 1992, *Journal of Applied Bacteriology*, **73**, 203–9; Vogel, R.F. *et al.*, 1993, *Journal of Applied Bacteriology*, **74**, 295–300.)

**Table 7.3** Factors affecting the competitive ability of starter strains of lactic acid bacteria

1. The initial number of indigenous lactic acid bacteria in the mix
2. The nature of indigenous lactic acid bacteria
3. Size of inoculum of starter lactic acid bacteria
4. The physiological state of starter lactic acid bacteria
5. The mix formulation.

A large number of species of lactic acid bacteria have been used as starters in sausage fermentations (Table 7.4), of which *Lb. plantarum*, *Lb. curvatus*, *Pd. damnosus* and *Pd. acidilactici* are currently most widely used. All of the commonly used lactobacilli are species which are of importance in natural fermentations, but attempts to use dominant indigenous strains as starters have tended to be unsuccessful.

The technology of starter cultures for sausage fermentations is crude compared with that of starters used in dairy fermentations. Despite this, there is increasing interest in the use of genetic modification to produce strains of improved performance, or with new technologically desirable attributes (Table 7.5).

**Table 7.4** Lactic acid bacteria commonly used as starter cultures in sausage fermentations

*Lactobacillus*
  *Lb. curvatus*, *Lb. jensenii*, *Lb. plantarum*, *Lb. sake*
*Pediococcus*
  *Pd. acidilactici*, *Pd. damnosus*, *Pd. pentasaceous*
*Lactococcus*
  *Lc. lactis* sub sp. *lactis*, *Lc. lactis* sub sp. *diacetylactis*

**Table 7.5** Possible genetic modification of lactic acid bacteria used as starter cultures in sausage fermentations

1. Enhancement of acid production at temperatures below 15° C
2. Ability to grow in higher NaCl concentrations
3. High level of nitrate reduction[1]
4. Production of desirable aromas[1]
5. Enhancement of competitive ability through utilization of special carbohydrates

[1]These features are attributes of *Micrococcus* and *Staphylococcus*.

Strains of *Micrococcus* and *Staphylococcus* used as starter cultures produce only very limited amounts of acid and are primarily added to reduce nitrate to nitrite and improve colour production. Various *Micrococcus* isolates have been used as starters, including a 'fermentative *Micrococcus*', but *M. varians* is now most common. In recent years *Staph. carnosus* has become very widely used, although *Staph. xylosus* is also available as a starter culture. *Staphylococcus carnosus* is considered to improve colour and aroma even in sausages made with nitrite, where nitrate reduction is of no significance. Many commercial meat starters now contain both lactic acid bacteria and *Staph. carnosus*, although lactic acid bacteria alone are widely used, especially in semi-dry sausages of low pH value. Equally *Staph. carnosus* is used alone in some types of dry sausage. Where starters contain both lactic acid bacteria and *Staphylococcus*, it is essential that the organisms are tolerant of each other or, preferably, grow synergistically. *Debaryomyces hansenii* may also be included in the starter rather than applied to the surface of the sausage. The yeast is considered to improve flavour and colour and contributes to inhibition of *Staph. aureus*. At the same time, nitrate reduction by the indigenous microflora is reduced and this can lead to serious defects in dry sausages made without a nitrate-reducing starter culture.

Bacteriophages are a major problem with starter cultures used in the dairy industry, especially in cheese manufacture. In contrast, there is little knowledge of the incidence of phages, or its practical significance, in sausage fermentations. A number of *Lb. plantarum* phages have been described and delay in acid production due to presence of such a phage has been demonstrated, when the homologous *Lb. plantarum* was used as starter. The industrial significance of this finding has been doubted and some microbiologists consider that phage infection is rarely, if ever, responsible for problems in practice. Phages capable of lysing starter strains of *Staph. carnosus* have been isolated in German and Italian factories. Products made with phage-infected cultures were of normal colour, taste and texture.

* Although *Staph. xylosus* is not a recognized pathogen, enterotoxin production has been reported in isolates from sheep milk. In contrast, *Staph. carnosus* is one of the very few species of *Staphylococcus* that has never been considered to be an actual or potential pathogen of humans and animals.

From the limited information available, it appears that phages are relatively rare in sausage fermentations and, when present, are of limited significance. It is possible that this is due to the relatively low rate of growth of starter bacteria in sausage fermentations reducing susceptibility to phage attack and to the physical nature of the sausage limiting spread of phage. In this context, it has been suggested that phage infection of *Staph. carnosus* may be responsible for a defect in which small discoloured spots appear below the sausage casing. Direct evidence is currently lacking.

A number of other micro-organisms have been proposed as starter cultures in sausage fermentations, including poorly defined species of *Vibrio*. Proposed starter strains have been identified with *V. costicola*, but doubts concerning the safety of these micro-organisms have precluded industrial use. Surprisingly, perhaps, the mycelium-forming bacterium *Streptomyces griseus* has been proposed as a starter bacterium and is claimed to improve flavour and aroma.

### (d) Technology of sausage fermentations

Starter cultures are usually supplied in freeze-dried form and require reconstitution in water before addition to the sausage mix. Reconstituted cultures can be used directly, but it is usual to resuscitate the organisms by holding the reconstituted starter for 18–24 hours at room temperature. An inoculum level of $10^6$–$10^7$ cfu/g mix is considered necessary, and levels of up to $10^8$ cfu/g mix are used in short, high temperature fermentations.

The temperature of fermentation varies according to the type of sausage. In general, higher temperatures are used where a fast pH fall is required, an increase of 5°C being considered to double the rate of acid formation, but increase the possibility of growth of pathogens, especially *Staph. aureus*, if the initiation of fermentation is delayed. The temperature of fermentation also appears to alter the relative quantities of lactic and acetic acid produced, lactic acid production being favoured by higher temperatures. In commercial practice, the temperature and length of fermentation can vary considerably. In general, however, dry sausages are fermented at 15–27°C for 24–72 hours, spreadable sausages at 22–30°C for up to 48 hours and semi-dry sliceable sausages at 30–37°C for a period as short as 14 hours and as long as 72 hours. Temperatures of less than 10°C, however, are used in fermentation of Hungarian salami, while semi-dry summer sausage of low pH may be fermented at temperatures as high as 40°C. Even higher temperatures have been used in the US, but these have not been entirely successful.

---

### BOX 7.3  **Sumer is icumen in**

In industrialized countries accurate control of temperature during fermented sausage manufacture is available, but elsewhere ambient temperatures must still be used. In many parts of Europe, before the advent of refrigeration, manufacture of sausages was suspended during the hottest summer months. In some cases the season dictated the type of fermented sausage produced, dry, fully ripened sausages being produced in winter and semi-dry, high acidity sausages of short shelf-life ('summer' sausages) being produced in the warmer months.

---

Control of relative humidity is important during fermentation, both to initiate drying and to prevent excessive surface growth of yeasts and moulds. At the same time it is necessary to avoid formation of a hard outer layer. Short fermentations at high temperatures are carried out at a relative humidity of 98%, but at lower temperatures a general rule is that the relative humidity in the fermentation chamber should be between 5 and 10% lower than that in the sausage interior (*ca.* 90%).

In modern practice sausages are fermented in enclosed chambers with a high level of control of temperature and humidity. Light smoking may be applied at this stage, but must not interfere with the course of the fermentation. In previous years, the inability to control temperature and humidity led, in some countries, to special precautions against spoilage of the sausage during fermentation. These precautions are now technologically redundant but may still be applied during manufacture of some traditional speciality products to obtain distinctive organoleptic qualities. In Scandinavia, for example, fermentation (and maturation) of some types of sausage is carried out under brine, while in Germany the *Schwitzreifung* process is sometimes still used. This involves fermentation at 25°C and high relative humidity, microbial growth being washed off the surface at frequent intervals.

The extent of acidification varies according to the product, but is greatest in semi-dry sausages, especially those made in the US where the pH value is often significantly below 5.0. Dry German sausages usually have a pH value in the range 5.0–5.5, but the extent of acidification is limited in other dry sausages, such as Italian salami. The presence of minor ingredients can affect the extent of acidification. Antioxidants, such as tertiary butylhydroquinone, reduce the pH fall, although the antioxidants themselves may have inhibitory activity against *Listeria monocytogenes* and other pathogens. Acid production is greatest in vacuum-filled sausages and is also greatest in large-diameter sausages, where oxygen is limited. Ammonia production, however, is also enhanced in large-diameter sausages and counterbalances the lowering of pH value by lactic acid production.

### 7.2.5 Drying and maturation

The extent of drying varies considerably and is a major factor in determining the physico-chemical and organoleptic properties of the sausage, as well as its storage stability. In the case of dry sausages, which are not subject to heat treatment, drying is a critical control point with respect to control of *Trichinella*.

In all cases it is necessary to control drying so that the rate of water loss from the surface of the sausage is equal to the rate at which moisture migrates from the sausage interior. In the case of semi-dry sausages, drying loss is less than 20% (by weight) and temperatures between 37 and 66°C are used at a relative humidity as low as 0%. At the higher temperatures drying is complete in a few hours, but a period of several days is required at lower temperatures. Rapid drying is only possible where the pH value is low and the correspondingly low solubility of proteins facilitates moisture loss. Drying at high temperatures can involve a single-stage process but in other cases the drying is conducted in several stages with decreasing relative humidity. Heating to inactivate *Trichinella* is usually the final processing stage before cooling; it is a critical control point and should be monitored accordingly. Cooling usually involves reducing the temperature to 1°C over a period of *ca.* 24 hours.

Drying of dry sausages is an extended process, the length being at least partly dictated by the diameter of the sausage. Drying is carried out at low temperatures, the final temperature being

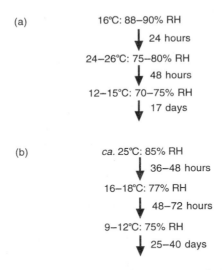

(a)

16°C: 88–90% RH

↓ 24 hours

24–26°C: 75–80% RH

↓ 48 hours

12–15°C: 70–75% RH

↓ 17 days

(b)

*ca.* 25°C: 85% RH

↓ 36–48 hours

16–18°C: 77% RH

↓ 48–72 hours

9–12°C: 75% RH

↓ 25–40 days

**Figure 7.2** Examples of temperature regimens during drying of fermented sausage.

typically in the range 12–15°C. It is common practice, however, to use higher temperatures in the initial stages, the temperature then being progressively lowered as drying continues (Figure 7.2). The relative humidity is also progressively lowered and is usually maintained at *ca.* 10% below that in the sausage.

Many types of semi-dry sausages are smoked during drying and smoking is also applied to dry sausages, which do not have a secondary mould or yeast microflora. Smoking is intended to inhibit mould growth by drying of the surface and by deposition of anti-microbial phenols, carbonyls and low molecular weight organic acids. Phenolic compounds also reduce the extent of fat oxidation, while smoking also has a significant effect on the organoleptic properties of the sausage. Although surface drying is desirable, it is essential that there should be no interference with the removal of water. Wood smoke, usually oak, is still widely used, conditions varying from 2 hours at 60°C and 57% relative humidity for semi-dry summer sausage produced in the US to a total of 160 hours at 8–18°C and 75–90% relative humidity for dry sausage (*chorizo*) produced in Spain. Liquid smoke has been introduced for some semi-dry sausages, but while this has some anti-microbial effect, inhibition of yeast and moulds is limited and vacuum packaging is often used to confer stability. The organoleptic nature of the sausage is inevitably changed and use of liquid smoke is uncommon in traditional markets.

In the case of dry sausages, especially those with surface mould or yeast growth, extensive chemical changes occur during drying, which may be considered as maturation. In some cases, drying is completed at a relatively early stage, maturation continuing up to consumption. Maturation affects the organoleptic properties of the sausage, imparting a characteristic aroma and flavour. The major reactions are lipolytic and proteolytic (see pages 340–343). Surface growth of moulds and yeasts modify the maturation process. Moulds have a particularly marked effect, with hyphae penetrating through the skin and deep into the meat mix. Metabolism of lactic acid results in a rise in pH value, while moisture content tends to be more constant, avoiding surface dehydration.

In some countries, fermented sausages are still dried by laying out in direct sunlight, or by hanging at low temperatures in natural caverns. In industrialized nations, sausages may be moved through a series of chambers maintained at constant temperature and humidity. Alternatively the sausages may remain in a single chamber in which temperature and humidity are changed for each stage of drying. In small-scale production of semi-dry sausages, the products remain in the same chamber throughout the entire production cycle of fermentation, drying and smoking. Microprocessors are now widely used in control of temperature and humidity. In all cases care must be taken to ensure that no localized areas of high moisture content remain in the sausage and an airflow of *ca*. 1 metre/second is required.

### 7.2.6 Packaging

Many traditionally made fermented sausages are minimally packaged, being placed in cardboard cartons to provide a measure of protection during transport and storage. Some types are placed in cloth bags or plastic shrouds, which offer individual protection. Vacuum packing of the whole sausage is applied to some semi-dry types. In some cases the packing is intended to provide protection during transport and is removed before retail display. Vacuum packing can lead to migration of moisture to the surface and rapid growth of yeast or mould after opening the pack.

Fermented sausages are now frequently sliced and pre-packaged before retail sale. Vacuum packaging is widely used and is effective in maintaining the colour of the sausage and in minimizing fat oxidation. Sausages should be sliced at low temperatures to avoid fat smearing and consequent poor appearance. The use of low temperatures also minimizes the possibility of fat contaminating the plastic film and causing problems of heat sealing. The use of high intensity display lighting can cause colour fading. Modified atmosphere packaging is also used but is unnecessary with respect to microbial stability.

### 7.2.7 Quality assurance and control

Manufacture of fermented sausages has been effusively described as an 'art' and the sausage maker as a 'true artist'. Such descriptions are misleading, although at least some craft elements are maintained, and modern processing requires a high level of technical control. The use of the HACCP system is recommended and no fewer than 19 control points have been identified. The most important are those directly concerned with control of pathogenic micro-organisms and, in the case of pork products, *Trichinella*.

The two main microbiological concerns with respect to safety are the possibility of growth and enterotoxin production by *Staph. aureus* during fermentation and the survival of pathogens such as *Salmonella* and *Listeria* (see pages 345–346). Acid production during fermentation is the critical control point for inhibition of *Staph. aureus*. The American Meat Institute (AMI) has defined a pH value of below 5.3 as limiting and many control systems are based on this pH value being obtained during fermentation. It is also necessary to take account of the possibility of growth of *Staph. aureus* before the limiting pH value is reached and the AMI consider the relationship between temperature and time taken for the pH value to fall to 5.3 to be a critical control point. The Institute has produced good manufacturing practice guidelines, which define correct processes in terms of time:temperature relationships.

American Meat Institute guidelines utilize a concept of 'degree' × hours which is the product of the the fermentation temperature minus 15 (°C) multiplied by the length of the fermentation (hours). Correct processes must satisfy the following conditions:

1. Less than 720 degrees × hours when the highest fermentation temperature is less than 32°C.
2. Less than 560 degrees × hours when the highest fermentation temperature is between 32 and 40°C.
3. Less than 500 degrees × hours when the highest fermentation temperature is greater than 40°C.

It must be stressed that when fermentation temperature is variable, calculations must be based on the maximum temperature and *not* the average temperature. The guidelines are equally applicable where fermentation is at a constant temperature.

The AMI Guidelines have been considered to be effective in practice. No account is taken of the contribution of the reduction in pH value before the target of 5.3 is obtained and it is likely that, in many cases, a substantial safety margin exists. However, the safety of some processes, where acidification is deliberately slow, is marginal.

Correct control of the fermentation stage eliminates the possibility of growth of *Staph. aureus* to critical levels and fermentation is a critical control point with respect to this organism. The AMI guidelines have been applied to other potential pathogens, on an empirical basis. The risk from pathogens such as *Salmonella*, however, involves survival as well as growth and thus control by fermentation is not absolute.

Temperature during fermentation and pH drop should be monitored. Management procedures should be in place to ensure that fermentations are not arbitrarily extended if insufficient fall in pH value has occurred. Staff should be experienced and aware of the importance of this stage in assuring safety.

Drying and maturation of sausages is an important stage with respect to ensuring character and quality. In the case of dry pork-containing sausages, the drying stage is critical for the control of *Trichinella* and control of the length and effectiveness of drying must reflect this. This requires control and monitoring of air humidity, air flow and temperature during drying. Reduced $a_w$ level may reduce the survival of some microbial pathogens, but elimination is not possible.

Drying of semi-dry sausages cannot guarantee the inactivation of *Trichinella* and heating to a minimum temperature of 58.3°C is required to ensure safety. Temperatures attained should be monitored to ensure correct heat treatment. This stage is not intended to eliminate vegetative microbial pathogens and any reduction in numbers that occurs is incidental. A full cooking process is applied to a small number of types of lightly dried fermented sausage with the intention of increasing stability and eliminating pathogens such as *Salmonella*. In such cases, control and monitoring must ensure correct heating and prevention of cross-contamination (see Chapter 5, pages 256–258).

End-product testing is applied to fermented sausages as a means of ensuring that compositional standards have been met. This involves determination of moisture and fat content and of levels of NaCl and curing agents. pH value of the end-product is of secondary importance to the pH value at the end of fermentation, but may be determined. In some cases $a_w$ level is also measured at the end of drying and maturation. Microbiological examination is usually restricted to ensuring that significant growth of *Staph. aureus* has not occurred.

## 7.3 CHEMISTRY

### 7.3.1 Nutritional properties of fermented sausages

The high fat, NaCl and nitrite content of fermented sausages leads to their being regarded as 'unhealthy' foods. In less developed countries, this must be set in the context of the continuing value of fermentation as a means of preservation of meat and the importance of fat as a source of dietary energy.

It has been suggested that lactic acid bacteria in fermented sausages may have a therapeutic role similar to that claimed for lactic acid bacteria in fermented milk products. At present, there appears to be no evidence for or against a therapeutic role for the lactic acid bacteria of fermented sausages, or for the expression of therapeutic properties in the sausage itself.

There is evidence that proteolysis in fermented sausages increases digestibility. Net protein digestibility of a pork and beef sausage fermented for 22 days, for example, increased from 73.8% to 78.7%, while protein digestibility increased from 92.0% to 94.1%.

## 7.3.2 Chemical and physical changes during manufacture of fermented sausages

### (a) Carbohydrate metabolism and pH value

Fermentation of carbohydrates commences soon after preparation of the sausage mix. As a general rule, *ca.* 50% of glucose is metabolized during active fermentation, *ca.* 74% of which is accounted for by organic acid formation. Lactic acid is predominant, but acetic acid and trace amounts of the intermediate pyruvic acid are also present in fermented sausages. Carbon dioxide accounts for *ca.* 21% of the fermentation products, the remainder consisting of $C_2$ compounds including ethanol. A further 18% of glucose is fermented during drying, of which 83% is converted to lactic acid. In addition to temperature, lactic acid production is affected by a number of other factors, including the composition of the microflora and the oxygen concentration. Complete oxidation to $CO_2$ and water is favoured by higher oxygen concentrations.

Production of lactic acid is accompanied by a fall in pH value. In the case of high-acid, semi-dry products that have not undergone significant maturation, the pH value is primarily determined by the quantity of lactic acid produced and the buffering capacity of the meat proteins. Ammonia, produced as an end-product of proteolysis, raises the pH value and, where proteolysis is significant, interaction between lactic acid, ammonia, water content and buffering capacity of meat proteins is the major factor determining pH value. In sausages ripened by moulds or yeasts, assimilation of lactate also tends to raise the pH value. A slight rise in pH value may also occur in the later stages of ripening, which cannot be attributed to either ammonia production or lactate assimilation. The main cause is the increased concentration of buffering substances resulting from drying, although an increase in the degree of dissociation of electrolytes is also involved.

Lactic acid production and fall in pH value have important consequences for the organoleptic properties of fermented sausages. Lactic acid does not have a major role as a flavour/aroma compound but, especially in high-acid semi-dry sausages, it plays an important role in imparting a 'tangy', acidic character. Under some circumstances, lactic acid may enhance saltiness, possibly by masking other flavour notes. Low pH values may also modify flavour by restricting proteolytic and lipolytic enzyme activity.

The consistency of fermented sausage is largely determined by pH value and $a_w$ level. Correct consistency cannot be obtained at a pH value above 5.4 unless the $a_w$ level is below 0.9. For this reason pH value is the major determinant of consistency in high-acid, semi-dry sausages, and $a_w$ level in low-acid, dry sausages.

Fall in pH value is accompanied by a decrease in the solubility of proteins and enhanced gel forming ability. This effect is enhanced by high levels of NaCl. Myofibrillar proteins exhibit the greatest decrease in solubility at low pH values and are probably of greater importance in determining consistency than sarcoplasmic proteins. It appears likely, however, that interactions occur in which NaCl-induced insolubilization of sarcoplasmic proteins affects the precipitation of myofibrillar proteins and enhances formation of a structured gel.

The behaviour of proteins during decrease in pH value probably explains the differences in consistency at different ripening temperatures, since acidification will occur at different rates. Marked differences in texture between the interior and exterior of mould-ripened fermented sausages may similarly be explained by the existence of a pH gradient resulting from assimilation of lactate by moulds growing at the sausage surface.

### (b) Metabolism of nitrogenous compounds

The importance of proteolysis during the maturation of fermented sausages is well established. The extent of proteolysis varies according to a number of factors, including the nature of the meat microflora and conditions during sausage processing. Changes in the total crude protein content typically involve a decrease of between *ca.* 20 and 45% over 14–15 days ripening, while increases of over 30% in non-protein nitrogen (NPN) content have been reported during ripening of dry sausage for 100 days. Non-protein nitrogen comprises free amino acids, nucleotides and nucleosides. The total quantity of NPN and its composition is of major importance in determining the aroma of fermented sausage. Proteolysis is primarily mediated by meat-derived enzymes, such as calpains and cathepsins. Proteolysis by micro-organisms appears to be of little significance under most circumstances, although levels of free amino acids are 10–11% higher when starters with proteolytic activity, such as *Micrococcus varians* are present. Micro-organisms are known to be of importance in influencing the composition of

**Table 7.6** Major amino acids produced in sausages fermented by different starter cultures

| *Pediococcus acidilactici* | *Pd. pentosaceous* | *Micrococcus varians* |
|---|---|---|
| valine | taurine | alanine |
| leucine | leucine | taurine |
| glutamine | glutamine | leucine |
| taurine | valine | glutamine |

NPN and the relative quantities of different free amino acids present. A comparison of the three starters, *M. varians*, *Pediococcus acidilactici* and *Pd. pentasaceous*, for example, showed greatest production by *Pd. pentasaceous* followed by *M. varians* and *Pd. acidilactici*. The amino acids produced also varies, reflecting different patterns of metabolism and each organism produced four amino acids, which accounted for 40–48% of the total free amino acid content (Table 7.6). Variations may be accounted for by differences in the amino acid requirements for growth of the different bacteria.

Although the overall level of free amino acids increases during fermented sausage manufacture, a marked fall may occur for individual amino acids. Falls in concentration of arginine, cysteine and glutamine can be of particular significance and may be accounted for by further metabolism (Table 7.7). Loss of amino acids such as histidine may occur by decarboxylation. Differences in the free amino acid composition of fermented sausages appear to have little effect on the sensory properties, despite the role of amino acids in aroma.

**Table 7.7** Possible pathways of metabolism of amino acids during sausage fermentations

| Amino acid | Pathway |
|---|---|
| *Arginine* | Metabolized to ornithine |
| *Cysteine* | Metabolized to pyruvate and sulphate, *or* converted to taurine |
| *Glutamine* | Converted to glutamic acid and $NH_3$ |

*Note*: From DeMasi, T.W. *et al.* (1990) *Meat Science*, **27**, 1–12.

Little is known of Strecker degradation reactions in fermented sausages. It has been suggested, however, that in the Swedish product, isterband, carbonyls may be formed as secondary degradation products of α-amino acids. Strecker degradation and, possibly, Maillard condensation reactions may account for the depletion of some amino acids as well as the formation of flavour compounds (cf. Chapter 4, pages 208–209, dry-cured hams).

Proteolysis is affected by a number of factors in addition to the type of starter culture used. The rate increases with higher ripening temperatures, up to the temperature at which enzyme inactivation occurs. Higher temperatures, however, also lead to increased acid production and faster lowering of the pH value, which leads to a reduction of proteolytic enzyme activity. Conversely low pH value has been reported to stimulate hydrolysis of myofibrillar proteins. Diameter of sausage has also been identified as a factor affecting proteolysis. This may be another pH-related effect, especially in the centre of large diameter sausages where the fall in pH value is particularly marked.

## (c) Metabolism of lipids

Free fatty acids and carbonyl compounds are recognized as being a major component of the flavour of fermented sausages. Formation is mediated by lipases and in many sausages intense lipolysis occurs during ripening. In well made sausages, lipolysis occurs in the absence of peroxidation and consequent development of rancidity and associated undesirable organoleptic effects.

Lipolysis is generally considered to be mediated by enzymes of microbial origin. *Micrococcus*, derived either adventitiously or from starter cultures, is the most important source. It has been recognized for some years, however, that many lactic acid bacteria also synthesize a lipolytic exoenzyme. Lipolytic activity of the lipase of lactic acid bacteria is significantly less than that of *Micrococcus*, but in many instances lactic acid bacteria are present in far greater

---

* Although possession of lipases by micro-organisms can be relatively easily demonstrated in pure culture, it is often more difficult to assess their role in lipolysis of foods. In the case of fermented sausages, strong circumstantial evidence is provided by the fact that the rate of hydrolysis decreases in the order linoleic, oleic, stearic, palmitic acids. This indicates specificity for lipolysis in position 3 of triacylglycerols, a characteristic of microbial lipases.

numbers. The role of microbial lipases has been questioned, since lipase from *Micrococcus* is of low activity at reduced pH values and low temperatures. Despite this a consensus of opinion considers microbial lipases to be of prime importance. Tissue lipases are also involved, especially where ripening conditions are controlled to limit lipolytic activity and lipases from moulds are of significance in some types of sausage.

As lipolysis proceeds, a continuing decrease in triacylglycerol fatty acids is matched by a corresponding increase in diacylglycerols, free fatty acids and, to a lesser extent, monoacylglycerols. True lipolytic activity is enhanced by higher temperatures, but apparent activity falls. This is due to temperature-dependent feedback inhibition of lipases by free fatty acids.

Oxidative changes in the fat of fermented sausages involves unsaturated fatty acids almost exclusively. Fat oxidation is often auto-catalytic and rate increases markedly with time. Oxidation may also be catalysed by lipoxygenase or by metals, such as copper. Oxidation leads to formation of lipid peroxides and carbonyl compounds with, in most circumstances, an increase in peroxide value. Increase of peroxide value is greatest when *Lactobacillus* is involved, but no correlation has been found between micro-organisms and carbonyl activity. Peroxide formation is reduced when micro-organisms of high catalase activity are present.

### (d) Pigment formation

Formation of haem-containing pigments in fermented sausages follows the same pathways as in other nitrite-containing meat products. The low pH value, however, has a strong influence, initially destabilizing myoglobin and increasing the rate of autoxidation to metmyoglobin. The haem group dissociates at the pH value of many fermented sausages and colour is primarily attributed to nitrosylmyoglobin. The use of rancid fats or the presence of large numbers of $H_2O_2$-producing bacteria adversely affect the colour through promotion of brown metmyoglobin formation.

### (e) Water activity ($a_w$) level

In most meat products, the $a_w$ level is determined by the addition or subtraction of water, by added solutes (especially NaCl) and, through decreasing the water content, the addition of fat. The $a_w$

level of freshly prepared sausage mix is depressed, the extent depending on solute concentration and fat content. This favours *Micrococcus*, or *Staphylococcus* in the initial stages of fermentation, but is insufficient to inhibit other micro-organisms with the possible exception of *Campylobacter*, which is highly sensitive to reduced $a_w$ level. Reduction of $a_w$ level during drying is largely a function of water loss and consequent increase in solute concentration.

The final $a_w$ level of fermented sausages is of obvious importance with respect to growth and survival of micro-organisms. There is also a powerful influence on the consistency of the sausage. Reduction of $a_w$ level during drying affects the course of maturation by reducing enzyme activity. This results from the enzyme molecule being unable to attain the optimal configuration for activity. Effects usually become significant at $a_w$ levels of 0.94 and below.

### 7.3.3 Chemical analysis

In general, analytical methods for determination of meat and fat content, NaCl and curing agents and pH value are the same as those used for other meat products. The high fat content can, however, cause problems. In determining pH value using a glass electrode, for example, it is necessary to clean the electrode regularly with a solvent to avoid erroneous readings.

Simple methods are available for determination of $a_w$ level. These are effectively hygrometers and determine $a_w$ level through measurement of equilibrium relative humidity. Although simple to use and of low cost, such methods are limited and unsuitable where definitive determinations are required. Instrumental methods, such as the Sina® meter are effective but expensive.

## 7.4 MICROBIOLOGY

### 7.4.1 Micro-organisms of public health significance

With the exception of moulds, some of which may be mycotoxigenic, fermented sausages will not support the growth of potentially pathogenic micro-organisms. Fermented sausages are generally considered to be low risk products, although there are two areas of concern: growth of toxin-producing micro-organisms, especially *Staphylococcus aureus*, in the meat before fermentation is established, and survival of bacterial pathogens such as *Salmonella* and *Listeria*.

## (a) Growth of Staphylococcus aureus during fermentation

Enterotoxin production by *Staph. aureus* has been a major problem in production of fermented sausages, although greatly improved control of fermentation has significantly reduced the scale of the problem in recent years. *Staphylococcus aureus* is a common contaminant of raw meat, although growth is usually suppressed, even at high storage temperatures, by competitive pressures exerted by the spoilage microflora. *Staphylococcus aureus* is, however, resistant to relatively high levels of NaCl and nitrite and is selectively favoured in the presence of these ingredients if the onset of fermentation is delayed. Growth and enterotoxin formation is rapid under these conditions, and while the organism itself subsequently dies, the enterotoxin remains for extended periods.

Microbiological examination of freshly made fermented sausage has been proposed as a means of ensuring the safety of fermented sausages, it being suggested that the product should contain less than $10^4$ cfu/g of *Staph. aureus*, immediately after fermentation. Where examination immediately after fermentation is not possible, the use of the thermonuclease test, which indicates the *likely* presence of enterotoxin, has been proposed as an alternative. It must, however, be appreciated that laboratory tests of any kind are secondary to control during manufacture, especially the rapid lowering of pH value by establishing a lactic fermentation and/or the addition of an acidulant, such as glucono-δ-lactone.

## (b) Survival of bacterial pathogens

Fermented sausages are raw meat products and the absence of bacterial pathogens cannot be guaranteed. Conditions in the sausage

---

* The outbreak of salmonellosis involving a salami-snack product was caused by *S. typhimurium* DT124 and had a number of unusual features. A total of 71 cases were confirmed, of which 55 were children aged less than 16 years. Epidemiologists investigating the outbreak felt that this factor, together with the proximity of the outbreak to Christmas, suggested that a sweet product, possibly a chocolate novelty was involved and initially the salami-snack was not suspected. A further unusual feature was the atypical symptomology, in that no less than 42% of persons affected reported bloody stools, an unusual symptom in salmonellosis. (Cowden, J.M. *et al.*, 1989, *Epidemiology and Infection*, **103**, 219–225.)

do not normally permit multiplication, but survival can be prolonged. *Salmonella* has been associated with food poisoning following consumption of salamis in Australia, Italy and, most recently, in the UK, where a salami-snack product was involved. The product had a very short processing and survival of *Salmonella* may have been enhanced. In all cases, however, the inability of *Salmonella* to grow in fermented sausages means that the infective dose must have been low. As with cheese, this may result in the protection of the organism from gastric acidity by the high fat content.

Little attention has been paid to the effect of method of acidification on survival of *Salmonella*. In salami-type sausages made under Australian conditions, acidification by starter cultures was significantly more effective in reducing numbers of *Salmonella* than acidification with glucono-δ-lactone. Indeed, growth occurred during acidification with glucono-δ-lactone where pre-exposure of *Salmonella* to acid conditions had resulted in adaptation. These findings suggest that caution is required when considering the use of glucono-δ-lactone in place of starter cultures.

*Listeria monocytogenes* has been isolated from fermented sausages and consumption of salamis has been identified, on epidemiological grounds, as a risk factor for listeriosis in the US. Despite this, the significance of the isolation of *L. monocytogenes* has not been fully evaluated.

The isolation of *L. monocytogenes* from fermented sausage has led to speculation that other pathogens, including *Campylobacter* and verocytotoxin-producing *Escherichia coli* (especially serovar O157:H7), may also present a risk. *Campylobacter* is highly sensitive to reduced $a_w$ levels and pH values so that extended survival appears unlikely. Verocytotoxin-producing *E. coli* is currently largely associated with consumption of undercooked beefburgers. The organism is most prevalent in beef, but can also be isolated from lamb, pork and poultry. Survival in fermented sausages would seem to be at least a possibility, but there is no evidence of the involvement of these products in morbidity due to the organism.

Although *Staph. aureus* is generally considered to present a risk only through growth during the early stages of fermentation, an outbreak of food poisoning has been reported in which growth and enterotoxin production occurred in the finished product. The sausage involved was of high $a_w$ level and contained relatively low concentrations of NaCl and curing ingredients. As a consequence, stability was highly dependent on the low pH value. Temperature abuse during storage led initially to mould growth and metabolism of lactate, which raised the pH value to a level at which growth and enterotoxin production could occur.

### (c) Mycotoxin production

Mycotoxigenic moulds have been isolated both from mould-ripened fermented sausages and sausages where mould contamination is undesirable. The significance of these findings has been questioned and it has been suggested that meat products in general, and fermented sausages in particular, are poor substrates for mycotoxin production even when mycelial growth is extensive. The evidence for this is conflicting. Some studies have shown that mycotoxin production is inhibited by low storage temperatures, reduced $a_w$ levels and, where appropriate, smoking. In contrast, some strains of *Aspergillus* have been shown to produce high levels of aflatoxins during growth on fermented sausages. The role of starter cultures has received less attention, but at low temperatures both *Lactobacillus* and *Pediococcus* have been shown to inhibit aflatoxin production.

To some extent, disputes over the ability of moulds to produce mycotoxins on fermented sausages have distracted attention from the more important issues concerning the selection of moulds for ripening. Safety can be assured by inoculating sausages with strains which are known to be non-toxigenic and it is then irrelevant whether or not adventitiously derived strains can produce mycotoxins. Similarly, with non-mould ripened sausages attention should be given to ensuring that conditions during storage and transport are such as to prevent the growth of moulds.

---

* In recent years, especially in the US, the intrinsic stability of some semi-dry fermented sausages has been reduced by lowering the concentration of NaCl and curing agents in the finished product in response to consumer concern over food additives. Such products are also of higher $a_w$ level than their traditional counterparts. Potential problems are caused by retailers, caterers and consumers failing to realize that considerably greater care is required in handling these products compared with traditional semi-dry sausages. It is seen as the responsibility of the manufacturer to ensure that labelling clearly indicates the relatively unstable nature of the products and clearly stipulates handling and storage conditions.

## (d) Production of biogenic amines

Species of *Lactobacillus* have been associated with production of the biogenic amines, histamine and tyramine, through decarboxylation of the amino acids, histidine and tyrosine. The presence of biogenic amines in significant quantities can lead to a critical increase in blood pressure, together with headache, flushing and, possibly, rashes. On some occasions, symptoms have been primarily gastrointestinal, including sudden onset of vomiting and diarrhoea with accompanying abdominal pain. Authenticated outbreaks of illness due to the presence of biogenic amines occurred following consumption of pickled (fermented) vegetables and cheese, notably Stilton and Swiss varieties. There have been no authenticated cases associated with consumption of fermented sausages, but the development of symptoms including headache and flushing has been circumstantially linked with biogenic amines. This interpretation should be treated with care, however, since high levels of nitrite can cause similar effects in sensitive persons.

Factors influencing histamine formation in mettwurst have been investigated. Histamine production largely occurs during the first two weeks of ripening. A large number of histidine-decarboxylating bacteria must be present, and levels of histidine must also be significantly higher than the norm. Such levels are associated with storage of the unprocessed meat at high temperatures and/or for excessive periods. Considerable protection against biogenic amine production is obtained by use of good quality meat as raw material and by the use of effective starter cultures of strains which do not have amino acid decarboxylating activity.

## (e) Survival of viruses

As raw meat products, fermented sausages have attracted a certain amount of attention with respect to survival of viruses. In some cases, concern has been expressed that fermented sausages could serve as vehicles for human pathogenic viruses, such as poliovirus and coxsackievirus. From an economic viewpoint, however, animal pathogenic viruses are of greater significance. This stems from the importance of fermented sausages in international trade and the risk of transmitting animal viruses into areas previously free of the associated disease or where the disease is considered to be under control. Particular concern exists over foot-and-mouth, swine fever and hog cholera viruses.

Only a limited number of studies have been made of viral inactivation in fermented sausages. In general, it appears that viruses are inactivated either during active fermentation or during maturation. In some cases, there is considerable variation between experiments. This may be due to differences in the strains of virus used or to technological differences between the sausages. At present there is insufficient data to distinguish patterns of survival or inactivation.

### 7.4.2 Spoilage of fermented sausages

Fermented sausages, as a group, are highly stable. This results from a combination of factors: low $a_w$ level and moisture content, reduced pH value, the presence of organic acids and high levels of NaCl and $NaNO_2$. In large-diameter sausages, the low redox potential further inhibits growth of aerobic spoilage organisms, while vacuum or modified atmosphere packaging has a similar effect in sliced pre-packed products. The stability of different types of fermented sausage varies, as does the relative importance of the different inhibitory factors. Dry fermented sausages are most stable and $a_w$ level is of prime importance. In contrast, spreadable, semi-dry sausages are least stable with a short shelf life and often requiring refrigeration during storage. The higher $a_w$ level means that other inhibitory factors, such as low pH value and $NO_2$ concentration, are of relatively greater importance in these sausages. In all cases, there is considerable interaction between inhibitors, which significantly increases the overall effectiveness of the system. A fall in pH value, for example, raises the limiting $a_w$ level, while nitrite is also more inhibitory at low pH values. Lactic and acetic acid have been assumed to play an important role in stability of fermented sausages but, in some cases at least, the effective antimicrobial activity of these acids is very limited.

Moulds and yeasts are normally the only micro-organisms capable of growth on fermented sausages. Growth is restricted to the skin of entire sausages; while hyphal development may appear extensive, significant penetration of the interior is rare. Several genera of mould may develop, including *Aspergillus*, *Cladosporium* and *Penicillium*, extensive growth often following temperature change and condensation of moisture on to the surface of the sausage. In the UK, it is long-established commercial practice to wipe the mould from the surface with vegetable oil, polish with talc or rice flour and, unless penetration of the interior is extensive, treat the

---

BOX 7.4 **The secret of the sands**

In the developed world, the original purpose of fermentation, preservation, tends to be overlooked since refrigeration and vacuum or gas packaging means that long storage lives can be given to meat products of many types. This is not the case in developing countries, where limited refrigeration facilities cause problems in distribution and a high level of spoilage loss of meat. Fermentation has been suggested as an alternative, but in hot countries problems are caused by high ambient temperatures and humidity during maturation. A systematic study of the problems in Egypt has shown that a sausage of satisfactory quality could be produced under uncontrolled, ambient conditions (average temperature 21.9°C, relative humidity 77.8%) by coating the sausage with a spice-based paste. This contained fenugreek, garlic, paprika and wheat flour. (Shehata, H.A., 1991, *Proceedings of the 37th International Congress of Meat Science and Technology*, Kulmbach, Germany, vol. 2, pp. 934–9.)

---

sausage as unspoiled. The possibility of mycotoxin production has been raised and mycotoxigenic species have certainly been isolated from fermented sausages. For this reason the practice of wiping mould from fermented sausages is now considered undesirable. Mycotoxin production does not, however, appear to have been demonstrated.

Yeast growth may be restricted on dry fermented sausages, although confluent growth may develop. A wide range of yeasts appear to be able to develop under some circumstances, including both oxidative and fermentative genera. There is no evidence of any adverse effects on product quality.

Bacteria are not normally involved in spoilage of correctly made fermented sausages, although growth can occur if a large quantity of condensation is present on the outer skin of the sausage or, in pre-packs, is trapped between the face of the sausage and the packaging film. Species of *Bacillus*, *Micrococcus* and, occasionally, 'coryneform' bacteria may be isolated, but the effect on quality appears to be limited. In other circumstances, loss of organoleptic quality can occur following the death of a significant proportion of the lactic acid bacteria responsible for fermentation. This usually involves the development of bitterness. The problem is not common and occurs only after extended storage, either of the whole sausage or of sliced pre-packed sausages. The mechanism of bitterness development has not been investigated but is assumed to result from the release of proteolytic enzymes following cell death and lysis, and the production of bitter-tasting peptides.

### 7.4.3 Microbiological examination of fermented sausages

Under the vast majority of circumstances microbiological examination of fermented sausages is unnecessary. Yeast and mould can be detected by visual examination and culture is not required, except on special occasions where identification of contaminants is required.

Where doubt exists over the fermentation stage, examination for *Staph. aureus* or for its thermonuclease may be required. In the former case, it must be remembered that *Staph. aureus* may be severely stressed immediately after, or even during, fermentation. This must be taken into account in choice of media and methods. Liquid enrichment is not appropriate, since it is not possible to assess the significance of small numbers of *Staph. aureus*. The widely used Baird–Parker agar medium is satisfactory under many circumstances, although in some situations a total *Staphylococcus* count, which includes coagulase-negative species, is preferred. In this case KRANEP medium is suitable. Improved recovery of *Staph. aureus* may be obtained by using a modification of Baird–Parker medium in a method involving 'solid repair' of stressed cells (Figure 7.3).

The reliability and sensitivity of staphylococcal thermonuclease determinations can be highly dependent on the method used. The

* Although themselves effectively immune from bacterial spoilage, fermented sausages can be of importance as a source of micro-organisms in cooked meat pre-packing operations. The major problem is often considered to be caused by lactic acid bacteria, including those which are capable of hydrogen peroxide formation, or slime production and thus of high spoilage potential in cooked cured meats. In pre-packing operations, however, fermented sausages also should be seen as potential sources of pathogens, such as *Salmonella* and *L. monocytogenes*. These pathogens are able to grow on uncured or mildly cured cooked meats although, in practice, lactic acid bacteria, also derived from the sausages, are likely to be inhibitory. Despite this, the management of meat pre-packing operations should be aware that fermented sausages are raw meats and should arrange production and sanitation schedules to avoid contamination from these products to cooked meats.

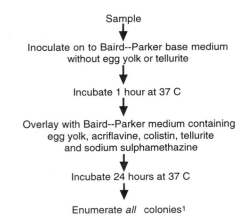

Sample

↓

Inoculate on to Baird--Parker base medium
without egg yolk or tellurite

↓

Incubate 1 hour at 37 C

↓

Overlay with Baird--Parker medium containing
egg yolk, acriflavine, colistin, tellurite
and sodium sulphamethazine

↓

Incubate 24 hours at 37 C

↓

Enumerate *all*  colonies[1]

**Figure 7.3**  Solid repair method for recovery of stressed *Staphylococcus aureus*. [1] The selectivity of the procedure is considered to be such that any colony developing in 24 hours may be accepted as *Staph. aureus* irrespective of appearance. (Method of Isigidi, B.K. *et al.*, 1989, *Journal of Applied Bacteriology*, **66**, 379–84.)

commercially available Staphynuclease™, a method based on antibody inhibition, is considered suitable. Assays for staphylococcal enterotoxins are not required on a routine basis. If they are considered necessary, account must be taken of difficulties caused by the high fat content of most fermented sausages, which leads to a low efficiency of extraction and poor recovery of the toxins.

Examination of fermented sausages for *Salmonella*, *Listeria monocytogenes*, etc. is only undertaken for specific purposes, including investigation of food poisoning. Under all circumstances methods must take account of the organisms being potentially stressed and resuscitation is necessary.

## EXERCISE 7.1

The alleged therapeutic properties of *Bifidobacterium* are exploited in 'bifid-amended' foods, to which cultures of the organism are added. Such foods are very popular in Japan and include a fermented sausage. It has been stated that fermented sausage is an 'ideal vehicle' for *Bifidobacterium* due to the high rate of survival during storage of the sausage, in gastric transit and in the presence of bile.

Consider the chemical and physical properties of fermented sausages and relate these to the survival of *Bifidobacterium* in the various circumstances. Do you consider that fermented sausages are likely to be a more favourable environment for *Bifidobacterium* than fermented milks, which are more common as a vehicle for the organism?

Fermented sausages contain high levels of fat, NaCl and nitrite and are thus perceived as 'unhealthy' foods. Discuss, from the viewpoint of consumer acceptance and marketing, the likely success of sausages as bifid-amended foods compared with perceived 'healthy' foods, such as skimmed milk and yoghurt.

(Further information on *Bifidobacterium* and its alleged therapeutic properties may be obtained from: Fuller, R., 1992, *Probiotics. The Scientific Basis*, Chapman & Hall, London; Varnam, A.H. and Sutherland, J.P., 1993, *Milk and Milk Products*, Chapman & Hall, London.)

## EXERCISE 7.2

It has been stated, on a number of occasions, that the manufacture of fermented sausages is directly analogous to the manufacture of cheese. To what extent do you consider this analogy to be valid? Discuss, with particular reference to differences in milk and meat and vegetable proteins, the alternative opinion that fermented sausages and cheese have nothing in common but a lactic fermentation and that fermented sausage manufacture is more closely analogous to manufacture of sauerkraut.

## EXERCISE 7.3

Production of organic acids is obviously the key technological role of lactic acid bacteria added as starter micro-organisms to sausage fermentations. For this reason, tests for acidifying activity often form the basis of quality assurance procedures for meat starter cultures. To what extent do you consider pure culture tests for acidifying activity to be of use in predicting the performance of starter cultures in the competitive environment of a sausage mix?

Design two tests for predicting starter culture performance in the presence of indigenous competitive bacteria. The first is required for selecting isolates from natural sausage fermentations for use as starter cultures. The test should be regarded as definitive, but should be as economical as possible in terms of labour, consumable materials and capital equipment.

The second test is to be used for monitoring the performance of existing starter strains in a factory laboratory. The test need not be definitive, but a high level of reproducibility is required. It must also be suitable for use in a laboratory equipped only to basic standards, by trained but non-graduate level personnel.

# 8

# FROZEN MEAT AND MEAT PRODUCTS

---

## OBJECTIVES

After reading this chapter you should understand
- The various types of frozen meat and meat products
- The basic principles of freezing
- Methods of freezing
- Precautions during freezing of whole meats
- Special requirements during formulation of frozen meat products
- Freezing as a manufacturing operation
- Effects of freezing on the structure and constituents of meat
- Chemical changes during frozen storage
- Microbiology of frozen meat products
- Quality assurance and control

---

## 8.1 INTRODUCTION

The freezing of meat on a commercial scale followed the development of efficient mechanical refrigeration during the 19th century. Freezing as a means of preservation was one of the factors which enabled meat to be imported from the new producing countries of Australia, New Zealand and the United States (cf. Chapter 2, page 66). Procedures were empirically based and frozen meat acquired a reputation for poor quality which remains today. Frozen meat was (and is) cheap and, despite higher refrigeration costs, it offered advantages in terms of long storage life to many sectors of the meat trade, mass caterers etc. Frozen storage was not, however, available to the vast majority of households or many retail outlets. Frozen foods developed as retail commodities when frozen storage became available on a domestic scale and the product range expanded from whole meat cuts to processed products, in parallel with the greater variety of non-meat frozen foods becoming available. In terms of consumption, however, the most important development was the availability of frozen chickens, which completely changed meat-eating habits in many countries.

---

**BOX 8.1　The idiot's lantern**

Growth in the frozen food market occurred at a time of increasing consumerism and greater availability of goods such as televisions and washing machines. Manufacturers of frozen foods confidently expected that the lure of the television would mean the demise of home cooking. In response a range of frozen 'television dinners' was produced which, after heating, were to be eaten off trays while watching the flickering electronic marvel. The early 1960s model of the happy family watching wholesome gameshows, while eating once-frozen meat and two vegetables, contrasts sharply with the 1990s couch potato, consuming beer and calorie-rich snacks while watching the more violent sports on satellite television.

---

Although frozen meats and meat products were well established in the catering and industrial sectors by the 1960s, growth of the retail sector tended to be dependent on growth in home ownership of freezers. In the UK, home ownership of large freezers grew considerably during the 1970s, resulting in increased sales of frozen foods. In retailing this was reflected in the appearance of dedicated freezer centres which, in some cases grew to become relatively large national chains. In the case of frozen meat products overall market growth was enhanced by new product development, which was also effective in adding value to basic, low profit products. Despite this the perception of frozen meats as being of low quality has persisted. Consumers who are happy to use frozen vegetables as a direct replacement for fresh are quite likely to reject high quality frozen meat products. The perception of the low quality of frozen meat products has been reinforced by the development of new, high-quality chilled products. For this reason sale of frozen meat products tends to be driven by convenience and price in industrial, catering and retail sectors of the food industry. Despite these constraints the market for frozen meat and meat products remains large and the industry is lively and innovative. To some extent, however, the advantages of long storage life, at least over a medium term, are being eroded by development of products with very long chilled lives. At the same time there is commercial pressure to develop fresh equivalents of products which currently require freezing for structural purposes (see page 122).

## 8.2 TECHNOLOGY OF FREEZING

### 8.2.1 The freezing process

The freezing process is deceptively simple. Meat (usually at chill temperatures) is cooled to 0°C, the water in the meat is converted to ice at 0°C and the material is then further cooled to its storage temperature (usually −18 to −20°C). The largest single factor in terms of heat to be extracted and thus the refrigeration capacity required is removal of latent heat during conversion of water into ice without change of temperature. For pure water, conversion of water at 0°C into ice at 0°C requires extraction of *ca.* 335 kJ/kg, compared with the *ca.* 155 kJ/kg required to lower the temperature from 37 to 0°C and the *ca.* 42 kJ/kg required to lower the temperature of ice from 0 to −20°C.

In meat, as in other foods, the actual situation is more complicated. The highest temperature at which ice crystals have a stable existence in a food is known as the freezing point. However, because of the nature of foodstuffs and the presence of water-soluble constituents, not all of the water freezes at this temperature. Under equilibrium conditions and at a temperature just below the freezing point, a certain fraction of water remains in the liquid phase. This fraction falls as the temperature is reduced and eutectic mixtures may separate from the unfrozen fluid. A small amount of water remains unfrozen at −20°C, typical quantities of water remaining unfrozen being: *ca.* 50% at −2°C, *ca.* 10% at −8°C, *ca.* 5% at −13°C and *ca.* 2% at −20°C. Completely freezing the water in meat usually requires a temperature of *ca.* −30°C. One consequence of this phenomenon is the difficulty of defining an endpoint to the freezing process.

A further difficulty lies in the fact that freezing occurs at different rates in different parts of the material being frozen. The rate is fastest at some point on the surface, while in the depth of the material there will be a point at which cooling is slowest. This point is known as the thermal centre and is usually the site at which measurement of temperature histories are made.

## 8.2.2 Freezing rate

The rate of freezing in early processes was low but later work by the eponymous Clarence Birdseye, amongst others, showed that the quality of the thawed product was improved by rapid freezing. This relationship does not always hold true and meat, especially of high fat content, is not particularly sensitive to freezing rate (Table 8.1). There is, however, strong evidence that drip losses in meat are greatest when freezing is slow. High rates of freezing lead to production of very small ice crystals, which are relatively highly reflective. Fast-frozen red meats appear to be paler in colour and poultry appears whiter. These are important factors in promoting consumer acceptability, although the intrinsic quality of the muscle is unchanged. In many cases, the benefits of quick freezing are lost on storage due to deteriorative chemical changes and, if storage temperatures fluctuate, the growth of large crystals at the expense of small.

It must be appreciated that rapid freezing is not merely a matter of product quality, but also of operating economics. Rapid freezing using low temperatures is thermodynamically less efficient and energy costs are accordingly higher. The cost of energy, however, is only a relatively small percentage of total operating costs: labour productivity and throughput in relation to floorspace are of significantly greater importance.

**Table 8.1**   Grouping of foods according to sensitivity to freezing rate

*Group 1: Freezing rate has no influence on final quality*
Products with high content of dry matter – peas, **high fat meats**, some ready-meals (including some **meat-containing** types)

*Group 2: Freezing rate should be >0.5–1.0° C/min*
Fish, **lean meat**, starch-containing ready meals

*Group 3: Freezing rate should be >3.0–6.0° C/min*
Strawberries, carrots, beans, gelatinous material such as egg products and sauces with a flour base

*Group 4: High rates of freezing advantageous*
Materials with relatively little dry matter – raspberries, tomatoes, cucumbers

The rate of freezing is difficult to define. This is a direct result of the variable freezing behaviour of meat and the difficulties in determining the endpoint. The freezing time may also be defined from the viewpoint of product quality and plant throughput. In the first instance the definition is referred to as the nominal freezing time and is based on the time during which the majority of ice is formed in the body. According to the International Institute of Refrigeration, the nominal freezing time is the time elapsing from the instant the surface of a body reaches 0°C to the instant that the thermal centre reaches a temperature of 10°C colder than the temperature of the initial ice formation. Freezing time from the viewpoint of plant throughput is referred to as the effective freezing time ($t_e$). This is defined as the time to lower the temperature of the product from the initial average value to a given value at the thermal centre.

### 8.2.3 Methods of freezing

*(a) Contact with cold solids*

The most common means of freezing by contact with cold solids is the plate freezer. Equipment of this type was used to produce the first 'quick frozen' foods and more modern versions are still widely used today. Plate freezers consist of a series of hollow, refrigerated plates, which may be of vertical or horizontal configuration (Figure 8.1). The space between the plates is variable and is controlled by a hydraulic or pneumatic ram. In operation the plates are opened for loading and unloading and closed for freezing. A slight pressure may be applied to ensure good contact between the plates and the material being frozen, which is in the form of parallel sided blocks. Modern plate freezers nearly always work on a semi-continuous basis.

Heat flux is perpendicular to the plates and the heat transfer coefficient between the surface of the food must be high and uniform. Packages should be completely filled with food and in uniform contact with the cooling plates. The plates must be kept free from ice and any other deposit which might interfere with heat transfer. The design and construction of the plates is of importance in allowing rapid and uniform heat transfer. The aluminium alloy extrusions now used are of very good performance.

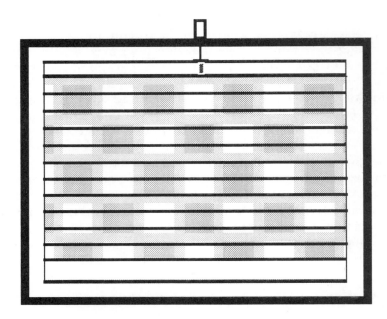

**Figure 8.1**   Simplified diagram of horizontal plate freezer.

Horizontal plate freezers are used for freezing ready-meals packed in rectangular plastic containers (inner packaging). The most suitable formulations are those which have a high liquid component, such as stews or meat in sauce. Vertical plate freezers are highly suited to freezing unpackaged meat and offal such as liver. The product is fed directly between the plates to form blocks, which may be compacted further by slight compression between the plates. At the end of the freezing cycle, it is necessary to warm the plates to release the frozen block and defrost and clean the plate surfaces. This requires use of refrigeration plant with a fast defrost cycle to minimize reheating of the frozen material. Minimal surface defrosting is necessary to minimize product loss and handling difficulties and to prevent the blocks freezing together in storage.

A further type of contact freezer consists of a continuous, horizontally mounted stainless steel bånd on which the product to be frozen is placed. The underside of the band is sprayed with refrigerated brines, freezing the product during its journey along the belt. At the end of the process, the product is broken off the belt by flexing over a roller, or vibrating with a pneumatic hammer.

This method of freezing is used for flat comminutes, such as burgers. Minced meat may also be frozen in this way.

## (b) Contact with a cooled liquid

Freezing by contact with a cooled liquid has the advantages of a high heat transfer coefficient and the ability to freeze irregular objects. Unlike plate freezers, individual (IQF) products can be frozen. Freezing may involve immersion in the cooled liquid or, alternatively, the liquid may be sprayed on to the surface of the product. Where food is unpackaged, the cooled liquid must be acceptable for food use, brines and syrups being widely used. These are not suitable for freezing meats, due to unacceptable flavour changes, and protective packaging must be used to prevent contact. Immersion freezing of poultry by immersion of film-wrapped birds in a brine is common practice, the rapid surface freezing producing small ice crystals, which enhance the whiteness of the muscle. Immersion freezing has also been used for individual pre-packaged joints of red meat. In either case, immersion freezing is usually an initial stage and the process is completed in a blast freezer. Whole sides, or primal joints, of meat may be frozen by spraying with cooled brine. Plastic shrouds are necessary to prevent contact between the brine and the meat.

## (c) Contact with cooled gas (air)

Freezing in cold air is a very early method, the original 'sharp' freezing involving refrigerated rooms with natural air circulation. Today forced air circulation (blast freezing) is used in even the smallest commercial installations. Cold-air freezing has many of the advantages of liquid freezing, but the heat transfer coefficients are significantly lower. Despite this, cold-air freezing is very widely used, being a relatively simple process and avoiding the inconveniences associated with brines and syrups.

Blast freezers may operate on either a batch or a continuous basis and normally employ air at $-20$ to $-40°C$. The heat transfer coefficient increases with higher air velocity. This does not lead to a proportionate decrease in freezing time because of the thermal resistance of packaging material and the effect of heat transfer by conduction within freezing food. A further factor is that at high air velocities, energy dissipation from the motors that drive the fans becomes a significant problem and further diminishes any advantage obtained through a higher heat transfer coefficient.

Batch-blast freezers are used only where throughput is low. The construction is basically simple, consisting of a well insulated chamber equipped with air coolers and fans. Design requires care and the chilled air must be directed to flow evenly over all product to be frozen. The product is usually placed on trays mounted on trolleys, which must be correctly positioned to maintain the correct airflow. In many cases guardrails are fitted inside the blast freezer to aid correct positioning. Incorrectly loaded trolleys can interfere with airflow and considerable care is required under some circumstances to ensure an even freezing time for the whole batch.

Continuous blast freezers are now almost universal in large-scale applications. The most common type is in the form of an insulated tunnel, through which product to be frozen travels on trolleys or on trays on a continuous belt or, in the case of beef sides, suspended from hooks. Chilled air flow can be parallel to product movement or perpendicular (cross-flow). Cross-flow freezers are now the most common type and have the advantage of minimizing chilled air loss at exit and entry ports. Thermal conditions are also easily controlled to maintain a high humidity and thus minimize freezer burn. The freezers consist of a series of unit coolers comprising air coolers, fans and associated ducting, mounted side by side along the length of the tunnel. Each unit cooler provides a blast of cold air in adjacent sections of the tunnel.

Blast-freezing tunnels are highly effective and are relatively economical in running costs. A problem, however, is the large amount of floor space required in relation to throughput. It is usual to configure the tunnel to make best use of available floor space. A special design is also available which reduces floor space requirements to a minimum. This consists of a flexible stainless steel conveyer belt, which carries products through air blast chambers while ascending and descending through vertical helices. Blast freezers are widely used for meat and meat products of all types and may be used alone or in conjunction with other types of freezer.

An alternative type of air freezer is the fluidized bed. This consists of a chamber with a porous floor, through which chilled air is blown. Product to be frozen is placed on the slatted floor, the airflow being upward through this 'bed', which may be stationary or vibratory. The air velocity is adjusted so that the volume of the bed is expanded, but individual pieces do not float. Fluidized bed freezers are designed on a 'plug-flow' basis, operating on a first-in, first-out principle giving a flat temperature profile. Fluidized bed freezers are efficient but are only suitable for small, regular-shaped pieces of food. Use with meat products is limited.

## (d) Two-phase freezing systems (cryogenic freezers)

In two-phase systems the cooling medium is either a subliming solid (usually carbon dioxide) or a boiling liquid (usually nitrogen). In the great majority of cases, the refrigerating effect is provided not by an on-site refrigeration plant but by the heat transfer medium itself as a consequence of its phase change. It must be appreciated, however, that there is no free ride in terms of energy usage and that a refrigeration plant at a remote site must originally be used to cool the heat transfer medium. Gas produced as a result of the phase change is ultimately vented to the atmosphere as waste. Two-phase freezing is often known as cryogenic freezing and the materials providing refrigeration are known as cryogens.

Solid $CO_2$ may be provided for small scale use as blocks or pellets, but it is most conveniently handled as cylinders of pressurized liquid. When released through nozzles to atmospheric pressure, a mixture of cold $CO_2$ gas and solid $CO_2$ 'snow' is formed. Solid $CO_2$ may be placed in direct contact with foods by means of special dispensing equipment. This type of freezing may be used to supplement mechanical systems at times of peak load or as a sole freezing source. Solid $CO_2$ finds a specialist use for deep chilling fresh turkeys to obtain a long shelf life at times of peak demand, such as Christmas (see Chapter 2, pages 67–68), and is used in a similar way for limiting temperature rise in mail order delicatessen meats delivered by non-refrigerated transport. Special boxes may be used, which are effectively miniature cold stores, in which $CO_2$ is contained in perforated chambers in the walls of the box. Chilling is then by cold $CO_2$ gas.

The boiling point of liquid $N_2$, at atmospheric pressure, is $-196°C$ and freezing by this medium is thus very rapid. Use of liquid $N_2$ as a freezing medium has been directly associated with improved quality in a number of meat products, including poultry pieces and steaks. The most striking perceived improvement was noted with

beefburgers, where the consistency of quality was also improved.
This is of very great importance in fast-food operations where stan-
dardization is highly desirable. The direct benefits of fast freezing
through use of cryogens are, however, restricted to fairly small
pieces of meat. With larger pieces, the heat transfer properties of
the meat become limiting. Problems also arise due to cracking as a
result of expansion of meat in the centre after the exterior has solidified. In addition to quality effects, manufacture of some meat
products is facilitated by very fast freezing. These are often further
processed poultry products, where the meat is combined with
other types of food. Examples are small pieces of chicken (nuggets)
surrounded by a layer of breaded cheese, or cheesebase, and
breaded minced chicken formed in a layer around cheesebase. A
further advantage of liquid $N_2$ is the very low water-holding
capacity of the very cold gas after evaporation, which means that
reduction in product weight by evaporation during freezing is
minimized. From an economic viewpoint, however, the low
moisture loss must be balanced against the relatively low cryogenic
efficiency (see below). Liquid $N_2$ tends to be expensive and its use
can be economically marginal, even where quality considerations
are of considerable importance. The increased use of liquid $N_2$ as
cryogen in the UK during the 1980s, for example, largely stemmed
from the relatively low cost, which resulted from fall in demand
from industrial users.

Early methods of cryogenic freezing with $N_2$ usually involved
immersion of the food in the liquefied gas. This has two major disadvantages. In the first place, it is not possible to control the rate
of cooling of a body immersed in liquid cryogen. Secondly, the use
of the refrigeration capacity of the cryogenic medium is inefficient.
The latent heat of evaporation of liquid $N_2$ at $-196°C$ is 200 kJ/kg
and gaseous $N_2$ warming from $-196°C$ to $-18°C$ absorbs a further
209 kJ/kg.

For this reason more modern plant exploits the refrigeration
capacity of $N_2$ gas. This may be achieved in two ways. In the first

---

\* The most striking evidence for improvement of quality of beefburgers by cryogenic freezing came from a large consumer trial. It was reported that a high percentage of consumers were able to differentiate cryogenically and air blast frozen
burgers, even when the burgers were eaten in a bun with salad and relish! Preference for cryogenically frozen burgers was strong, superior texture and juiciness
being the main reasons.

system, $N_2$ is sprayed on to food passing along a conveyer belt. Cooling rate is controlled by varying the volume of liquid $N_2$ sprayed on the food. Some surplus is inevitable and this may be recirculated in liquid form, or evaporated by $N_2$ gas returned from the 'warm' end of the freezing tunnel. The very cold gas produced, together with that which evaporates when liquid $N_2$ contacts the food being frozen is used to effect further cooling. A common arrangement is to employ a three-stage cooling process. Food entering the freezer moves first through a 'pre-chilling' section, where the cooling medium is gaseous $N_2$. Liquid $N_2$ is sprayed on to the food in the second, 'freezing', stage and the process is completed by passing through an 'equilibration' stage where gaseous $N_2$ is again the cooling medium.

In the second system, cooling is entirely by $N_2$ gas and there is no contact whatsoever with liquid $N_2$. In this system liquid $N_2$ is used to cool thermostatically controlled chambers. In continuous freezers of this type, food passes sequentially through a number of chambers on a conveyer belt. Batch freezers are also available. Although not suited to large-scale production, batch freezers of relatively small capacity are logistically well suited to production of frozen ready-meals in situations where large number of varieties are produced but production runs are short.

---

### BOX 8.2  Global warming

In 1967, the multinational chemical manufacturer DuPont developed a cryogenic freezing technique based on 'R12' (dichlorodifluoromethane). This compound boils at $-30°C$ at atmospheric pressure and is an excellent heat transfer medium, which can be used for cryogenic freezing by either immersion or spraying. 'R12', however, is too expensive to be allowed to vent to waste and freezing equipment was designed to facilitate recovery and recycling. Under good operating conditions losses of 'R12' amounted to no more than a few per cent of product weight. The full benefits of using 'R12' as a cryogen were realized in the early 1970s and 150 000 tons of food were frozen by this method in 1973. During the 1980s, however, the potential role of 'R12' in depletion of stratospheric ozone and contribution to global warming became apparent. For these reasons use of 'R12' in freezing of foods is no longer considered acceptable.

## 8.2.4 Thawing

Thawing is the reverse process of freezing. It is ironic that while the frozen food industry must overcome problems leading to premature thawing, deliberate thawing can cause difficulties in other parts of the food industry. In theory, the same quantity of heat must be put in to warm the meat from, for example, $-18°$ up to $5°C$ as was extracted when cooling from $5°$ down to $-18°C$. In practice, thawing by heat transfer from a warmer source is fundamentally less efficient than freezing. This largely stems from the fact that the thermal conductivity of water is much lower than that of ice. This means that thawing is slowed by the heat having to pass through an increasing thickness of water, rather than through an increasing thickness of ice. Inefficiency is also due to the fact that, in practice, it is difficult to obtain a large temperature difference between the heating source and the meat undergoing defrosting. This is because of the need to limit microbial growth during thawing, especially where water is used as the heating medium. Equally, very high temperatures are unacceptable due to discolouration and possibly other undesirable changes.

It is a striking fact that, in otherwise well managed processing operations, the thawing of frozen meat before further processing can be a shambles. Conditions are usually least satisfactory where frozen meats are used only occasionally and in smaller operations, especially catering. There is, however, a tendency to overlook thawing as a food processing operation and cases are known where this stage has been omitted from otherwise comprehensive HACCP flow sheets.

Water is widely used as a means of thawing and is relatively efficient in terms of heat transfer. However, unless the meat is in impermeable packaging, there will be leaching of nutrients into water and microbial growth may result. Running water is more satisfactory but may not entirely overcome the problem. In any

---

* Failure to thaw frozen meats adequately, especially poultry, is considered to be a major contributory factor in sporadic outbreaks of *Campylobacter* and *Salmonella* food poisoning and has also been involved in some larger, point source outbreaks. Large joints and whole birds, especially turkeys, are most commonly involved. With these products, defrosting is slow and remaining frozen parts in the centre can be very difficult to detect. Failure of an inexperienced cook to thaw burgers adequately, however, was believed to be contributory to undercooking and an outbreak of *Escherichia coli* O157:H7 infection in northern England.

case the higher water content at the surface of the meat will also lead to increased microbial growth rate. Thawing in air may involve no more than placing the frozen meat in air at ambient temperature. With large objects, this can lead to significant microbial growth in the outer layers before the inner part is defrosted. Low temperatures are unsatisfactory because of the excessively slow rate of thawing, but it is often considered that a satisfactory compromise is slow circulation of air at 5-7°C. Rapid thawing in warm air is commonly used and the application of saturated steam in a vacuum chamber has also achieved some popularity. In vacuum, the saturation temperature of steam is only 20°C, but heat transfer is very good and the method is efficient.

Microwave ovens are widely used for thawing frozen foods on a domestic and small catering scale and the principle has been adopted on an industrial scale. Continuous equipment is available using either microwave or radiofrequency heating. Equipment can be calibrated to permit rapid and complete thawing with only minimal temperature rise in the meat. Product being defrosted should ideally be of even shape and size, and there can be problems of localized overheating (runaway thawing). These have been much reduced in modern equipment. Some use has also been made of electrical resistance heating.

Commercial requirements usually demand the fastest possible thawing time. In the case of whole meat it is necessary to appreciate that the optimum rate for thawing is linked to the freezing rate. Few, if any, problems result from fast thawing of fast-frozen meat. Slow thawing of fast-frozen meat leads to growth of large ice crystals with associated leakage through cell membranes and collection of water (drip) in the extracellular spaces. Equally, fast thawing of slow-frozen meat does not permit fluid in the extracellular spaces to be readsorbed, leading to immediate high drip losses. As a corollary, the adverse effects of slow freezing can be minimized by slow thawing, which allows at least some fluid to be readsorbed. Matching thawing rate with freezing rate does, of course, require that the conditions of freezing are known to those responsible for the thawing process.

With carcasses or primal joints, complete thawing before butchery is often neither necessary nor desirable. Only a very limited temperature rise is required, the purpose being to raise the temperature sufficiently to facilitate further butchery. Problems are much less severe, apart from the limited temperature rise required, since sufficient ice remains to maintain a high thermal conductivity.

## 8.3 TECHNOLOGY OF FROZEN MEATS AND MEAT PRODUCTS

Freezing is not a process intended to enhance quality, or to change fundamentally the physical nature or eating quality of food. With some significant exceptions (see pages 372–373), freezing is a means of preservation and, as such, the technology amounts to a damage limitation exercise, the intention being to induce as few changes as possible. This means, of course, that positive quality determinants are largely independent of the freezing process. Basic quality management should be the same as that applied to non-frozen equivalent products. Within these constraints there are certain fundamental rules which must be obeyed if the prime objectives of freezing – to preserve without compromising organoleptic quality – are to be met (Table 8.2). Equally, it must be recognized that these rules are not in any sense specific to frozen meats or to frozen foods as a whole. They merely reflect good manufacturing practice.

### 8.3.1 Carcass meat and retail cuts

Although chilled meat is invariably considered to be of better quality, there is still a considerable international trade in carcass meat. This is primarily lamb, with rather less beef and only small quantities of pork. In addition substantial quantities of deboned rabbit are exported from China. Whole frozen chickens and turkeys are produced in large numbers, together with smaller numbers of ducks, despite a continuing trend to chilled birds. Production is

**Table 8.2**  Conditions necessary for the successful freezing of meats

---

1. Meat must be of initial good quality with respect both to microbiological and chemical status.
2. Pre-freezing processing must be in accordance with good manufacturing practice.
3. Meat must be frozen as soon as possible
4. Where delays are unavoidable, the meat must be protected from contamination and refrigerated to minimize microbial growth.
5. Freezing should be in accordance with predetermined parameters.
6. Frozen meat must be stored at correct temperatures **throughout** its storage life.

---

---

### BOX 8.3  The year of the pig

Although red meat consumption in the UK continues to fall, the Meat and Livestock Commission have been remarkably successful in developing new frozen ready-meals for caterers. Public house catering was targeted initially and will be followed by airline catering and in-store restaurants. Ideas are generated by consultants and catering colleges as well as by competitions for the general public. Basic ideas are then adapted to suit the requirements of the caterer. Pork is by far the most popular meat and is used to take up flavours in a manner analogous to poultry. Indeed pork from modern lean carcasses is almost considered to be a white meat. (*Frozen and Chilled Foods*, **May 1994**, 27.)

---

primarily for domestic markets. Imported carcass meat is usually broken down into retail cuts at the importing country but meat may also be butchered before freezing. There is a substantial domestic trade in frozen chicken portions and meat, and added-value variants have been developed.

The most important aspect of freezing red meat is the time of freezing in relation to onset of rigor mortis. If meat is frozen before ATP and glycogen levels are depleted, post-mortem glycolysis is suspended. On thawing, however, the meat undergoes severe contraction with associated toughening and loss of large quantities of drip (thaw rigor). For this reason meat has normally been frozen post-rigor. Despite this, freezing as soon as possible after slaughter is preferred and pre-rigor freezing is possible if precautions are taken to avoid thaw rigor. The ice matrix physically prevents shortening if ATP and glycogen levels are depleted while the muscle is still frozen. This can be achieved either by raising the temperature of the meat from $-20°$ to $-1°C$, or by holding meat at $-12°C$ for 20 days or at $-3°C$ for 5 days. As with matching thawing times to freezing times (see page 367), it is necessary for the user to be aware that the meat has been frozen in the pre-rigor state.

A more satisfactory solution is the use of electrical stimulation to initiate early onset of glycolysis and rigor and thus permit freezing shortly after slaughter. Use of electrical stimulation has consequences for optimum freezing rate. Based on pH value, drip loss, water-holding capacity and shear strength, the quality of nonstimulated beef was best at freezing rates of 2.5 cm/h, while that of electrically stimulated beef was best at 1.8 cm/h. Freezing of meat requires relatively large amounts of energy and hot deboning is common practice ahead of electrical stimulation to avoid the cost of freezing bones. Thaw rigor is not a problem with poultry, where post mortem glycolysis is very rapid and completed during progress along the processing line.

Freezing of carcass meat and retail cuts presents no special problems. Plate freezers may be used for deboned meat, but blast freezing is most common for primal joints and carcasses. A high freezing rate is generally desirable in terms of both quality and economics and very slow freezing must be avoided to prevent cold shortening before the muscle is frozen. Where blast freezing is used, it is usual to use plastic film wraps to prevent dehydration and protect from contamination during subsequent handling. Vacuum packaging may be employed with primal joints to minimize oxidative deterioration.

Retail cuts of red meat and poultry portions may also be frozen using blast freezers. Cryogenic freezing, however, leads to improved quality with small pieces. Brine immersion freezing has been used with larger cuts. This method is common, in conjunction with blast freezing, for freezing whole poultry. Mince can be frozen in final packaging using plate or blast freezers. Alternatively fluidized bed freezers may be used, the mince being packaged after freezing.

### 8.3.2 Manufacturing meat

Substantial quantities of meat from lower value parts of the animal are frozen for use in meat products. Mechanically recovered meat and meat surimi may also be frozen. General considerations are the same as for meat destined for retail sale, although presentation is unimportant and retention of functional properties is of prime importance. For convenience manufacturing meat is often comminuted before freezing and is particularly prone to loss of functionality through protein denaturation. Similar problems exist with mechanically recovered meat and meat surimi. Enhanced lipid oxidation also affects the functional properties of proteins. This factor limits the effective frozen storage life of some types of mechanically recovered poultry meat to less than three months, for example.

In recent years there has been considerable interest in use of cryo-protectants to minimize damage to functional properties. Addition of sucrose and sorbitol to minced pork before freezing was found to improve gelling properties during subsequent use in cooked meat products, but use of these compounds is limited by excessive sweetness. Various non-sweet alternatives have been successfully used including hydrolysed starch and non-fat dried milk powder. The most effective cryoprotectant currently available, however, is probably Polydextrose®, a non-sweet dextrose polymer. This is particularly effective when used in combination with polyphospates. Use of Polydextrose® and polyphosphates makes the freezing of salted pre-rigor beef feasible. This has not been possible in the past due to marked loss of functional properties on thawing.

### 8.3.3 Raw and cooked uncured meat products

A wide range of frozen uncured meat products is available. In many cases sales are in bulk packs, the main customers being caterers. Frozen uncured meat products are effectively of two types. The first consists of products such as sausages, pies and burgers, which are the direct equivalent of fresh products except that minor recipe modifications may be necessary. The second type consists of products such as restructured meat, where freezing is a pre-requisite of mechanical stability.

### (a) Frozen equivalents of fresh products

In most cases the manufacturing technology of fresh and frozen meat products is identical. An exception is pies where, although the filling is cooked, the pastry is usually raw. In other cases, it is necessary to modify the recipe to minimize the effects of a long period of frozen storage. In many cases the level of herbs and other flavouring is increased to compensate for loss during storage and also to mask incipient rancidity. Frozen sausages, for example, often contain monosodium glutamate as flavour enhancer and hydrolysed vegetable protein to increase overall flavour strength. Lipid oxidation is a problem in many frozen meat products and antioxidants are often added, where permitted. Addition of antioxidants is forbidden in some countries, although there may be attempts to overcome this by flavouring with herbs such as rosemary, which have antioxidant properties (cf. Chapter 5, page 266). Sodium chloride is a recognized pro-oxidant in frozen foods, and some protection from oxidation may be obtained by use of encapsulated NaCl.

Depending on the physical nature of the product, a greater or lesser quantity of water will be released on thawing the frozen product. For this reason fresh sausages (British type) are almost invariably of the 'skinless' variety to avoid accumulation of liquid between the meat and the casing. A water-binding agent to readsorb water may also be included in products such as sausages. Some types of protein concentrate or skimmed milk powder are suitable.

Meat products are often individually blast frozen, although plate freezers are used for some ready-meals. Cryogenic freezing has become popular and its advantages in terms of quality have been clearly demonstrated in burgers (see page 364) and some other products. Use of cryogens has also made it possible to freeze some types of emulsion sausage, which are of unacceptable quality when frozen using slower freezing methods.

## (b) Products depending on freezing for mechanical stability

The largest group of products of this type are restructured meats. In the current context these are products which have been manufactured from relatively large pieces of meat and formed into a roll. The physical nature of such products differs from re-formed or comminuted meat products but they occupy a similar market position. Despite restructured meats being a frozen product, they are generally regarded as being of superior quality to comminuted or re-formed meat and there may even be preference over whole steaks.

Manufacture of restructured products normally involves tumbling or massaging the meat pieces followed by moulding into a roll. Poultry meat is very popular, the soft texture allowing easy formation of exudate and thus assisting binding. Removal of connective

---

* In the UK frozen meat products, such as sausages and burgers, provide examples of the use of redundant preservatives in that sulphite, in one form or another, is often present. Sulphite is associated with a number of adverse effects, including destruction of thiamine and adverse reactions in consumers, especially asthmatics and is considered by many to be undesirable even in foods where a positive preservative role is acknowledged.

tissue is required with red meats, especially beef, and better quality is obtained by use of individual muscle blocks. Sodium chloride is added to enhance protein extraction during tumbling and polyphosphates are added to improve water binding. Binding between the meat pieces is not sufficiently strong to enable the product to be handled in chilled form and so freezing is necessary. Restructured meats may be retailed as a whole roll or sliced. In the latter case the slices are often coated and breaded. Restructured meats can also be cut into novelty shapes, such as dinosaurs for the children's market.

The second type of product in which an ice matrix is required for mechanical stability comprises two-component products in which meat (usually chicken) is combined with cheese, cheesebase or some type of sauce. Examples are small pieces of chicken enrobed with flavoured cheesebase, or a comminuted chicken mix moulded around cheesebase or a thick, starch-based sauce. Freezing and formation of an ice matrix imparts sufficient strength for mechanical stabilization of the non-meat component and to maintain the integrity of the entire product.

### 8.3.4 Cured products

Raw cured products, such as bacon, present no specific difficulties in freezing. Sodium chloride, however, is a strong pro-oxidant and the frozen product often deteriorates very rapidly. Polyphosphates are effective in minimizing rancidity but not eliminating it. Packing in nitrogen can also be used, but frozen bacon is not a premium product and the extra cost is difficult to justify.

Cooked cured meats have rarely been frozen in the past. The growth of salad and sandwich bars, however, has created a demand for long-life, pre-prepared ingredients, such as sliced ham. There is now growing production of frozen ham, which is sliced, cut into strips or diced and individually frozen. Cryogenic freezing

---

* A novel technology involving treatment of meat at pressures of *ca.* 150 atmospheres has been applied to experimental production of restructured steak. The meat is subjected to pressure before processing and this results (it is claimed) in improved water holding and protein solubilization properties. Pressure treatment is most effective when applied to pre-rigor meat and its use permits restructured steaks to be made without use of NaCl or binding agents.

---

### BOX 8.4  **Pizza to go**

In recent years in the UK there has been a rapid growth in sandwich bars and pizzerias serving the take-away and home (or office) delivery market. In many cases food preparation areas are minimal and there is a considerable reliance on pre-prepared ingredients. At the same time a considerable variety in terms of toppings or fillings is offered. Sales of some types are small and ingredients must have a long life to minimize wastage. This has created a demand for a new type of frozen catering product. In addition to simple products, such as sliced ham, a wide range of combinations is produced, including ham strips with sweetcorn and peppers and pepperoni slices with cream cheese and herbs.

---

gives markedly better quality, especially in terms of reduced drip on thawing. The ham is packed in bags of low moisture permeability and stored at $-23°C$, this temperature apparently maintaining quality more effectively than $-18°C$.

### 8.3.5 Packaging of frozen meat and meat products

A number of the quality problems suffered in the past by frozen meat and meat products has been attributed to poor packaging techniques. It is particularly important to protect frozen meat from dehydration, potentially leading to freezer burn, and this requires a film of low water permeability. Equally, considerable mechanical strength is required since the rigid nature of frozen meat can easily lead to damage to the packaging.

The packaging of frozen retail cuts of meat has been greatly improved by the introduction of films such as Sarlyn®. These films are of low water permeability and adequate physical strength and are shrink-wrapped on to the meat, eliminating any airspace between the product and the film. This minimizes formation of metmyoglobin and maintains levels of oxymyoglobin for extended periods in the dark. Pigments in frozen meats are highly prone to light-mediated oxidation but an ultraviolet impermeable layer can be incorporated into packaging films, which markedly improves appearance during prolonged display in the light. Alternatively, translucent or opaque packaging may be used. Bulk, individually frozen products, such as burgers, sausages or chicken pieces, are packed loose in polythene or other plastic bags, which require a considerable degree of mechanical strength.

## 8.3.6 Quality assurance and control

A considerable part of quality assurance procedures with frozen meats relate to production of the items to be frozen. These are not usually specific to frozen meats and are essentially the same as for the chilled equivalent. In the case of restructured meats, which are almost invariably frozen, overall quality assurance and control procedures are largely the same as those for uncooked comminutes.

Specific quality assurance procedures for frozen meats relate to the need to freeze the items as soon as possible after production, to freeze under the correct conditions and to minimize temperature fluctuations during storage and transport. In large operations production is continuous employing a high level of automation. Quality assurance largely involves monitoring the performance of equipment, especially with respect to temperatures. A contingency plan must be available in case of breakdown and food awaiting freezing should be refrigerated if delays occur. A high level of control is required in small operations where production is on a batch basis. Problems tend to result from delays in freezing and incorrect loading of freezers, particularly of the blast type. Formal quality management systems should be used to control these problems and a high level of supervision is necessary.

Minimizing temperature fluctuation during storage and distribution is a matter of providing suitable cold stores and refrigerated transport. Temperatures should be monitored at all stages. Problems at manufacturing level again tend to be greatest in small operations, particularly if operating over capacity. Mishandling at manufacturing level, however, tends to be minimal and insignificant in comparison with that occurring at retail and domestic level. Overfilling display cabinets is by no means restricted to small shops and examples of the practice are common in large supermarkets.

---

* Freezers do break down and failure can result in substantial financial loss at any stage from manufacture to consumption. Some multiple retailers operate a contingency plan in which totally defrosted food is discarded but food which still contains ice crystals may be sold for immediate use to caterers. Judgement of the suitability of partially defrosted food for use is subjective and decisions should only be taken by experienced and responsible personnel.

Frozen meat products should be subject to the same chemical and microbiological analysis as their chilled equivalents. Microbiological analysis of unprocessed frozen meat is common practice, particularly when the meat is imported.

## 8.4 CHEMICAL AND PHYSICAL PROCESSES

### 8.4.1 Ice formation and related events

The major changes affecting the quality of frozen meat tend to be those affecting tissue structure. To a large extent these are due to crystallization and size and location of ice crystals and they involve changes in water-holding capacity. This leads to drip production with resultant unpleasant appearance, loss of juiciness after cooking and enhanced microbial growth rate due to surface wetness. Changes affecting quality occur at all stages in the life of frozen meat, from initial freezing to final thawing.

### (a) Ice formation during freezing

During freezing of meat, the local freezing rate at the border of the cut meat is much higher than in the centre. A fairly steep temperature gradient is established across the meat and creates mechanisms such as nucleation, dendritic growth of ice crystals towards the centre, water migration through the sarcolemma and increases in solute concentration. During slow cooling, nucleation does not occur in the intracellular fibres. Nuclei in the extracellular spaces grow at the expense of water migrating out of tissues. The concentration of solutes increases in the extracellular spaces, drawing further water by osmosis from the unfrozen intracellular space. Extracellular nucleation also occurs at fast cooling rates. Release of water through the sarcolemma becomes the controlling step, however, and intracellular supercooling is attained leading to intracellular nucleation.

Extracellular and ultimately intracellular crystals formed at the refrigerated border grow dendritically towards the centre of the meat piece. The whole freezing process may be considered to occur in four stages (Table 8.3). The pre-cooling stage is considered to extend from the start of cooling at an initial temperature $t_i$, until the refrigerated border reaches nucleation temperature, $t_n$. When this occurs there exists in the product a temperature distribution which leads to a supercooled zone ($\delta$) extending inward from the

**Table 8.3** Stages of ice formation during the fast freezing of meat

1. Pre-cooling
2. Nucleation and subsequent growth of thermal dendrites
3. Partial remelting of thermal dendrites and consolidation of dendritic structure
4. Columnar growth

border. When nucleation takes place close to the refrigerated border the temperature of the nuclei is raised by heat of crystallization and approaches the equilibrium temperature $T_e$. As a result an unstable condition is generated on the interface of each nucleated crystal, leading to the formation of thermal dendrites. These grow laterally until contact is made between adjacent dendrites. Growth also occurs along the direction of the thermal gradient and is very rapid (Figure 8.2), being at the expense of the existing supercooling in the $\delta$ thickness zone. As a consequence of the effective elimination of supercooling, no subsequent nucleation occurs. Penetration by the thermal dendrites depends on freezing conditions. Under normal commercial conditions, however, it is unlikely that penetration would exceed 2 cm, a distance covered by dendrites in *ca.* 10 seconds.

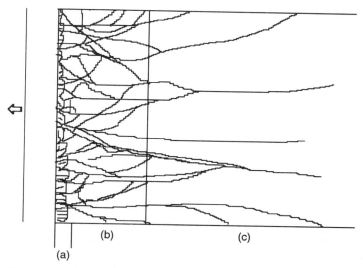

**Figure 8.2** Ice morphology in a sample of meat frozen under a temperature gradient: a = nucleation zone; b = free dentritic zone; c = cellular dendritic zone. Arrow indicates direction of heat flow.

Experiments with model systems have shown that the freezing front tends to retreat slightly after thermal dendrites cover the originally supercooled area. It seems likely that the heat removed by the external refrigeration system is used for advancing a consolidation front, while the remelting of thermal dendrites is a consequence of the flux coming from the zone in which $T > T_e$. Consolidation of the thermal dendrite zone requires conduction through the solid structure, lateral growth of the original dendrites and concentration of solutes. Under any conditions, thermal supercooling must eventually cease although constitutional supercooling can be established. This permits cellular or, more usually, cellular dendritic growth at a later stage.

In contrast to single component model systems, meats are multicomponent consisting of an intracellular and extracellular space. Events in the two follow a different course, while the whole situation is complicated by the fact that heat flow out of the meat may either follow (longitudinal) or be at right angles to the meat fibres. After nucleation thermal dendrites advance laterally and along the thermal gradient. This is followed by partial remelting of the extracellular dendrites and simultaneous consolidation of the residual dendritic structure. Consolidation of the extracellular dendritic structure leads to an increase in the solute concentration in the extracellular space and migration of water from the interior of the fibres. At a high rate of heat removal (fast freezing), supercooling and nucleation within the cells occurs followed by production of dendrites. When heat flow is longitudinal the dendrites advance towards the centre of the meat, but when heat flow is at right angles dendrite advance is lateral.

Following consolidation of the zone of free dendritic growth, subsequent development involves formation of a cellular dendritic structure. As noted above, this involves a selective process, with growth of large crystals at the expense of small. Average crystal

---

* During the growth of cellular dendrites, events are dictated to a large extent by a selective mechanism. This favours crystals orientated so that most rapid growth occurs in the plane of the thermal gradient. At the same time, large columns predominate over small. The combined effect of these two factors is that the number of columns decreases along the direction of growth. This pattern of crystal development, in which the larger and more favourably orientated crystals grow at the expense of the smaller, appears to be common in the material world and has been observed in metal alloys and several other situations.

size also increases with increase in the local freezing time ($t_c$) according to the equation:

$$D = c + d\log t_c$$

where $D$ is the average crystal diameter and $c$ and $d$ are constants.

Changes in the water-holding capacity during freezing lead to drip on thawing. These result from water migration into the extracellular space and distortion of the myofibrillar structure. Dehydration of the fibres and a marked increase (10-fold at $-20°C$) in solute concentration, together with myofibrillar distortion, leads to protein denaturation, which indirectly affects the water-holding capacity. Denaturation increases considerably with slower freezing rates and almost entirely involves the myosin head; the actin filaments and the myosin tail are largely unaffected. Denaturation, especially at low freezing rates, also involves cleavage of the actin–myosin bonds. Denaturation alone cannot, however, account for differences in drip at different freezing rates.

The relationship between the amount of drip and the freezing rate is relatively complex. At fast freezing rates, where both intracellular and extracellular crystals are formed, the amount of drip is strongly dependent on the freezing rate. Where the rate is very fast and a large number of very small intracellular crystals are present, drip production is very small. Intracellular crystals become larger with slower rates of freezing and there is a concomitant increase in drip production. Maximum drip production under fast-freezing conditions occurs when a single crystal occupies most of a fibre. Drip production in fast-freezing conditions is considered to result from distortion of the myofibrillar structure with increasing crystal size.

At slow-freezing rates crystals are located exclusively in the extracellular space. Drip production is effectively independent of freezing rate. Dehydration of the fibres leads to denaturation of the myosin head and loss of water-holding capacity. The reduced water-holding capacity means that the fibres are unable to reabsorb all of the water from the extracellular space when the meat is thawed. This water is then released as drip. Protein denaturation is also responsible for the loss of functional properties. Cryoprotectants bind water in the intracellular space and limit dehydration even during very slow freezing.

*(b) Changes during frozen storage*

Changes to the crystalline structure of ice in frozen meat occur during storage and are accompanied by further protein denaturation. Extent of change is directly related to the length of storage. Changes to the ice structure tend to involve recrystallization and an increase in the average crystal size. Large crystals increase in size at the expense of smaller ones and the number of crystals falls as the average size increases. The process is driven by differences in free energy between crystals of different size and the overall tendency of the system to minimize free energy. Water transport between crystals of different sizes involves a melting process at the small crystals, with diffusion and solidification on the large crystals. Recrystallization occurs to some extent at constant temperature but is greatly accelerated where temperature fluctuations occur, leading to melting of the small crystals.

Protein denaturation continues during storage, which tends to enhance the effects of initial freezing. During 15 weeks of storage of fast-frozen meat, containing both intracellular and extracellular crystals, denaturation of the myosin head increased from 19% (immediately after freezing) to *ca*. 60%. This was independent of the storage temperature. Cleavage of actin–myosin bonds continued, although this effect was temperature dependent and denaturation of the myosin tail also occurred. A direct relationship exists between myosin tail denaturation and storage temperature, the rate increasing significantly as storage temperatures increase.

In slow-frozen meat the majority of the myosin head denaturation (40%) occurs during freezing. Denaturation during storage is slow and only a further 20% occurs over 40 weeks. Myosin tail denaturation and cleavage of actin–myosin bonds follow a similar pattern in both fast- and slow-frozen meat, while actin and sarcoplasmic proteins appear unaffected by frozen storage.

Dissociation and denaturation of the myosin head appears to be a continuous process, which largely occurs during freezing where freezing rates are slow and during storage when freezing rates are high. Myosin tail denaturation appears to be the consequence of long periods at high solute concentration and is accompanied by formation of insoluble aggregates. The overall effect of changes during storage is an increase in drip production as storage time and temperature increase, with a corresponding loss of functional properties.

## (c) Changes during thawing

At the onset of thawing, the reduction in water-holding capacity has already been determined by freezing rate and frozen storage. Exudate production after thawing, however, also depends on thawing rate, although the actual water-holding capacity is not affected. Under equilibrium conditions during storage, the $a_w$ level in the intracellular and extracellular spaces will be similar, irrespective of the size and number of ice crystals. Melting of ice in the extracellular space during thawing raises the $a_w$ level, leading to migration of water through the sarcolemma towards the still frozen intracellular space, where the $a_w$ level is higher. Water reaching the intracellular space is reabsorbed by the partially dehydrated fibres. This results in the establishment of a series mechanism, involving the melting of ice in the extracellular compartment, at a rate proportional to the thawing rate, and the reabsorption of water in the intracellular space. The difference in the $a_w$ level between the extracellular and intracellular compartments is dependent on thawing rate, being greatest at high thawing rates. Under these conditions the overall process is controlled by the ability of dehydrated fibres in the intracellular space to reabsorb water. Water which is not absorbed accumulates in the extracellular space and is ultimately released as drip.

## 8.4.2 Colour of frozen meat

The haem pigments of frozen meat are the same as those of fresh meat (see Chapter 1, pages 27–29). Myoglobin in frozen meat is susceptible to oxidation to metmyoglobin and there is a strong association with fat rancidity. Under any circumstances, frozen meat can appear darker than fresh. This stems primarily from concentration of the pigments during the freezing process and is offset in rapidly frozen meat by the reflectance of the small ice crystals. This can make the meat appear unacceptably light in colour.

Dehydration at the surface also concentrates pigments and favours formation of metmyoglobin. In extreme cases this results in 'freezer burn', which is a consequence of sublimation of ice from unprotected surfaces and severe dehydration (see pages 362–374).

Freezer burn is apparent after thawing and affects appearance, eating quality and functional properties. Severe dehydration is necessary, although 'freezer scorch', which is seen as small white to grey areas, is an earlier manifestation of dehydration. Freezer scorch is reversible and usually disappears when the meat is thawed.

### 8.4.3 Deteriorative changes in fat

Deteriorative changes in fat generally occur slowly in frozen meat. Two types of change may occur, oxidative rancidity and lipolysis, the former generally being of greater importance.

### (a) Oxidative rancidity

Lipid oxidation during frozen storage of meats is a well established phenomenon that can lead to significant loss of organoleptic quality. There are two freezing-related factors which affect the rate of oxidation. In the first place, the lowering of temperature reduces the rate of chemical reactions, while in the second, reactions proceed more rapidly due to the concentration of reactants in the remaining free water. Lipid oxidation virtually ceases at $-30°C$, when effectively 100% of water is frozen, but at temperatures above *ca.* $-20°C$ the increase in reaction rate due to concentration is greater than the decrease due to low temperature. Lipid oxidation occurs most rapidly at *ca.* $-2°$ to $-4°C$, when most of the water is frozen and enzyme activity is fairly high. A rapid onset of rancidity may be expected during storage in domestic 'one-star' (storage at *ca.* $-6°C$) and even 'two-star' (storage at *ca.* $-12°C$) freezers.

Under standardized conditions, resistance to oxidation is determined by the degree of unsaturation (e.g. beef > chicken white meat > chicken dark meat). For a number of years, there was considerable debate concerning the relative importance of phospholipids and triacylglycerols in development of oxidative rancidity during frozen storage. Results of many studies were directly contradictory, some indicating that triacylglycerols were primarily involved, while others showed the triacylglycerols to be largely unaffected, but that significant losses of phospholipids occurred. To a large extent these discrepancies resulted from differences in time scales and it became apparent that lipid oxidation in frozen meat is a two-stage process. The first stage, which occurs over as little as 3 months, involves oxidation of phospholipids, while tria-cylglycerols are oxidized during the second stage, which commences after storage for 5–6 months.

The lability of phospholipids to oxidation is partly a consequence of their high content of unsaturated fatty acids, especially linoleic and arachidonic acids. These fatty acids exist in close association with cell membranes and are thus susceptible to processes such as comminution. Comminution disrupts the membranes, leading to exposure to oxygen, enzymes, haem pigments and metal ions, which can lead to rapid rancidity. For this reason rancidity can be a major problem in frozen minced meat where antioxidants are not usually permitted.

Sodium chloride is a powerful pro-oxidant and is involved in rapid onset of lipid oxidation in frozen bacon and, to a lesser extent, in other products such as burgers. A number of theories have been proposed to explain the catalytic effect of NaCl, but most are unable to account for all observed phenomena. Mechanisms have still not been fully elucidated, but it seems probable that the catalytic effect of NaCl is derived from enhancement of the pro-oxidative effect of chelatable iron atoms. The mechanism involves NaCl displacing Fe ions from binding macromolecules which, in the absence of NaCl, limit the ability of Fe ions to participate in oxidative reactions. Polyphosphates, which limit the pro-oxidative effect of NaCl, appear to stabilize Fe binding.

## (b) Lipolysis

Rancidity due to enzymic lipolysis occurs in frozen meat, although its significance is usually less than oxidative rancidity. Both lipases and phospholipases are involved, resulting in release of free fatty acids. The free fatty acids are then subject to oxidation. The rate of oxidation of free fatty acids derived from triacylglycerols is greater than the rate of oxidation of the original triacylglycerols. This is not true, however, of free fatty acids derived from phospholipids. In this case, generation of free fatty acids inhibits subsequent oxidation. This phenomenon has not been adequately explained.

## 8.4.4 Chemical analysis

There are no analytical procedures which are applied to frozen meat products and not their chilled equivalents. Purchasers of frozen meat may require determination of peroxide value (oxidative rancidity) or free fatty acids (lipolytic rancidity), although the predictive value of these tests is debatable.

There have been anecdotal reports that immunological methods for determination of meat species cannot be used with frozen meats. In general terms, there appears to be no justification for these reports, although there can be an occasional difficulty due to weak false–positive reactions with pork.

In some circumstances, a purchaser may wish to know if meat has been frozen at any stage in its history. Disruption of membranes, including those of the mitochondria, occurs during freezing and results in delocalization of some enzymes. This can be exploited in tests for differentiation between frozen meat which has subsequently been thawed and meat which has never been frozen. Various enzymes have been chosen as markers, including mitochondrial cytochrome oxidase and β-hydroxyacyl-CoA dehydrogenase.

## 8.5 MICROBIOLOGY

Micro-organisms do not grow below *ca.* $-10°C$ and considerations of spoilage are normally relevant only to handling before freezing or during thawing. In these contexts, frozen meats behave like their unfrozen counterparts, although growth rates are often faster after thawing. This results from the release of drip. In the past, considerable quantities of carcass meat were imported at temperatures of $-5$ to $-10°C$. At these temperatures psychrotrophic moulds such as strains of *Cladosporium*, *Geotrichum*, *Mucor*, *Penicillium*, *Rhizopus* and *Thamnidium* grow slowly and, during prolonged storage, cause spoilage through development of 'whiskers' or 'spots' of various colour depending on the species of mould.

Very little meat is stored at these temperatures in modern commerce and mould spoilage is largely of historic importance. Despite this, the topic is still much discussed and ghostly fleets of ships carrying cargoes of black, green and white spotted (and even luminescent) carcasses continue to sail across the pages of some food microbiology textbooks. Mould and yeast growth may also be a problem with re-formed meats held at $-5°C$, but these are not considered as true frozen meats.

Some death of micro-organisms may be expected during freezing and subsequent storage. This is largely a consequence of the lowering of $a_w$ level and the effect is most marked at low freezing rates. The practical consequences are at most limited and frozen storage cannot be seen as a means of reducing the load of pathogens.

Microbiological analysis of frozen meat products largely reflects that applied to the chilled equivalent. Microbiological analysis of imported frozen meat may also be carried out to ensure that temperature control and hygiene standards have been adequate. Standard methods may be used. With the exception of frozen poultry, where examination for *Salmonella* may be used as a means of monitoring the success of control programmes, it is not usual to test for specific pathogens. If analysis is required, however, pre-enrichment (resuscitation) is a necessity.

## EXERCISE 8.1

You have recently been appointed as meat technologist at an abattoir using electrical stimulation to permit rapid chilling and freezing of lamb. Over the last 12 months complaints have been received of thaw rigor affecting a significant number of carcasses. The abattoir operates a 7-day week and it appears that the problem affects animals killed during weekend working. Your predecessor formulated three hypotheses: animals being held for a longer period in the lairage; problems with the operation of the electrical stimulation apparatus due to observed power supply difficulties; problems with the operation of the electrical stimulation apparatus due to the less experienced operatives employed at weekends. Consider these possibilities from a theoretical viewpoint and determine what further information is required. Draw up a systematic plan for investigation and resolution of this problem, bearing in mind the need for a rapid solution, the 'hire and fire' attitudes of your employers and the difficulties of carrying out investigations in a situation where interruptions to throughput are not tolerated.

## EXERCISE 8.2

You are employed as a chemist by a small frozen foods manufacturer which wishes to develop a new range of 'natural' ready-meals. The chief technologist wishes to examine natural antioxidants and has produced a range of trial dishes. There is considerable pressure to develop the products as rapidly as possible and the chief technologist has asked you to devise accelerated tests to determine the relative effectiveness of the antioxidants. To what extent do you consider accelerated tests to be valid in assessing antioxidants in frozen meats? Depending on your opinion of accelerated tests, write a short report either describing the most suitable approach and any precautions to be taken in interpretation of results, or describing why accelerated tests are not valid for this purpose.

# 9

# DRIED MEATS, INTERMEDIATE MOISTURE MEATS AND EXTRACTS

---

## OBJECTIVES

After reading this chapter you should understand
- The various types of dried and intermediate moisture meats and meat extracts
- The basic principles of air- and freeze-drying and the application to meats
- Formulation of intermediate moisture meats
- Manufacture of meat extracts
- Uses of dried meat and meat extracts in other foods
- Quality assurance and control
- Chemical changes during drying and storage
- Microbiology of dried meats, intermediate moisture meats and meat extracts

---

## 9.1 INTRODUCTION

Drying of meat is a very old process which, for obvious reasons, was most widely practised in hot countries where sun-drying was possible. Large quantities of meat are still sun-dried in Africa using traditional techniques that have been unchanged for a vast number of years. Improvement of these techniques, without changing the underlying technology, is seen as being of great potential value in raising nutritional standards in parts of Africa today (see pages 394–395). Modern hot air-drying techniques have been applied to meat in temperate climates. There are inevitable textural changes and air-dried meat is almost entirely used in other products, such as canned soups in industrialized countries. A new generation of dried meat-based snack products is now, however, under development.

---

### BOX 9.1 **The armies of the night**

Hot air-drying of meat was practised on a large scale in the UK during the Second World War, when much of the fundamental technology in use today was developed. The intention was to have available, in the event of extreme emergency such as invasion, a large stock of meat which was stable without dependency on refrigeration and hence power supplies. Retention of nutritional properties was seen as one of the prime objectives. Large quantities of meat were successfully dried and subsequently used in military rations. None was released direct to the public, but a quantity was used in processed meat products.

---

Freeze-drying, an alternative means of drying, was originally seen as offering an alternative means of preservation to freezing. In contrast to hot air-drying, large pieces such as steaks can be dried, the quality after reconstitution being considered very good. Freeze-drying is, however, an expensive process and its use is very restricted. Small quantities of freeze-dried meat are used in what are effectively dehydrated ready-meals, where rapid rehydration is required, and use is also made in specialist applications such as meals for long-distance walkers.

Traditional intermediate moisture meat products, such as fully dried fermented sausages, are still made in large quantities. A new generation of intermediate moisture meats was envisaged, stemming from the development of moist, temperature-stable intermediate moisture meat meals for the US manned space programme and, perhaps oddly, the success of intermediate moisture pet foods. Very little progress has been made with new intermediate moisture meats for human consumption as a consequence of continuing problems with acceptability and palatability. At present the greatest likelihood of commercial success for products of this nature lies in the innovation-hungry snack industry.

Meat extracts have been prepared since the 18th century, although commercial-scale processing developed in the 19th century with the growth of beef production in South America. Meat extract was a means of utilizing the meat of animals slaughtered for their hides and fat. Corned beef manufacture was originally a by-product of meat extraction processes. Meat extracts are primarily used in other products but also form the basis for consumer products, including stock cubes and beef beverages.

## 9.2 TECHNOLOGY

### 9.2.1 Drying as a generalized process

Drying, by any means, is a method by which microbial growth is arrested by deprivation of moisture. Many chemical reactions are also retarded. Drying is not a sterilization process and means must be provided to preserve the equilibrium and prevent foodstuffs regaining moisture before deliberate reconstitution. Removal of moisture is not the only criterion, however, since the material dried must retain the capability of returning to a condition resembling the original after reconstitution.

Although conventional drying involves the extraction of all or part of a solvent, which is usually water, the inhibition of micro-organisms is not determined by water content alone, but by the availability of the water to the micro-organisms. This is usually expressed as water activity ($a_w$) level, which is the ratio of the vapour pressure above a solution to that above pure water at the same temperature:

$$a_w = P/P_o$$

The basis of intermediate moisture foods is the lowering of $a_w$ by solutes sufficiently to inhibit microbial growth (see pages 401–402). The relationship of $a_w$ level to microbial growth is illustrated in Table 9.1.

### 9.2.2 Air-dried meat

#### (a) Principles of air-drying

In air-drying, hot air is the vector supplying the external exposed surface of the meat with energy and also the vector which removes

* Water activity level effectively indicates the difficulty further water has in moving out of solution and gives a simple measure of what may be a complex situation. The term 'activity of water content' has been applied to very low moisture conditions when erratic molecular migration may occur. This concept is generally thought of as concerning the availability of water for microbial conversion as well as enzymatic and chemical reactions. This can be used as a guide to the effectiveness of a particular condition of dryness in retarding the elements of quality loss.

**Table 9.1** Relationship between $a_w$ level and microbial growth

| $a_w$ level | Microbial growth |
|---|---|
| 0.90 | Lower limit of growth for most bacteria |
| 0.86 | Lower limit for staphylococcal enterotoxin production |
| 0.85 | Lower limit for growth of many yeasts |
| 0.84 | Lower limit for staphylococcal growth |
| 0.80 | Lower limit of growth for most fungi |
| 0.70 | Lower limit of growth for most xerophilic fungi |
| 0.60 | Extreme limit for xerophilic fungi and osmotolerant yeast |
| 0.55 | DNA disordered |

the resulting water vapour. The air-drying process is consequently controlled by the properties of the moist air surrounding the meat (external process variables), together with the properties of the product itself (internal process variables). The relative importance of external and internal processes can be predicted by a drying curve. In parallel with this competition the thermal and mass transfers take place at their own individual rates. In an air-drying process, the thermal transfers are many times faster than the mass transfers.

The air-drying process is fundamentally a diffusion-like one. The external driving force is the difference between the water vapour partial pressure at the surface of the meat and the water vapour partial pressure of the air remote from the meat surface. This can be expressed by two boundary conditions:

$$F_m = h_p \, (P_{vsurf} - P_{v\delta}) \text{ for moisture evolution}$$

$$Q = h_T \, (T_{surf} - T_\delta) \text{ for temperature evolution}$$

where $F_m$ is the mass of water evaporated, $Q$ is the energy transferred per unit time and square area, $h$ is heat capacity, $P_v$ is partial pressure, $T$ is temperature (°C), $_{surf}$ is the descriptor for the meat surface and $_\delta$ the descriptor for dry air.

The air-drying process is often considered to consist of three stages: heating, constant-rate drying and falling-rate drying. The main condition in the early part of the process is free water movement with a large quantity of fresh air. As drying continues, resistance to transfers becomes the limiting factor and higher temperature rather than faster air flow is required for efficiency. Air of lower relative humidity is required at the end of drying to obtain the required low moisture content in the meat. Factors such as case hardening and shape distortion due to shrinkage can complicate the general situation and lead to reduction in drying rates.

### Heating

During the heating stage, the temperature of the wet material rises to the wet bulb temperature of the drying air. This is a very short stage in relation to overall drying time and heat transfer is the main effect.

### Constant-rate drying

Constant-rate drying is an isoenthalpic regimen depending only on boundary conditions. A surface phenomenon is involved initially, during which free water is removed with a constant mass flux. Evaporation of water occurs at the extreme outer surface and the stage extends as long as free water is available from the interior of the meat. The process is very similar to evaporation of a water film and the limiting factor is the rate of moisture evaporation at the outer surface. This is dictated by the characteristics of the drying air.

### Falling-rate drying

Falling-rate drying is usually the longest stage, during which free water at the outer surface becomes increasingly scarce. The temperature of the meat in contact with air reaches the hygroscopic moisture threshold or reaches a zero moisture content for non-hygroscopic material (material without bound water). This latter state is rare in foods. The prevailing conditions mean that the vaporization 'drying front' at the extreme surface moves into the meat. Water from the interior moves towards the drying front under capillary forces, while bound water (hygroscopic material) and water vapour are removed from the meat by diffusion. The reduction in the drying rate is a consequence of the sudden and often dramatic decrease in the effective area of surface transfer, which results from the shortage of free water. The process during falling-rate drying is governed by resistance to heat and mass transfer and the rate of drying depends on the rate of moisture migration from the interior to the exterior.

### (b) Methods of air-drying

In recent years there has been a revival of interest in solar drying in countries of ambient high temperature and long hours of sunshine.

Two basic types of solar drier are available: the direct and the indirect. In the direct type meat is placed behind a transparent cover and energy is primarily provided by radiative transfer to the meat, which acts as a solar collector, and to parts of the drying cabinet. This type of drying therefore differs from other types of hot air-drying, convective air acting only to remove water vapour. In contrast, in the indirect type of solar drier the construction is designed to enhance the convective effect. Heated air both supplies energy and removes water vapour. The characteristics of the air, however, are closely linked to the diurnal cycle of the sun, via the efficiency of the solar cycle.

Where solar drying is not possible it is necessary to use heated air. Various types of drier are available, operating on both a batch and a continuous basis. All types are suitable for drying a range of foods and none are specifically designed for meat. The simplest is the cabinet drier in which meat to be dried is placed on trays within an enclosure fitted with a heater, a circulatory fan and a means of controlling the airflow. Cabinet driers are not suited to large scale production, but find an application where production runs are short.

Belt driers are widely used and represent a relatively simple means of continuous drying. They consist of a stainless steel continuous belt which carries trays of meat through a tunnel comprising a number of sections. The airflow, air temperature and humidity in each section is independently controlled. In some cases the belt is perforated and hot air is blown upwards through the drying product. Dryers of this type have similar characteristics to overdraught driers. The simplest type of belt drier has only a single belt, but others have a number of belts in vertical array. These travel at different speeds so that meat being dried passes quickly and in a thin layer through the earliest and hottest part of the machine, but more slowly and in thick layers through later and cooler parts.

Overdraught driers are now very widely used for general drying purposes and may be of either the batch or the tunnel type. Meat is arranged in shallow multilayers with high velocity airflow directed across and between surfaces. Large volumes of air are required and, where possible, recirculation is used to reduce costs. Recirculation must be controlled, however, to avoid raising the humidity sufficiently to interfere with drying. Design of overdraught driers is relatively complex, requiring large-capacity fans, air heaters and ducting between the heaters and the drying chamber(s) which must be arranged to provide very even heat distribution. Dampers are installed to control air recirculation and flow through intake and exhaust ducts.

Meat is dried on trays, which are placed on wheeled trolleys. In the continuous type of overdraught drier the trolleys pass through a tunnel, airflow either being cocurrent (in the same direction as trolley movement), countercurrent (in the opposite direction to trolley movement), combined cocurrent and countercurrent, or transverse (perpendicular to trolley movement). Air recirculation is not possible with cocurrent driers due to the low drying capacity of the air at the exit end of the tunnel and fuel consumption is significantly higher than that of countercurrent driers. Recirculation is possible with this type, which is also of simple design and widely used. The combined design is often considered most effective. This may consist of two parallel drying tunnels. Undried meat enters the first, 'wet' tunnel, which is of the cocurrent type, where it is partly dried and then passes to the second, 'dry' tunnel which is of the countercurrent type. An alternative combined design, which requires less floor space, consists of a single tunnel with heaters at each end and a central exhaust fan. The first half of the tunnel is thus effectively a cocurrent drier and the second half effectively a countercurrent drier, through which the trolleys move in stages.

In all combined overdraught driers, the second stage must be longer than the first to allow for the slower drying rate. Alternatively a finishing drier may be used to complete drying. These are usually batch driers consisting of a bin or deep trays through which air at *ca*. 55°C is blown for several hours. The transverse type of overdraught drier has a very high evaporation rate and is of good thermal efficiency. Trolleys must fit very tightly, however, to maintain the correct airflow. Transverse driers are economical with floor space and a very high level of control is possible, but drier design is more complex and the need for several heating units means that power consumption is high. Where transverse driers are used, it is usual practice to dry the meat to *ca*. 20% moisture and then transfer to a bin drier to complete drying.

Fluidized bed driers are very popular for general food dehydration and are successfully used for meat. The general design is the same as that of fluidized bed freezers (see Chapter 8, pages 362–363), except of course that hot air rather than cold is blown through the bed. The airflow must also be capable of removing large volumes of moist air. Fluidized bed drying is usually a two-stage process involving two driers arranged as a cascade. Two plug-flow driers may be used or, alternatively, the first drier may be of the well-mixed type. Particle mixing is perfect in the well-mixed type and initial drying of meat is rapid. This type of drier is also highly suited to handling sticky, partially dried material. Meat pieces leaving the drier are not, however, of equal moisture content and plug-flow driers must be used for final drying. A further disadvantage of well-mixed driers is that the residence time is very uneven and some pieces can remain in the drier for extended periods. This can result in significant loss of quality.

### (c) Production of air-dried meat

Traditional dried meats are not usually cooked before drying but are cut into strips and, in the case of charqui in South America and biltong in South Africa, lightly salted. The meat is placed on racks, or merely on the ground, and dried in the sun. Drying may be relatively slow, allowing microbial growth to take place, although the increasing NaCl content diminishes growth rates as the water content falls. Meat produced in this way is also subject to environmental contamination.

Modifications have been made to traditional dried meat products to improve quality and nutritional status. This may be illustrated by reference to the Sudanese dried meat, sharmoot. This is a part of the staple Sudanese Asida meal. In this, reboiled sharmoot is cooked with powdered dried okra to form a thick sauce and is served on a jellied substance prepared from boiled, fermented sorghum meal. Traditional production of sharmoot is compared with a proposed 'improved' method in Figure 9.1. It is notable, however, that the 'improved' method involves use of imported hot-air ovens, requiring electrical supplies for operation. A more viable method for improving the production of traditional dried meats in non-industrialized countries may well involve use of high performance solar driers, several designs of which are available (section 9.2.2(b)).

---

* There is considerable interest in drying foods in a bed of inert particles. This permits a high rate of drying at low air temperatures. Several materials have been used to form the inert bed, but activated alumina appears to be most successful for drying meat.

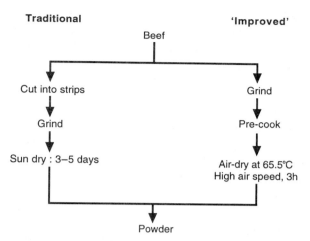

**Figure 9.1** Traditional and improved methods for production of the Sudanese dried meat, sharmoot.

In industrialized countries, meat is cooked before drying. This is not necessary from the strictly technical viewpoint but greatly improves quality by minimizing case hardening. The usual process is to cut meat into slices and then cook, mince or cut it into small cubes to provide a large surface:volume ratio, and dry it (Figure 9.2). The extent of pre-cooking is important, since overcooking

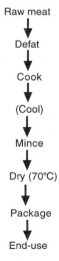

**Figure 9.2** General method for industrial-scale production of dried meat.

degrades connective tissue, giving dry granules of unsatisfactory texture. Undercooked meat is slow to dehydrate and rehydrate and has a dry and brittle texture. Water is normally used for cooking. Leaching of nutrients and flavour compounds occurs and these may be returned to the dried product along with any fat which has rendered out.

Care must be taken to prevent contamination of the meat after cooking, and handling at this stage is an important control point. Cooked meat may be transferred direct to the drier, but if this is not the practice the meat should be cooled and refrigerated before drying.

A high fat content (>35% dry weight) in the meat significantly reduces the drying rate, while above 40% rendering-out occurs. Below 35%, however, fat content has little effect on drying rate and continuous drying at a fixed drying time is used. Fatty meat is usually trimmed before drying and defatting may be used, but low-grade meat from a lean carcass is preferred.

Various types of equipment are suitable (section 9.2.2(b)) but dried meat is not a premium quality product and choice is largely dictated by drying economics. Continuous belt ovens are very popular, although there is increasing interest in fluidized bed driers. Temperature of drying is of considerable importance with respect to quality of the dried product. A temperature as high as 80°C can be tolerated for *ca.* two hours from the onset of drying, but a temperature as low as 50°C can cause damage when the water content is low. A temperature of 70°C can be tolerated throughout the drying process without significantly lowering quality. The size of the meat particle is not of itself important in determining drier performance in the range 0.3–0.8 cm diameter. Where the meat is dried on trays, however, particle size influences density of loading, which has a direct effect on drying time. The final water content of the meat should be 4–8%.

Dried meat is sometimes compressed to reduce volume and facilitate handling. This is achieved by pressing into cans or moulds to form blocks, at a pressure of *ca.* 20 000 kPa. There is a tendency for fat to render out at this pressure and pressing should be at 0°C to minimize this phenomenon. Dried meat, whether pressed or not, must be protected from moisture and should be packed in material of low water permeability. Product for export to hot climates may be packed in cans and flushed with nitrogen to minimize fat oxidation.

Dried meat snacks are made from high quality meat of low fat content which is cooked, cut into very thin flakes and air dried. Details of technology are proprietary.

---

### BOX 9.2  Pecos Pete

The Western Trading Post company of Clay Cross, Derbyshire, UK, was awarded a £90 000 grant by the Department of Trade and Industry for development of a 'Western-themed snack food'. These are based on beef jerky and are expected to be competitive with potato crisps. The products, Pecos Pete's steak flakes, are promoted by not only the eponymous Pete but also Black Bart, a former Wells Fargo worker who, disaffected, has taken to robbing his former employees. (*The Guardian*, August 2nd 1994.)

---

Dried meat of any type requires moisture-impermeable packaging. Film containing an aluminium foil layer is ideal for consumer packs of dried beef snacks. A layer of high mechanical resistance to puncture must be incorporated. Dried meat for incorporation into other products is usually packed in heat-sealed plastic bags of very low permeability to moisture, often protected by a cardboard outer. Air is evacuated unless storage is only for very short periods.

### 9.2.3 Freeze-dried meat

#### (a) Principles of freeze-drying

Freeze-drying at its simplest involves two physical processes: freezing and sublimation (Figure 9.3). Ice is formed from the natural liquid content of the meat (or other food) and the frozen segregate is then extracted directly as vapour. Sublimation is carried out under carefully controlled conditions of temperature and pressure to leave the fundamental structure restorable. The basic dehydrating factor is freezing, extraction of the frozen moisture as vapour being secondary.

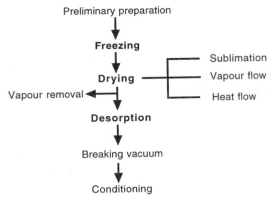

**Figure 9.3** Main stages in freeze-drying.

Freezing dictates the form of ice and accessibility for extraction. Very rapid freezing and thus production of a large number of small crystals is not desirable, since true sublimation is difficult or slow from within cells. Slow freezing produces large crystals which sublimate very readily, but may damage the structure. It is thus necessary to use an intermediate rate of freezing which balances ease of sublimation with minimal structural damage.

Sublimation commences at $-20°C$ in a vacuum of less than 1.33 mbar, provided that sufficient heat is available from the surroundings (benign heat). In commercial practice, heat is normally supplied by radiant heaters to the outer surface of the meat. As sublimation proceeds, the outer dried layer acts as an increasingly effective insulator of the remaining inner core of ice crystals. Heating the inner core through the insulating material is thus the rate limiting factor. Various means of increasing heat flow have been investigated. In the accelerated freeze-dried (AFD) technique, heated expanded metal plates are pressed against the product. These supply heat by conduction while allowing vapour to escape.

Vapour flow from sublimation at the core of the meat can also cause problems and severe restriction may occur due to the outer layer of porous dried material. The extractive diffusion coefficient decreases with decreasing pressure until the mean free path of gas molecules at intermediate pressures becomes comparable with the pore diameter. Wall collisions then begin to compete with collisions in the gas stream itself in limiting flow by diffusion. This condition is known as transition flow and is the dominant vapour flow in the freeze drying of all types of food. Vapour flow may be increased by pressure cycling. This involves periodically increasing (two-thirds cycle) and decreasing (one-third cycle) the vacuum, which leads to 'vapour flash' at the surface of the product. This raises the thermal conductivity of the gas and causes a small increase in the temperature of the ice interface. The overall drying time of cooked meat can be reduced by up to 34% in this way.

Water vapour must be removed from the sublimation chamber. The usual means has been freezing on to a refrigerated coil, but better results are obtained by use of desiccants. This creates a high and sustained vapour pressure driving force.

A certain amount of water remains unfrozen and cannot be removed by sublimation. This residual water together with incompletely sublimed ice, which is usually present under practical conditions, is reduced by desorption (secondary drying). This involves heating under a vacuum of 0.13–0.67 mbar to a surface temperature of not greater than 60°C. Radiant heating is most commonly used, but microwave heating has been introduced and has a number of advantages, including lack of surface browning and ease of control. One of the main benefits of the use of desiccants to remove water vapour is the ability to use significantly lower temperatures for desorption.

Desorption marks the end of the freeze-drying process proper and when completed the vacuum is broken, usually under a nitrogen atmosphere. Depending on drier design, significant temperature gradients may exist in the cabinet leading to uneven drying. Conditioning is employed to achieve even distribution of residual water and involves combining product from all parts of the drier and storing for 24 hours in an airtight cylinder.

## (b) Types of freeze-drier

Most freeze-driers are of similar basic design, differences lying in the method of supplying heat and removing water vapour. Dryers

* Microwave heating has also been proposed as a means of supplying heat to sublimate ice from the frozen material. Despite a number of potential advantages there are continuing problems of localized overheating leading to melting of ice rather than sublimation. More alarmingly it has been reported that under conditions in the primary drying phase, microwaves may ionize the air with a consequent risk of explosion.

may be either batch or continuous, but in each case the basic construction is a vacuum chamber fitted with a means of applying heat to the frozen meat and of removing sublimated water vapour (see above). Meat is dried in shallow trays carried in racks on wheeled trolleys. Where conduction heating is used, heating plates are positioned in the vacuum chamber to intersperse with the shelves of the trolley, which is positioned by guide rails. Batch driers are loaded and sealed before the vacuum is pulled, but in the continuous type a vacuum must be maintained throughout the operating period. This means that trolleys must enter and leave the drying chamber through a pressure lock at each end. The chamber is of considerably greater length than that of batch driers, to permit adequate residence time for drying to be completed. The chamber of continuous freeze-driers is divided into sections to allow for different conditions during primary drying and desorption.

### (c) Production of freeze-dried meat

In contrast to air-drying, either raw or cooked meat can be freeze-dried successfully. In many cases the meat is intended for eating after reconstitution and minimal heating. Cooked meat is invariably used for this purpose, the cooking process being required to inactivate vegetative pathogens. Some types of meat sausage have also been freeze-dried for special purposes, and meat dishes such as stews are freeze-dried for military and expedition purposes. In such cases, the prime objective is light weight and ease of preparation rather than organoleptic properties.

Muscle and fatty tissue cannot be dried simultaneously and fat causes problems by melting and interfering with vapour flow by blocking pores in the dried part of the meat. For this reason lean meat is selected for freeze-drying, excess fat being trimmed off. The meat is moulded into blocks and frozen. Blast freezers are normally used and it is common practice to freeze the meat to *ca.* $-20°C$ to ensure the majority of the water is converted to ice and then allow the temperature to rise before extraction of the frozen water. Care must be taken not to allow the ice to melt or the crystalline structure to change. Temperature rise may occur during operations after freezing and transfer from the freezer to the drying cabinet. Alternatively blast freezers may be used in controlled cycles to bring the meat to the desired temperature before drying. Evaporative cooling may be used, but this effectively involves drying from the liquid phase to extract latent heat. This leads to an unacceptable level of protein denaturation in raw meat, but evaporative cooling is acceptable with cooked meat where proteins are already denatured.

After freezing, the meat is cut into slices 10–15 mm thick. Cut surfaces should be at right angles to the direction of the muscle fibres, so that the sublimation front moves parallel with the grain during drying and remains parallel to the direction of heat flow. This assists heat and vapour transfer.

After drying, the meat may be cut into pieces, if used as an ingredient in a dried ready-meal, for example. Moisture-impermeable packaging is essential and it is usual to pack in an atmosphere of nitrogen.

### 9.2.4 Intermediate moisture meat products

#### (a) Principles and formulation

In air-dried or freeze-dried meats, the reduction in $a_w$ level necessary to eliminate microbial growth is achieved by physical removal of water. Water removal is inevitably accompanied by changes leading to toughening (see page 406) even when freezedrying is employed. This problem may be solved, or at least minimized, by simply maintaining the $a_w$ at a higher level. The usual approach is to reduce the $a_w$ level sufficiently to prevent bacterial growth and to minimize growth of yeasts and moulds. This objective may be achieved by lowering the $a_w$ level to below 0.85, either by physical removal of some of the water or by the use of ingredients which lower the $a_w$ level of the native water (humectants), or by a combination of the two. Growth of yeasts and moulds is controlled by addition of an antimycotic agent, usually sorbic acid. Formulation of intermediate moisture foods is

---

* A modification of freeze-drying is reversible freeze-dried compression. The moisture content of the meat or other food is reduced to 90% by freeze-drying and then compressed into bars at a pressure of 69 000 kPa. The remaining moisture keeps the meat elastic during compression. The bars are then dried by radiant heat in a vacuum chamber and packaged in an inert atmosphere. By these means a highly stable product is obtained with an ambient temperature storage life of 5 years. Foods of this type are used as field rations by the US military consisting of separate bars of, for example, pepperoni, meat stew, granola dessert and an orange drink. Reconstitution is very rapid. The compressed food 'groans, rumbles and quivers' before returning to normal shape and size.

based on hurdle (combination) technology and it is possible to use an $a_w$ level higher than 0.85 if the pH value is lowered to 5.0.

For many purposes use of humectants to lower the $a_w$ level is the preferred approach. This minimizes toughening and results in a fully moist product. Traditionally salt or sugars have been used as humectants, but in many products these have an unacceptable effect on flavour. Most modern formulations are based on glycerol as the major humectant. Glycerol has a number of advantages in being soluble, relatively stable, non-volatile and almost free of colour and odour. A mixture of glycerol with NaCl, glycine and lactic acid is considered highly effective in lowering the $a_w$ level and pH value and in maintaining organoleptic quality. There are, however, problems with the impact of glycerol on flavour, this being a major factor in the low level of acceptability of products of this type.

### (b) Types of intermediate moisture meat products

Intermediate moisture meat products are prepared from comminuted or similarly processed meat to permit distribution of humectants and other ingredients. A high level of flavouring is added to mask the taste of the humectants and antioxidants are also usually present. Cooking is required to inactivate vegetative pathogens, which may survive in the final product even though growth is not possible. The main type of intermediate moisture product is a variety of cooked emulsion sausage, which is usually processed in-pack in a pouch of very low permeability to water to prevent take-up of water during storage. Such products have had very little commercial success.

A further type of intermediate moisture meat product, which has also had little commercial success, is an ambient temperature spreadable meat product, resembling pâté. In some cases this was presented as a snack product, packaged in a sealed plastic tray together with savoury biscuits and a spreader.

The snack market, currently consisting mainly of starch-based products such as potato crisps (chips), is large and growing and there is much interest in development of meat-based snacks (see page 397). Attempts are being made to develop intermediate moisture snacks, using a combination of meat and starch-based ingredients. Extrusion cooking is used for processing the snacks, which may be 'puffed' to enhance eating quality. Such products are seen as a means of obtaining higher prices for low grade meat as well as gaining entry to a largely new market for meat products. In view of the development effort, it is ironic that the only successful meat-based snack at present is a traditional intermediate moisture product, a small salami stick. On a wider scale it has been envisaged that the novel technology, when fully developed, may be used for making temperature-stable, high protein foods for areas where nutritional status is low.

### 9.2.5 Meat extracts

Meat extracts are now largely produced as a by-product of corned beef manufacture and other meat canning operations. The basic procedure is to cook the meat to be canned at 95–100°C for *ca.* 30 minutes. The cooking water is filtered to remove fine solids and coagulated protein and concentrated, initially under vacuum but in the final concentration stage in an open pan. Boiling in an open pan is considered essential to develop the characteristic flavour of meat extract when beef is used, but chicken extract may be concentrated entirely under vacuum. At the end of the process, fat is allowed to separate by gravity and is solidified by cooling to facilitate removal.

Unconcentrated extract from a single cooking cycle is low in solids content and it is common practice to cook batches of meat in the same fluid, to increase the concentration of solids and reduce energy costs during evaporation. Extraction of solids is also increased with length of boiling for periods of up to 1 hour and by higher temperatures.

Where the final product is to be described as a 'stock', coarse-ground bones are added and cooked at a minimum temperature of 115°C in a pressurized kettle. Bones (with attached raw meat) and liver are used as raw materials in the manufacture of bone extract and liver extract, respectively.

A number of types of meat extract are produced, using basically similar technology. In many cases, however, there is no common terminology and the same description may be used in different contexts by different manufacturers. The most common extract, often known simply as meat extract, or extract number 1, is invariably a by-product of corned beef. The extract contains *ca.* 17% moisture and *ca.* 50 kg of meat is required to produce 1 kg of extract. The low moisture content and the NaCl content of 5% means that this type of extract is shelf stable. In the US, regulations require that the fat content should not exceed 0.6% and that the minimum nitrogen content should be 8%. The nitrogenous component should consist of at least 40% meat bases and 10% creatine plus creatinine (creatinine content alone is usually *ca.* 7%).

Although meat extracts are traditionally liquid or semi-liquid products, drying has been used for a number of years to reduce bulk and increase stability. Roller-drying was widely used but it is now being replaced by spray-drying, which results in a generally superior powder and a general absence of burnt flavour notes. Some more conservative users consider some of the characteristic strong flavours of roller-dried extracts to be lacking in the spray-dried equivalents.

### 9.2.6 Use of dried meat and meat extracts as ingredients

Dried meat and meat extracts find only limited direct use as consumer products. Both types of product, however, are used as ingredients in a wide range of consumer products. In contrast, intermediate moisture meats are primarily seen as consumer products.

Air-dried meat granules are widely used in canned and dried soup mixes and have the advantage to the processor of avoiding the need for refrigeration and (in the case of frozen meat) thawing. Other products of these types in which air-dried meat is utilized include canned and dried meat sauce mixes, such as bolognese sauce, and some canned meat products, such as mince-and-onion pie. Air-dried meat is also an ingredient in some types of stock cube and beef drink.

The higher cost of freeze-drying means that the correspondingly higher price of the consumer product must be justified in terms of either markedly higher quality or greater convenience through faster reconstitution and shorter cooking. Justification on the basis of higher quality is difficult with dried meat products, which are not usually considered as premium items. However, freeze-dried meat is used in dehydrated ready-meals and also in some specialist foods designed for hill walkers.

Meat extracts find very wide ingredient use within the food industry to enhance the meat flavour of both fresh and canned meat products as well as some use in flavouring snacks. To some extent, especially with snacks, synthetic meat flavours have replaced meat extracts. These are heterocyclic thiols and disulphides, which are associated with the characteristic aroma of cooked meat (see Chapter 5, page 259).

Meat extract is the basis of a distinct range of consumer goods which encompasses stock cubes, gravy browning and beef drinks. The latter, such as Bovril™, may be referred to simply as meat extract, although a number of other ingredients are present.

Stock cubes are made from beef or chicken stock (see above), which is roller- or spray-dried. The resulting powder is pressed into blocks, which are wrapped in a foil-based material of low water permeability. Finely ground dried meat is present as well as a small quantity of animal fat and flavour enhancers such as monosodium glutamate and sodium 5'-ribonucleotide. In modern products starches are present to increase viscosity after reconstitution and maltodextrin may be present as a bulking agent. Soup cubes and powder are essentially similar products, which contain a higher proportion of dried meat powder together with additional materials (according to variety) including dried vegetable powder, further flavouring, colouring and antioxidants.

Gravy browning is available as powder or as granules. Powder contains ground dried meat, meat extract and a starch as thickening agent, colouring (usually caramel), flavour enhancers, etc. Granules are prepared by instantizing spray-dried browning. The whole product may be dried and instantized in a single operation, or the main ingredients co-aggregated during instantization. Preparation of gravy from powder requires boiling, but granules are merely dissolved in hot water. This requires use of a starch which gels with relatively little heating.

Beef drinks may indeed be consumed as a hot beverage but are also used in cooking and as a sandwich spread. Products are packed as low-moisture pastes and contain both meat extract and dried meat, as well as hydrolysed vegetable protein, yeast extract, caramel, etc.

---

BOX 9.3  **The medium is the message**

The meat scientist with a microbiological bent may well be better acquainted with meat extract through its use in bacteriological media than through personal consumption. Early commercial production of meat extract was based on Justin von Liebig's development of *extractum carnis*. Liebig's Extract of Meat Company was subsequently formed, from which the acronym LEMCO was derived. A new improved product, OXO, was marketed for human consumption in 1899, the manufacturers becoming Oxo Ltd on the outbreak of war in 1914. Oxo Ltd later created Oxoid who, at the time, were devoted to the sale of LEMCO meat extract for bacteriological purposes (as well as cattle gland extracts as pharmaceuticals) and an improved version LAB-LEMCO was developed. In the early 1920s, solubilized dried beef was introduced for convalescent patients, a similar product, peptone, being used in bacteriological media. Although most of the media in common use today are in ready-mixed dehydrated form (developed by Oxoid during the 1940s), peptone and LAB-LEMCO are still widely used as ingredients.

---

## 9.3 CHEMISTRY

### 9.3.1 Changes during processing of dried meat

*(a) Air-drying*

Conventional air-drying has relatively severe effects on meat. A considerable degree of protein denaturation occurs during the combined process of heating and water removal. The extent of protein denaturation is probably the reason underlying the markedly better quality of dried meat which has been cooked prior to drying: it is thought that denaturation in the presence of the original moisture is less severe than during drying. The extent of denaturation depends on conditions during drying and also on the pH value of the meat. Denaturation and accompanying loss of water-holding capacity are markedly more severe with meat at a pH value close to the isoelectric point of proteins than at a higher value.

A considerable change in distribution of salts occurs during air-drying. This is a consequence of salts being carried to the surface with migrating water and results in a high concentration near the surface. The high concentration of salts in turn has an adverse effect on protein stability.

### (b) Freeze-drying

The effect of freeze-drying is very much less than that of air-drying. Despite this, some denaturation of protein occurs, this being highly dependent on pH value. There is relatively little change in the concentration of salts during sublimation, any effect on proteins being further limited by the low temperature.

## 9.3.2 Changes during storage of dried and intermediate moisture meat

### (a) Dried meat

Both air-dried and freeze-dried meat are subject to deteriorative changes during storage. Various factors affect the extent of change including the initial quality of the meat, processing conditions and temperature, oxygen level and moisture content during storage. Lipid oxidation is the major problem, requiring use of antioxidants (where permitted) and special packaging. Freeze-dried meats can be particularly prone to lipid oxidation because the structure provides a large interior surface area for interaction with oxygen. Lipid oxidation is accompanied by development of rancid flavours and destruction of oxidizable nutrients, including essential fatty acids and some vitamins. Lipid oxidation is also often accompanied by metmyoglobin formation. Under some circumstances, haematin pigments may degrade with formation of green, yellow or colourless bile pigments and globin (see also Chapter 1, page 29).

Non-enzymic browning is also a problem with dried meats. The Maillard reaction is involved, leading to formation of dark pigments and loss of nutritive value of amino acids and proteins. There are also changes in the physical properties of proteins leading to a significant hardening of texture. The Maillard reaction also has implications for flavour and can lead to changes regarded as being both desirable and imparting a typical character to dried meats.

### (b) Intermediate moisture meats

Both lipid oxidation and non-enzymic browning occur during storage of intermediate moisture foods. Protein solubility is reduced by cross-linking associated with lipid oxidation and there is also some promotion of cross-linking by glycerol and other polyols used as humectants. This appears to be an indirect effect caused by peroxides and aldehydes produced during degradation of glycerol in the presence of oxygen.

Some degradation of protein occurs during storage of intermediate moisture foods. Haemoproteins and collagen appear to be most affected. Degradation of collagen occurs to the greatest extent and affects the texture of the meat after cooking.

### 9.3.3 Composition of meat extract

The composition of meat extract varies according to the method of extraction and subsequent processing and the meat species. Average levels of constituents of beef extract are summarized in Table 9.2. Levels of non-volatile constituents are summarized in Table 9.3 and volatile constituents, as determined by gas–liquid chromatography, are listed in Table 9.4. Creatine and creatinine are present at high concentrations and are important quality parameters. The combined concentration of creatine and creatinine should not be less than 10% and it is often considered that the presence of these two compounds accounts for the superior quality of meat extract when compared with the cheaper sub-

**Table 9.2** Average composition of meat (beef) extract

| | |
|---|---|
| Moisture | 19.41% |
| NaCl | 4.76% |
| Creatinine | 7.02% |
| Soluble organic compounds | 43.70% |
| Water-insoluble compounds | 1.70% |
| Ash | 24.0% |
| Thiamin | 10.0 µg/g |
| Riboflavin | 35.0 µg/g |
| Niacin | 1200 µg/g |
| Pyridine | 5.0 µg/g |
| Pantothenic acid | 25.0 µg/g |
| Vitamin $B_{12}$ | 0.5 µg/g |

*Note*: Data from Pearson, A.M. (1993) In *Encyclopaedia of Food Science, Food Technology and Nutrition*, (eds R. Macrae, R.K. Robinson and M.J. Sadler), pp. 2954–9, Academic Press, London.

**Table 9.3** Non-volatile constituents of meat extract (% dry weight)

| | |
|---|---|
| Amino acids | 2.20 |
| Peptides | 6.33 |
| Guanidines | *ca.* 10.10 |
| Purines | 2.60 |
| Protein | 10.92 |
| Lactic acid | 14.60 |
| Other organic acids | *ca.* 3.24 |
| Carnitine | 3.30 |
| Choline | Trace |
| Urea | 0.11 |
| Ammonia | 0.42 |
| Inorganic matter | *ca.* 27.85 |
| Colouring matter | 18.30 |

*Note*: Data from Pearson, A.M. (1993) In *Encyclopaedia of Food Science, Food Technology and Nutrition*, (eds R. Macrae, R.K. Robinson and M.J. Sadler), pp. 2954–9, Academic Press, London.

**Table 9.4** Volatile constituents of meat extract

| | |
|---|---|
| **Hydrogen sulphide** | **Methyl mercaptan** |
| Ethyl mercaptan | Dimethyl sulphoxide |
| **Acetaldehyde** | Propionaldehyde |
| Isobutyraldehyde | Acetone |
| Methyl ethyl ketone | Methanol[1] |
| Ethanol[1] | |

[1]Present only in some samples at low concentrations.
*Notes*: 1. Compounds printed in **bold** are those producing major peaks when detected by gas–liquid chromatography.
2. Data from Pearson, A.M. (1993) In *Encyclopaedia of Food Science, Food Technology and Nutrition*, (eds R. Macrae, R.K. Robinson and M.J. Sadler), pp. 2954–9, Academic Press, London.

stitute, yeast extract. Meat extract also contains relatively large quantities of inosine and inosinic acid. These compounds are flavour enhancers and are thought to make an important contribution to flavour and aroma. Meat extracts have a relatively high content of amino acids. Involvement in the Maillard reaction appears to be related to the characteristic flavour of meat extracts.

## 9.4 MICROBIOLOGY

### 9.4.1 Microbiology of dried meat

Microbial growth cannot grow in dried meat of any type unless the water content and $a_w$ level are raised very considerably. Drying is not, however, a sterilization process, although some inactivation of vegetative micro-organisms is likely. Dried meat will therefore contain viable micro-organisms, as both endospores and vegetative cells. Numbers depend on the load on the initial meat, any heat treatment applied and any contamination after heat treatment and the extent of inactivation during drying and storage. In general rod-shaped bacteria are more sensitive to dehydration than coccal, while endospores are largely unaffected.

*Staphylococcus aureus* is present in small numbers in dried meats and is normally considered to be of no significance. Dried meat has, however, been implicated as a cause of stapyhlococcal intoxication. A traditional dried meat product was involved and it is assumed that *Staph. aureus* grew during drying. Favourable conditions would be created if ambient temperatures were above 25°C and the rate of drying slow, leading to an extended period during which the $a_w$ level permitted growth of *Staph. aureus* but not its competitors.

Microbial growth can occur during reconstitution and, if prolonged, refrigeration is required. There have been reports of food poisoning due to growth of *Bacillus subtilis* and *Staph. aureus* in meat during reconstitution, but these appear to be apocryphal and evidence is lacking.

### 9.4.2 Microbiology of intermediate moisture meat

The microbiology of intermediate moisture meat is more complex than that of dried meat in that stability is dependent on a number of factors and the synergistic relationship between them. In some cases, the margin of safety is not great and cases are known, with products manufactured on an experimental basis, where high storage temperatures have led to a breakdown of the antimycotic system and growth of mould.

It is essential that intermediate moisture meats are correctly formulated and that the composition is sufficiently homogeneous to avoid localized areas where growth of micro-organisms can occur. It is also essential that the preservative system should remain stable during storage. Particular concern has been expressed over the possibility of moisture migration leading to localized areas of high $a_w$ level. It has been suggested that this could result as a consequence of a heterogeneous structure or variations in storage temperature.

Many pathogens survive for extended periods in intermediate moisture foods and the products must be cooked before consumption. The minimum safety margin with respect to *Staph. aureus* is not great and there have been suggestions that enterotoxin can be elaborated by non-growing or very slowly growing cells. There is no evidence to substantiate this theory.

### 9.4.3 Microbiology of meat extracts

Depending on the processing received, meat extracts contain a mixed microflora which is often dominated by *Bacillus* spp. and the more heat-resistant vegetative micro-organisms, such as *Enterococcus*. Stability depends on $a_w$ level and NaCl content and a number of types used as ingredient require refrigeration. The level of heat-resistant endospores can be a problem when meat extracts are used in canned products, such as soups.

Consumer meat extracts (beef drinks) are concentrated to a low $a_w$ level and have a relatively high NaCl content. The products are hot filled and stable at room temperature. There have been allegations of *Staph. aureus* food poisoning associated with beef drinks, following growth of the organism during manufacture and survival of the enterotoxin to point of consumption. The allegations were made many years ago and refer to products made under poor conditions with no relevance to modern practice.

## EXERCISE 9.1

Problems of malnutrition result not only from failure to produce adequate food but also from losses due to spoilage. This is of considerable significance in Africa and Asia where high ambient temperatures lead to rapid spoilage. Solutions applicable in the industrialized nations, primarily involving refrigeration, are often not appropriate to developing countries and considerable attention has been given to enhancing traditional preservation technologies (cf. pages 391–392).

As a food science graduate you have been employed by an international relief agency to examine the best means of minimizing spoilage losses of beef in an equatorial African country. Although there is interest in traditional preservation technology you have been told to keep an open mind and not to dismiss 'high-tech' solutions without due consideration. After a preliminary study you are asked to draw up a detailed comparison of the following methods of extending storage life: refrigeration and/or freezing, salting, irradiation, traditional fermentation, solar drying, use of humectants to lower $a_w$ level. Base your deliberations on the availability and maintenance of suitable technology, suitability for operation by poorly educated operators, acceptability after processing, cost, effectiveness, safety and nutritional consequences.

## EXERCISE 9.2

You are employed as a technologist by a company which has been awarded a contract to manufacture freeze-dried meals for the military. The contract is valuable, but specifications are very strict and the overall profit margin small. A number of the meals are variants on meat stews and problems have been encountered in manufacture due to the liquid freezing in a glassy state, making sublimation very difficult. This has consequences for both product quality and production costs. Summarize the methods which could be used to overcome this problem while minimizing changes to the product and increase in production costs. Prepare a briefing for both your management and the military authorities explaining why action is necessary and presenting (and justifying) what you consider to be the most acceptable solution.

# Further reading

## GENERAL

### Chemistry

Ledward, D.A., Knight, M.K. and Ledward, D.A. (eds) (1992) *The Chemistry of Meat-based Foods*, Royal Society of Chemistry, London.

Price, J.F. and Schweigert, B.S. (eds) (1987) *The Science of Meat and Meat Products*, Food and Nutrition Press, Westport, Ct.

### Microbiology

Brown, M.H. (ed.) (1982) *Meat Microbiology*, Elsevier Applied Sciences, London.

Dainty, R.H. and Mackey, B.M. (1992) The relationship between the phenotypic properties of bacteria from chill-stored meat and spoilage processes. *Journal of Applied Bacteriology (Supplement)*, **73**, 103–114.

Pearson, A.M. and Dutson, T.R. (eds) (1986) *Advances in Meat Research, Meat and Poultry Microbiology*, Macmillan, London.

Varnam, A.H. and Evans, M.G. (1991) *Foodborne Pathogens: An Illustrated Text*, Wolfe Publishing, London.

### Technology

Girard, J.P. (1992) *Technology of Meat and Meat Products*, Ellis Horwood, Chichester.

Pearson, A.M. and Tauber, F.W. (1985) *Processed Meats*, 2nd edition, AVI, Westport.

## Quality control and assurance

Pearson, A.M. and Dutson, T.R. (eds) (1994) *Quality Attributes and their Measurement in Meat, Poultry and Fish Products*, Blackie Academic and Professional, Glasgow.

Sutherland, J.P., Varnam, A.H. and Evans, M.G. (1986) *A Colour Atlas of Food Quality Control*, Wolfe Publishing Ltd, London.


## 1 INTRODUCTION

Kinsman, A.K. and Kotula, A. (1994) *Muscle Foods*, Chapman & Hall, London.

Lawrie, R.A. (1985) *Meat Science*, 4th edition, Pergamon Press, Oxford.

Offer, G., Knight, P., Jeacocke, R. *et al.* (1989) The structural basis of water-holding, appearance and toughness of meat and meat products. *Food Microstructure*, **8**, 151–70.

Shahidi, F. (ed.) (1994) *Flavour of Meat and Meat Products*, Blackie Academic and Professional, Glasgow.


## 2 CONVERSION OF MUSCLE INTO MEAT

Faustman, C. and Cassens, R.G. (1990) The biochemical basis for discolouration in fresh meat: a review. *Journal of Muscle Foods*, **1**, 217–43.

Hood, D.E. and Tarrant, P.V. (eds) (1981) *The problems of Dark Cutting in Beef*, Martinus Nijhoff, The Hague.

Koohmaraie, M. (1992) The role of $Ca^{2+}$-dependent proteases (calpains) in *post-mortem* proteolysis and meat tenderness. *Biochimie*, **74**, 239–45.

Mead, G.C. (1989) *Processing of Poultry*, Chapman & Hall, London.

Ouali, A. (1990) Meat tenderization: possible causes and mechanisms, A review. *Journal of Muscle Foods*, **1**, 129–65.

Ouali, A. (1992) Proteolytic and physicochemical mechanisms involved in meat texture development. *Biochimie*, **74**, 251–65.

Pearson, A.M. and Dutson, T.R. (eds) (1988) *Edible Meat By-products*, Chapman & Hall, London.

Romita, A., Valin, C. and Taylor, A.A. (eds) (1987) *Accelerated Processing of Meat*, Chapman & Hall, London.

## 3 UNCOOKED COMMINUTED AND RE-FORMED PRODUCTS

Hafs, H.P. and Zimbelman, R.G. (1994) *Low-fat Meats: Design Strategies and Human Implications*, Academic Press, Orlando, Fl.

Pearson, A.M. and Dutson, T.R. (eds) (1987) *Restructured Meat and Poultry Products*, Van Nostrand Reinhold, New York.

## 4 CURED MEAT

Cassens, R.G. (1990) *Nitrite-cured meat. A Food Safety Issue in Perspective*, Food and Nutrition Press, Trimbull, Ct.

Gardner, G.A. (1983) Microbial spoilage of cured meats, in *Food Microbiology: Advances and Prospects*, (eds T.A. Roberts and F.A. Skinner), pp. 179–202, Academic Press, London.

## 5 COOKED MEAT AND COOKED UNCURED MEAT PRODUCTS

Ockerman, H.W. (1989) *Sausage and Processed Meat Formulations*, Chapman & Hall, London.

St Angelo, A.J. and Bailey, M.E. (eds) (1987) *Warmed Over Flavour of Meat*, Academic Press, Orlando, Fl.

Xiong, Y.L. (1994) Myofibrillar proteins from different muscle fiber types: Implications of biochemical and functional properties in meat processes. *Critical Reviews in Food Science and Nutrition*, **34**, 293–320.

## 6 COOKED CURED MEATS

Roberts, T.A. and Jarvis, B. (1983) Predictive modelling of food safety with particular reference to *Clostridium botulinum* in model cured meat systems, in *Food Microbiology: Advances and Prospects*, (eds T.A. Roberts and F.A. Skinner), pp. 179–202, Academic Press, London.

## 7 FERMENTED MEAT

Campbell-Platt, G. and Cook, P. (1994) *Fermented Meats*, Blackie Academic and Professional, Glasgow.

## 8 FROZEN MEAT AND MEAT PRODUCTS

Honikel, K.O. (1990) The meat aspects of water and food quality, in *Water and Food Quality*, (ed. T.M. Hardman), pp. 277–304, Elsevier, London.

Spiess, W.E.L. and Schubert, H. (eds) (1990) *Engineering and Food. Vol. 2. Preservation Processes and Related Techniques*, Elsevier, London.

## 9 DRIED MEAT, INTERMEDIATE MOISTURE MEATS AND MEAT EXTRACTS

Brennan, J.G. (1990) Dehydration of foodstuffs, in *Water and Food Quality*, (ed. T.M. Hardman), pp. 277–304, Elsevier, London.

# Index